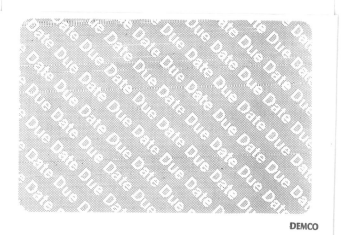

Non-chemical Weed Management

Principles, Concepts and Technology

Non-chemical Weed Management

Principles, Concepts and Technology

Edited by

Mahesh K. Upadhyaya

*Faculty of Land and Food Systems, University of British Columbia,
Vancouver, British Columbia, Canada*

and

Robert E. Blackshaw

Agriculture and Agri-Food Canada, Lethbridge, Alberta, Canada

CABI is a trading name of CAB International

CABI Head Office
Nosworthy Way
Wallingford
Oxfordshire OX10 8DE
UK

Tel: +44 (0)1491 832111
Fax: +44 (0)1491 833508
E-mail: cabi@cabi.org
Website: www.cabi.org

CABI North American Office
875 Massachusetts Avenue
7th Floor
Cambridge, MA 02139
USA

Tel: +1 617 395 4056
Fax: +1 617 354 6875
E-mail: cabi-nao@cabi.org

A catalogue record for this book is available from the British Library,
London, UK.

Library of Congress Cataloging-in-Publication Data

Non-chemical weed management : principles, concepts and technology /
edited by Mahesh K. Upadhyaya & Robert E. Blackshaw.
    p. cm.
   ISBN 978-1-84593-290-9 (alk. paper) -- ISBN 978-1-84593-291-6 (ebook)
1. Weeds -- Control. 2. Pesticides -- Risk mitigation. I. Upadhyaya, Mahesh K.
II. Blackshaw, Robert E. III. Title.

   SB611.N66 2007
   632′.5--dc22

ISBN-13: 978 1 84593 290 9

                                                          2007014230

Produced and typeset by Columns Design Ltd, Reading, UK
Printed and bound in the UK by Biddles Ltd, King's Lynn

Contents

Contributors   vii

Preface   ix

1 Prevention Strategies in Weed Management   1
P.J. Christoffoleti, S.J. Pinto de Carvalho, M. Nicolai, D. Doohan and M. VanGessel

2 Understanding Weed–Crop Interactions to Manage Weed Problems   17
B.D. Maxwell and J.T. O'Donovan

3 Cultural Weed Management   35
R.E. Blackshaw, R.L. Anderson and D. Lemerle

4 Cover Crops and Weed Management   49
J.R. Teasdale, L.O. Brandsæter, A. Calegari and F. Skora Neto

**5 Allelopathy: A Potential Tool in the Development of Strategies for
Biorational Weed Management**   65
L.A. Weston and Inderjit

6 Biological Control of Weeds Using Arthropods   77
B. Blossey

7 Bioherbicides for Weed Control   93
M.A. Weaver, M.E. Lyn, C.D. Boyette and R.E. Hoagland

8 Mechanical Weed Management   111
D.C. Cloutier, R.Y. van der Weide, A. Peruzzi and M.L. Leblanc

9 Use of Non-living Mulches for Weed Control   135
A.C. Grundy and B. Bond

10 Thermal Weed Control   155
J. Ascard, P.E. Hatcher, B. Melander and M.K. Upadhyaya

11 Soil Solarization and Weed Management   177
O. Cohen and B. Rubin

12 Non-chemical Weed Management: Synopsis, Integration and the Future   201
M.K. Upadhyaya and R.E. Blackshaw

Index   211

Contributors

Randy L. Anderson, *USDA-ARS, Brookings, SD 57006, USA (randerson@ngirl.ars.usda.gov)*

Johan Ascard, *Swedish Board of Agriculture, Alnarp, Sweden (johan.ascard@telia.com)*

Robert E. Blackshaw, *Agriculture and Agri-Food Canada, Lethbridge, AB, Canada, T1J 4B1 (blackshaw@agr.gc.ca)*

Bernd Blossey, *Cornell University, Department of Natural Resources, Ithaca, NY 14853, USA (bb22@cornell.edu)*

Bill Bond, *Henry Doubleday Research Association, Coventry, UK (bbond@hdra.org.uk)*

C. Douglas Boyette, *United States Department of Agriculture, Agricultural Research Service, PO Box 350, Stoneville, MS 38776, USA (DBoyette@msa-stoneville.ars.usda.gov)*

Lars O. Brandsæter, *Norwegian Institute for Agricultural and Environmental Research, Ås, Norway (lars.olav.brandsaeter@bioforsk.no)*

Ademir Calegari, *Instituto Agronômico do Paraná, Londrina, PR, Brazil (calegari@pr.gov.br)*

Saul Jorge Pinto de Carvalho, *University of São Paulo, ESALQ, Department of Crop Science, CP 9, PO Box 09, Piracicaba, SP 13418-900, Brazil (sjpcarvalho@yahoo.com.br)*

Pedro J. Christoffoleti, *University of São Paulo, ESALQ, Department of Crop Science, CP 9, PO Box 09, Piracicaba, SP 13418–900, Brazil (pjchrist@esalq.usp.br)*

Daniel C. Cloutier, *Institut de Malherbologie, 102 Brentwood Rd, Beaconsfield, Québec, QC, Canada, H9W 4M3 (assistant@ciuss-scm.ca)*

Oded Cohen, *Department of Geography and Environmental Development, Ben Gurion University of the Negev, Beer Sheva 84105, Israel (odedic@gmail.com)*

Douglas Doohan, *The Ohio State University, Department of Horticulture and Crop Science, Ohio Agricultural Research and Development Center, Wooster, OH 44691, USA (doohan.1@osu.edu)*

Andrea C. Grundy, *Weed Ecology and Management, Warwick HRI, University of Warwick, Wellesbourne CV35 9EF, UK (andrea.grundy@warwick.ac.uk)*

Paul E. Hatcher, *School of Biological Sciences, The University of Reading, Reading, UK (p.e.hatcher@reading.ac.uk)*

Robert E. Hoagland, *United States Department of Agriculture, Agricultural Research Service, PO Box 350, Stoneville, MS 38776, USA (bob.hoagland@ars.usda.gov)*

Inderjit, *Centre for Environmental Management of Degraded Ecosystems (CEMDE), University of Delhi, Delhi 110007, India (inderjit@cemde.du.ac.in)*

Maryse L. Leblanc, *Institut de Recherche et de Développement en Agroenvironnement, 3300, rue Sicotte, C.P. 480, Saint-Hyacinthe, Québec, QC, Canada, J2S 7B8 (maryse.leblanc@irda.qc.ca)*

Deirdre Lemerle, *EH Graham Centre for Agricultural Innovation, New South Wales Department of Primary Industries and Charles Sturt University, Wagga Wagga, NSW 2678, Australia (deirdre.lemerle@dpi.nsw.gov.au)*

Margaret L. Lyn, *United States Department of Agriculture, Agricultural Research Service, PO Box 350, Stoneville, MS 38776, USA (mlyn@srrc.ars.usda.gov)*

Bruce D. Maxwell, *Land Resources and Environmental Science Department, 334 Leon Johnson Hall, Montana State University, Bozeman, MT 59717, USA (bmax@montana.edu)*

Bo Melander, *Aarhus University, Faculty of Agricultural Sciences, Department of Integrated Pest Management, Research Centre Flakkebjerg, Slagelse, Denmark (Bo.Melander@agrsci.dk)*

Marcelo Nicolai, *University of São Paulo, ESALQ, Department of Crop Science, CP 9, PO Box 09, Piracicaba, SP 13418-900, Brazil (marcelon@esalq.usp.br)*

John O'Donovan, *Grains and Oilseeds Team, Agriculture and Agri-Food Canada, Lacombe, AB, Canada, T0H 0C0 (O'DonovanJ@agr.gc.ca)*

Andrea Peruzzi, *Sezione Meccanica Agraria e Meccanizzazione Agricola (MAMA), DAGA – University of Pisa, via S. Michele degli Scalzi, 2 – 56124, Pisa, Italy (aperuzzi@agr.unipi.it)*

Baruch Rubin, *RH Smith Institute of Plant Sciences and Genetics in Agriculture, Faculty of Agricultural, Food and Environmental Quality Sciences, The Hebrew University of Jerusalem, Rehovot 76100, Israel (rubin@agri.huji.ac.il)*

Francisco Skora Neto, *Instituto Agronômico do Paraná, Ponta Grossa, PR, Brazil (skora@iapar.br)*

John R. Teasdale, *United States Department of Agriculture, Agricultural Research Service, Beltsville, MD 20705, USA (John.Teasdale@ars.usda.gov)*

Mahesh K. Upadhyaya, *Faculty of Land and Food Systems, University of British Columbia, Vancouver, BC, Canada, V6T 1Z4 (upadh@interchange.ubc.ca)*

Rommie Y. van der Weide, *Applied Plant Research, PO Box 430, 8200 AK Lelystad, The Netherlands (Rommie.vanderweide@wur.nl)*

Mark VanGessel, *University of Delaware, Plant and Soil Sciences Department, Research and Education Center, Georgetown, DE 19947, USA (mjv@udel.edu)*

Mark A. Weaver, *United States Department of Agriculture, Agricultural Research Service, PO Box 350, Stoneville, MS 38776, USA (mark.weaver@ars.usda.gov)*

Leslie A. Weston, *College of Agriculture and Life Sciences, Cornell University, Ithaca, NY 14853, USA (law20@cornell.edu)*

Preface

The increased availability and acceptability of highly effective and selective synthetic herbicides in the decades following World War II diverted the focus of weed researchers and managers away from non-chemical weed management. Herbicides became the predominant option for weed control, with the ecological and social consequences of herbicide use being ignored or downplayed. An over-reliance on herbicide use led to the widespread development of herbicide-resistant weeds and concerns about potential negative effects on human health and the environment.

The sustainability of our food production systems and the health and environmental consequences of pesticide use are rapidly becoming important global issues. Organic farming is increasing in popularity in many parts of the world due to an increasing demand for pesticide-free food. Weeds pose a serious problem in organic farming. Several weed management options that were once labelled 'uneconomic' or 'impractical', and their technology development practically discontinued, are now being revisited.

The unavailability of a comprehensive book on non-chemical weed management has been a problem for weed science students and instructors around the world. We feel that this book, which deals with the principles, concepts, technology, potential, limitations and impacts of various non-chemical weed management options, will fill this gap. The book consists of chapters on prevention strategies in weed management, exploitation of weed–crop interactions to manage weed problems, cultural methods, cover crops, allelopathy, classical biological control using phytophagous insects, bioherbicides, mechanical weed control, non-living mulches, thermal weed control and soil solarization. The final chapter is a synopsis and integration of all the information presented in the various chapters. We expect that this book will serve as a valuable source of information on non-chemical weed management options and will stimulate research in this area.

Since protection of the environment is a global concern, specialists from around the world have been selected to write these chapters, with an international focus wherever possible. While different options for non-chemical weed management are covered in different chapters, an optimal integration of these alternatives is necessary in order to achieve weed management objectives. The need for a more holistic way of thinking in weed management cannot be over-emphasized.

While the academic level of this book is aimed at upper-level undergraduate courses in weed science and vegetation management, it could also be used for some graduate level courses, and as a supplementary text or reference book for agroecology and organic agriculture courses. Weed scientists and vegetation management professionals working for academic institutions or government agencies, agri-business consultants, organic farmers and other environmentally conscious producers will find this publication to be a valuable resource. The learning objective for students using this book is to understand the principles, concepts, technology, potentials and limitations of various non-chemical weed management options and to think holistically by considering the entire agro- or natural ecosystem involved while managing weed problems. The options described in this book indeed have a variety of impacts on different aspects of ecosystems.

Lastly, we would like to thank all the authors of this book for their hard work in writing chapters in their areas of specialization, peer reviewers for their critical and constructive comments, and our families for their cooperation, patience and encouragement.

Mahesh K. Upadhyaya   Robert E. Blackshaw
Vancouver   *Lethbridge*
British Columbia   *Alberta*
Canada   *Canada*

1 Prevention Strategies in Weed Management

P.J. Christoffoleti,[1] S.J.P. Carvalho,[1] M. Nicolai,[1] D. Doohan[2] and M. VanGessel[3]

[1]*University of São Paulo, ESALQ, Department of Crop Science, PO Box 09, Piracicaba, SP 13418-900, Brazil;* [2]*The Ohio State University, Department of Horticulture and Crop Science, Ohio Agricultural Research and Development Center, Wooster, OH 44691, USA;* [3]*University of Delaware, Department of Plant and Soil Sciences, Research and Education Center, Georgetown, DE 19947, USA*

1.1 Introduction

Prevention has been a cornerstone of weed management throughout history. The importance of prevention to early farmers can be inferred from religious references and other historical documents. For instance the biblical parable of the tares in Matthew 13, in which an enemy invades a farmer's field at night to sow tares (probably Persian darnel (*Lolium persicum*)) undoes the farmer's sound weed control practices. In addition several quotations from ancient sources recorded in *Farm Weeds of Canada* illustrate the respect for weed prevention apparent during the medieval period (Clark and Fletcher, 1906). Muenscher (1955) in his classic book *Weeds*, written before the discovery of selective herbicides, recommended three fundamental objectives that farmers should strive for in weed control; prevention, eradication and control. Muenscher defined prevention as the exclusion of weeds from areas not yet infested or preventing spread from infested to clean fields.

Prevention is a pillar of integrated pest management (IPM) (Norris *et al.*, 2003) and arguably the most cost-effective approach that a grower can take. However, preventive management is complex, involving integration of a group of practices and policies that avoids introduction,

infestation, or dispersal of certain weed species to areas free of those species (Rizzardi *et al.*, 2004). Preventive management is a very efficient technique for any property size, from a small vegetable crop seedbed to large areas devoted to major field crops.

Many government agencies have laws and regulations prohibiting the movement of weed propagules. Seed purity laws are designed to ensure the purity of crop seeds and prevent the spread of weed seeds. Species that are regulated by government statutes are usually designated as exotic or noxious, carrying requirements to mitigate introduction or dispersal, and requiring owners of infested properties to eradicate or prevent propagule production. Generally these regulated weed species are not indigenous to the protected region.

Private landowners may practise elements of weed prevention on their own farms and individual farmers evaluate the risks associated with new weed problems according to their experiences (Slovic, 1987; Pidgeon and Beattie, 1998). Farmers in many countries perceive weeds as familiar, controllable, not catastrophic, and caused by natural forces rather than human failure (Pidgeon and Beattie, 1998). This attitude can be attributed largely to herbicide availability. Prevention also requires management that many farmers are unwilling to practise, particularly if the land is rented or leased.

At the agroecosystem level, seed or propagule dispersion from field to field and from farm to farm needs to be recognized as an important factor that affects the whole agricultural system and should be included in comprehensive weed management planning (Thill and Mallory-Smith, 1997; Woolcock and Cousens, 2000). Preventive management at the farm and at the landscape/ecosystem level require awareness of the processes and practices that contribute to species introduction and proliferation. Integration of preventing new weed infestations, controlling isolated weed patches in the area, as well as preventing seed production are important components of any weed prevention strategy. In some special cases, mainly in areas where exotic weeds are present, eradication should be considered (Woolcock and Cousens, 2000). Prevention should be implemented at all crop production stages, from the acquisition of machinery, seed, water and fertilizers, to crop harvest and processing. The practices of weed prevention are similar to the weed management elements of best production practices; however, they differ because of a requirement for management that is more intensive and the need to amortize costs over the longer term.

Practices that contribute to weed dispersal and which are amenable to prevention are described in this chapter. Attention is given to preventive actions during crop seed purchase, transport of harvested material, cleaning of machinery and equipment, and irrigation, livestock farming and soil management.

1.2 Dispersal of Weeds by Environmental and Ecological Processes

Adaptations for efficient dispersal within and between ecosystems are characteristic of many weeds. In the absence of human activity, weeds rely upon the same natural processes for dissemination as do other plants: dispersal by wind and water, adhesion to fur or feathers, and through food webs. However, farming, trade, and human migration usually amplify the impact of these dispersal adaptations.

Many weed species are disseminated by wind; some as whole plants that shed seed as they move across the landscape, while others rely on the wind to move only the seed. Russian thistle (*Salsola kali*) and kochia (*Kochia scoparia*) are species specially adapted to tumbling with the wind and disperse seeds as the plants move across the landscape. Seed of many species have adaptations to aid in wind dispersal. Examples include the pappus, which is common to many species of the Asteraceae as well as the winged fruits (samaras) of many woody species. However, there is very little information on the distances that weed species may be dispersed by the wind. Research with horseweed (*Conyza canadensis*) identified seeds at altitudes of 140 m, which implies that seeds can travel hundreds of kilometres while remaining aloft in the wind (Shields *et al.*, 2006).

Seeds of some weeds possess special modifications to provide greater buoyancy for efficient transport in open water channels including rivers, streams, and irrigation and drainage channels (Dastgheib, 1989; Lorenzi, 2000). Because rivers may traverse ecosystems, they are conduits for long-distance dispersal of plants. Water corridors are regularly disturbed by natural processes such as flooding and ice movement, and weeds dispersing in the water are thereby provided with an ideal habitat to colonize along the banks and shorelines.

Seed transportation by animals is used by many plants (Harper, 1977). The efficiency of animals in dispersing seeds depends on the specific animal and plant species involved (Couvreur *et al.* 2005; Mouissie *et al.* 2005). Dispersion of weed seeds through adherence of fruits or seeds to fur and feathers, or by ingestion, has been mainly attributed to birds and mammals. Birds may contribute to both short- and long-distance dispersion. The range of some large vertebrates and various endemic plant species have overlapped, resulting in distribution across ecosystems during annual migrations. Invertebrates and small mammals such as rodents regularly play a role in short-distance dispersion within the agroecosystem.

Dispersal by environmental and ecological processes is complex, and prevention practices intended to minimize dispersal by these means are generally unlikely to be effective. However, at the farm and community levels, preventing seed production of newly introduced species will eliminate the opportunity for dispersal by these natural processes. The remainder of this chapter will address the role of humans in opportunities to practise prevention within that dimension.

1.3 Dispersal of Weeds by Human Activities

Because weeds are adapted to the disturbance regimes of agriculture and human activity, it should come as no surprise that nearly every human activity plays a role in their spread and distribution. A complete treatment of the role that humans have played and continue to play in weed dissemination is beyond the scope of this chapter and we will restrict our discussion to those activities with direct impact upon food and fibre production.

Plant introductions

Several of the most highly competitive or trouble-some weeds have at one time been introduced to a new area as a potential food or medicinal crop, livestock feed, fibre or ornamental plant. *Panicum miliaceum* was initially introduced in Canada in the mid-1800s for grain production, but it adapted to the new habitat and became an aggressive weed in North America over the last 25 years (Bough *et al.*, 1986). *Cynodon dactylon*, *Sorghum halepense* and *Digitaria* spp. were introduced to many countries as pasture species and subsequently have become significant weeds (Kissmann, 1997; Zimdahl, 1999). Many other species can be mentioned as examples of intentional human dispersion (e.g. *Ageratum conyzoides*, *Linaria vulgaris*, *Nicandra physalodes*, *Opuntia stricta*, *Pistia stratiotes*, *Salvinia* spp. and *Sagittaria montevidensis*) (Kissmann and Groth, 2000).

Despite laws that have been passed by many national governments to prevent the introduction of invasive plants, this problem continues. Laws and regulations are for the most part reactive; therefore, prohibitions are generally not in place until after introduction and widespread distribution of a weed in the new habitat has already occurred. Efforts to develop predictive models that will enable governments to regulate

species before they are introduced are under way in Australia and in Hawaii, USA. However, enhanced regulation alone will not prevent the introduction of new noxious species. Prevention requires individual responsibility – not only of regulators but also of individual property owners, and those involved in the discovery and commercialization of novel plants.

Crop seeds

The production, selection and use of quality seed has direct implications for weed prevention as well as crop yield. Several problematic weed species can be traced back to the use of weed-infested crop seed. Some of these examples include *Echinochloa* spp. with rice (*Oryza sativa*), *Lolium persicum* or *Agrostemma githago* in small grains, and *Vicia sativa* in lentils (*Lens culinaris*) (Harper, 1977; Lorenzi, 2000). Many countries have laws establishing purity standards for commercial crop seeds. These laws also identify species that are unacceptable at any number in commercial lots (Thill and Mallory-Smith, 1997; Zimdahl, 1999). Some species are specifically targeted by seed laws due to their aggressiveness, difficulty of obtaining control, or difficulty in removing the weed seed from the harvested crop seed.

Even in those countries where seed purity laws are enforced there are generally no laws that require farmers to use commercial seed. Dastgheib (1989) analysed the influence of different sources of weed infestation of wheat (*Triticum aestivum*) in the Fars province of Iran, and observed that the use of 'saved seeds', produced on-farm, contributed 182,000 weed seeds/ha, representing 11 species (Table 1.1). This practice of 'saving seeds' continues to be a problem globally. For instance in Utah, USA, small grain seeds showed a decline in wild oat (*Avena fatua*) contamination from 1958 until 1988, yet an appreciable number of the farmers surveyed were continuing to plant wild-oat-infested crop seed (Thill and Mallory-Smith, 1997).

A similarity between certain weed and crop seeds in shape and size makes it very difficult to distinguish between species during the seed-cleaning process. Some of the best-known examples of this phenomenon are *Camelina*

Table 1.1. Various sources contributing weed seeds, based on studies attempting to quantify the input of weed seeds.

Seed source	Estimated number of seeds/ha[a]	Number of species collected	Reference
Irrigation water	48,400	34	Wilson, 1980
Irrigation water	10,000–94,000	137	Kelley and Bruns, 1975
Irrigation water	92	4	Dastgheib, 1989
Dairy farms	91,000–1,000,000[b]	na	Cudney *et al.*, 1992
Dairy farms	3,400,000[c]	48	Mt Pleasant and Schlather, 1994
Sheep pasture	9,900,000	92	Dastgheib, 1989
Cattle pens	5,300,000[d]	23	Rupende *et al.*, 1998
Wheat 'saved seed'	182,000	11	Dastgheib, 1989

[a] Based on authors' estimates or 22 t/ha for manure as a fertilizer source.
[b] Seven dairies were sampled.
[c] Twenty farms were sampled: four farms had no detectable seeds, and only one farm had >200,000 seeds per tonne of manure. Value presented is mean of 16 farms, with weed seeds at 75,100 seeds per tonne.
[d] Four farms sampled.
na = not available.

sativa in flax (*Linum usitatissimum*) seed lots, *Echinochloa* spp. in cultivated rice seed (Baker, 1974), soybean (*Glycine max*) seed infested with balloonvine (*Cardiospermum halicacabum*) and *Polygonum convolvulus* in wheat seed (Lorenzi, 2000; Rizzardi *et al.*, 2004). Such problems pose great challenges for both the producer of commercial seed and the farmer using 'saved seed'.

These examples illustrate the limitations of a regulatory approach to seed purity and the ongoing need for educational activities that will help farmers appreciate the long-term impacts of using weed-free crop seeds. Recent initiatives in developing countries to preserve biodiversity and improve quality of 'saved seed' through social network seed-cooperatives should be expanded to train farmers on the risks associated with planting crop seeds contaminated with weed seeds (Seboka and Deressa, 2000).

Machinery

Agricultural machinery disperses weeds during field preparation, cultivation and harvesting. Cultivation disperses weeds over short distances, while harvesters can transport seeds over greater distances in the field; and both can move propagules extensive distances if the machinery is moved from one location to another (Thill and Mallory-Smith, 1997; Bischoff, 2005). This includes the purchase of previously owned machinery, since farm equipment is often sold at great distances from the location of original use.

Dispersion by harvesters has been mentioned in several studies (Ghersa *et al.*, 1993; Thill and Mallory-Smith, 1997; Blanco-Moreno *et al.*, 2004). Seed dispersal associated with harvesting is dependent on the number of weed seeds remaining with the plant at time of harvest and varies by seed shape and size. Examples where harvesters are implicated in weed dissemination to previously uninfested fields include *Rottboellia exaltata* spread among soybean fields (Zimdahl, 1999) and the spread of jointed goatgrass (*Aegilops cylindrica*) in the wheat-producing region of the USA (Donald and Ogg, 1991).

Few studies have attempted to quantify the amount of propagules dispersed by machinery because so many variables are involved. In one of the few studies on dispersal by machinery, Ghersa *et al.* (1993) studied *Sorghum halepense* dispersal from isolated clumps (about 2 m diameter) in a maize (*Zea mays*) field due to machine harvesting. At harvest, 40 to 60% of the seeds had naturally dispersed. The combine dispersed 50% of the remaining seeds within the first 5 m from the clumps and the remainder dispersed uniformly over 50 m at a rate of 1% per metre

($\sim$5 per m^2). The dispersal potential of combine harvesters has been reported to exceed 18 m (Blanco-Moreno *et al.*, 2004) and 50 m (Ghersa *et al.*, 1993). Modifications to harvesters can reduce the in-field spread of weed seeds in some situations, in particular the collection of chaff where the majority of the weed seeds are located (Matthews *et al.*, 1996; Shirtliffe and Entz, 2005). Shirtliffe and Entz (2005) reported that 74% of

wild oats seed dispersed from a combine were distributed in the chaff.

Cultivation disperses species that propagate by seeds and/or vegetatively. Bischoff (2005) observed that cultivation promoted dispersion of *Lithospermum arvense* and *Silene noctiflora* seeds short distances (1 to 2 m) (Fig. 1.1). Guglielmini and Satorre (2004) observed that 50% of the vegetative parts of *Cynodon dactylon*

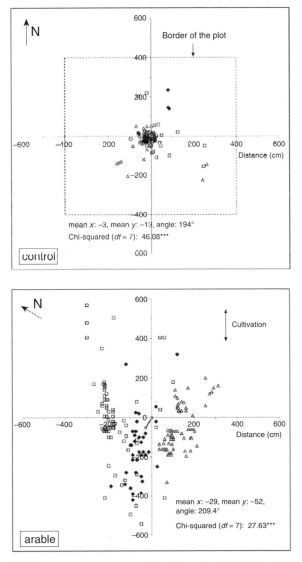

Fig. 1.1. Distribution of *Lithospermum arvense* seedlings from mother plants, with different symbols for different patches; *y* axis represents north–south direction in the control (upper panel) and direction of cultivation in the arable treatment (lower panel) (Bischoff, 2005).

dispersed by chisel-ploughing were between 0.9 and 2.0 m from the initial patches. Similar distances have been reported for other species, although distances do vary with size of seed and depth of seed in the soil profile (Rew and Cussans, 1997). Soil conditions at the time of cultivation influence the amount of seed, rhizomes or tubers dispersed, with moist soil resulting in more propagules being moved (Mayer *et al.*, 2002).

Fields infested with weed species that farmers are trying to prevent from spreading require additional management. Weed seeds and vegetative propagules imbedded in soil and debris inside the machine or adhering to surfaces should be removed by thoroughly cleaning before using the machine at the new location. Careful cleaning of tractor wheels, parts of implements used in soil preparation and seeding, as well as horizontal surfaces of harvesters, are important (Thill and Mallory-Smith, 1997; Rizzardi *et al.*, 2004).

Infested areas can be sprayed with a desiccant to kill the weeds and increase the seed shed of the weeds before harvesting, or harvest can be delayed to allow for greater seed shed. When the infestation is too dense or the risk of transporting undesirable seed is too great, the area should not be harvested.

Transportation of plant parts

Harvest operations that remove the entire plant require additional caution for preventing weed seed spread. Weed propagules are more likely to be harvested when the entire crop plant is removed for uses such as straw, silage, or further processing (e.g. sugarcane, machine-harvested vegetables). For many species, immature seeds can ripen after harvest and are capable of producing viable offspring. Because agricultural products are marketed globally, the likelihood of transporting weed seeds with the commodity over great distances is high.

Plant dissemination through straw of cereals has occurred with long-distance transport of humans and goods. During the colonization of the Americas, several weed species were introduced with the material transported in the ships. Straw was a common packing material or bedding that was discarded at the final destination. This brought weed species to both sides of the Atlantic Ocean (Baker, 1974).

Transport of raw commodities contributes to local and long-distance dispersal of weeds. The relative small seed size of most weed species makes them prone to be blown by the wind when plant material is transported in uncovered

Fig. 1.2. Trucking bales of sugarcane (*Saccharum officinarum*) straw, providing an opportunity for weed seeds to be blown along the roadside.

vehicles (Fig. 1.2). The diversity of weeds alongside roads and highways is often greater than in fields a short distance from the road. In Brazil, *Ricinus communis* dispersion is believed to be related to truck traffic (Kissmann and Groth, 1999). *Sorghum halepense* was introduced in the southern region of Brazil by the rail transport of flax, lucerne (*Medicago sativa*), sunflower (*Helianthus annuus*) and sorghum (*Sorghum bicolor*) from Argentina (Kissmann, 1997). Velvetleaf (*Abutilon theophrasti*) infestation of dairy farms in Nova Scotia, Canada, was directly associated with long-distance shipments (2000 km or more) of contaminated maize from fields in south-western Ontario and the US 'corn belt' (LeBlanc and Doohan, 1992).

In sugarcane (*Saccharum officinarum*)-producing areas of Brazil, an increase in weed density, especially that of *Cyperus rotundus* and *Cynodon dactylon,* has been observed. This phenomenon has been attributed to the transport of baled sugarcane straw and its disposal after processing for either ethanol or sugar (Medeiros, 2001). In the case of ethanol production, the crop residues are deposited in fields following processing. During sugar extraction, crop materials are decanted into discharge tanks. This liquid suspension, as well as the filter cake, is a potential reservoir of weed seeds and propagules. These materials, known as vinasse, are usually transported and returned as organic matter to the fields, sometimes via irrigation canals. Similarly, discarded plant materials from vegetable-processing facilities can contain weed seeds. These materials are often fed to cattle or returned to the fields as organic material. As described later in this chapter, animal digestion often does not destroy all weed seeds.

Transporting recently harvested commodities with coverings will minimize the dispersion of weed seeds along roadways and rail systems. However, this will not influence the final dispersal of alien weeds in a new area when the transported crop is fed to livestock. Once weed-seed-contaminated commodities arrive at the farm or the processing facility, grinding, palletizing or fermentation processes can be applied to reduce or eliminate viable weed seeds.

Composting is an approach to reduce weed seed density in organic matter prior to land application. Techniques for composting are beyond the scope of this chapter, but there are a number of references available (Cooperband, 2002; Anonymous, 2005). It is generally recommended that the temperature of the composted material should reach 60°C and be held at that temperature for 7 days to kill weed seeds and vegetative propagles. However, it is very difficult to reach this temperature near the surface of compost piles, and thus only weed seeds in the interior of the pile are killed. As a result, compost piles need to be turned or mixed to ensure that all seeds are exposed to the internal temperatures.

Transportation of plants, plant parts and associated soil is often regulated in a highly variable manner. The individual farmer needs to be zealously cautious when purchasing plant materials for planting or feeding to livestock in order to be certain that new weeds are not introduced.

Transportation of soil

Transport of soil, intentionally or otherwise, contributes significantly to the dispersal of weeds. Soil that adheres to roots of transplant seedlings and nursery stock is a source of weed seeds and vegetative propagules which may be transported to new areas. For example, the installation of new fruit orchards or coffee (*Coffea arabica*) crops demands the acquisition and transport of a great quantity of young plants. The nurseries that produce these plants must be rigorous regarding the cleanness of saleable material, especially with the substrate used for plant development. To prevent the introduction of weeds with nursery seedlings, use of media with little or no soil is recommended. However, this may be practical only with container-grown plants.

Animals and manure

The dispersion of weeds promoted by animals can be divided into two methods: dispersion by adherence and dispersion by ingestion. Dispersion by adherence occurs due to fixative structures present on seeds and fruits (i.e. thorns, awns and hooks) that allow them to stick in an animal's fur, promoting their dispersion throughout new areas. Weed species that are disseminated in this manner include *Bidens* spp., *Cenchrus echinatus, Cynodon dactylon, Desmodium tortuosum, Digitaria sanguinalis, Hypericum perforatum* and *Xanthium strumarium* (Radosevich and Holt, 1984). Livestock may also play a role in spreading weeds to and from rivers and streams. Paths to watering holes create the disturbance needed for weed establishment and provide a corridor between watering holes and pastures.

Dastgheib (1989) compared the weed seed contributed to fields by irrigation water, using 'saved seeds', and fertilizing with sheep (*Ovis aries*) manure. From these three sources, the author observed that the manure had the highest contribution of seeds, with 9.9 million seeds per hectare; approximately 54 and 107,000 times more than saved seeds and irrigation water, respectively (Table 1.1).

Manure from dairy cows (*Bos* spp.) was collected from various sites on seven farms in California, USA, and 2 to 21.7 thousand seeds per tonne of manure were recovered (Cudney *et al.*, 1992). Some of the differences between the sites were attributed to the quality of feed used for milking cows versus non-lactating cows. Mt. Pleasant and Schlather (1994) analysed the presence of plant seeds in bovine manure from 20 farms in New York, USA. They documented viable seeds of 13 grass species and of 35 broadleaved species, with *Chenopodium album*, found in manure from more than half of the analysed farms. Number

of seeds per tonne of manure ranged from 0 to 400 thousand, with an average of 75 thousand seeds per tonne. In Zimbabwe, 6 grass species and 17 broadleaved species were identified in cattle manure (Rupende *et al.*, 1998). One tonne of manure was estimated to contain 66 thousand seeds.

The percentage of seeds that pass through the animal's intestinal tract and remain viable varies according to the weed species, and to the animal species that has eaten them (Thill *et al.*, 1986). Weed seeds passing through pigs or cattle had higher viability than seeds passing through horses or sheep, while poultry was the most efficient at destroying weed seeds (Table 1.2) (Harmon and Keim, 1934). Neto *et al.* (1987) and Stanton *et al.* (2002) also compared farm animals for their ability to reduce weed seed viability and reported sheep to be more effective in destroying seeds than cattle. Although animals differ in their ability to reduce weed seed viability, complete destruction of weed seeds is often not achieved (Blackshaw and Rode, 1991; Gardener *et al.*, 1993; Mt. Pleasant and Schlather, 1994; Wallander *et al.*, 1995).

Composting manure is a common practice for many livestock farmers, and dramatic declines in weed seed viability have been documented with increased composting time and/or adequate moisture levels (Table 1.3) (Grundy *et al.*, 1998; Rupende *et al.*, 1998; Eghball and

Table 1.2. Percentage of viable seeds passed by various animals. Percentage is based on the total number of seeds fed.

Kinds of seeds	Percentage of viable seeds passed by					
	Calves	Horses	Sheep	Pigs	Chickens	Mean
Field bindweed						
Convolvulus arvensis	22.3	6.2	9.0	21.0	0.0	11.7
White sweetclover						
Melilotus alba	13.7	14.9	5.4	16.1	0.0	10.0
Pennsylvania smartweed						
Polygonum pennsylvanicum	0.3	0.4	2.3	0.0	0.0	0.6
Red sorrel						
Rumex acetocella	4.5	6.5	7.4	2.2	0.0	4.1
Velvetleaf						
Abutilon theophrasti	11.3	4.6	5.7	10.3	1.2	6.6
Whitetop						
Cardaria draba, Lepidium draba	5.4	19.8	8.4	3.1	0.0	7.3
Mean	9.6	8.7	6.4	8.8	0.2	6.7

From Harmon and Keim (1934).

Table 1.3. Effect of time of removal from compost windrow on weed seed viability averaged for 1997 and 1999 studies.

Weed	Control	Time of removal			
		Day 14	Day 21	Days 42–50[a]	Days 70–91[a]
		Viable seed (%)			
Green foxtail					
Setaria viridis	86	4	1	0	0
Redroot pigweed					
Amaranthus retroflexus	78	6	4	0	0
Pennycress					
Thlaspi arvense	11	5	4	1	4
Wild buckwheat					
Polygonum convolvulus	47	32	15	4	0
Wild oats					
Avena fatua	72	13	1	0	0

[a] The earlier date is for 1997 and the later date for 1999 data. Adapted from Larney and Blackshaw (2003).

Table 1.4. Average weed seed viability after ensiling in a silo, fermentation in the rumen, or both, 1986–1989.

Species	Control	Ensiling in a silo	Rumen	Silo and rumen
		Viable seed (%)		
Green foxtail				
Setaria viridis	96	0	17	0
Downy brome				
Bromus tectorum	98	0	0	0
Foxtail barley				
Hordeum jubatum	87	0	0	0
Barnyardgrass				
Echinochloa crus-galli	97	0	0	0
Flixweed				
Descurainia sophia	92	5	7	5
Kochia				
Kochia scoparia	94	10	15	10
Redroot pigweed				
Amaranthus retroflexus	93	6	45	4
Lambsquarters				
Chenopodium album	87	3	52	2
Wild buckwheat				
Polygonum convolvulus	96	30	56	16
Round-leaved mallow				
Malva pusilla	93	23	57	17
Pennycress				
Thlaspi arvense	98	10	68	10

Adapted from Blackshaw and Rode (1991).

Lesoing, 2000; Larney and Blackshaw, 2003). At five dairy farms in California, USA, the amount of seed in compost ranged from 323 to 4128 viable seeds per tonne of manure (Cudney *et al.*, 1992). Lethal temperatures for manure composting were dependent upon the weed species and the length of time the manure was allowed to compost. Tompkins *et al.* (1998) reported ≤1% viable seed after 2 weeks in a windrow of typical beef feedlot manure for 9 out of 12 weed species. *Amaranthus retroflexus* had the highest viability (3.5%) after 2 weeks. No viable seeds were detected after 4 weeks. Eghball and Lesoing (2000) reported no viable seeds after 1 week in beef feedlot manure that had adequate moisture, while dry manure required at least 3 months to destroy all the seeds. In the same study, composted dairy manure had destroyed all seeds except velvetleaf (*Abutilon theophrasti*), which was <17% viable after 3–4 months of composting. Anaerobic digesters are also able to reduce the viability of weed seeds (Katovich *et al.*, 2006). However, neither an animal digestive tract, composting, nor anaerobic digesters consistently eliminated all weed seeds.

Providing a weed propagule-free diet is one method of eliminating weed seeds associated with the digestion system and the manure. Good weed control can be used to minimize weed growth and weed seed contamination of forages. Mowing or removing weeds at or prior to the bud stage is necessary in order to ensure that no seeds are produced. However, some weed species cut early in the flowering stage were capable of producing viable weed seeds (Gill, 1938; Derscheid and Schultz, 1960).

The nutritional content of some weed species is high and, as such, farmers may make a conscious decision to allow some weeds in pastures, hay, silage or feed (Mueller *et al.*, 1993; Stanford *et al.*, 2000), or weeds may not be controlled in order to increase the tonnage of a pasture. Statutes that govern contents and quality of livestock feedstuffs may also play a role in the proliferation of certain species. For instance the 'mixed feed oats' classification that is used in the USA and in Canada may allow for a wild oat content of up to 50%. This statute has effectively enhanced the general distribution of wild oats throughout these countries, as elevators typically add wild oats to the grade, up to the maximum allowable level.

The farmer can take specific actions to reduce the viability of weed seeds that are harvested with the crop. Grinding or pelleting feed can reduce weed seed viability, but it does not consistently kill all seeds (Zamora and Olivarez, 1994; Cash *et al.*, 1998). Seed viability after processing depends on how finely ground the material is and the plant species involved. Grinding reduced viability of sulphur cinquefoil (*Potentilla recta*) seeds by 98%, but reduced spotted knapweed (*Centaurea macu-*

losa) by only 15% (Zamora and Olivarez, 1994).

The fermentation process of producing silage from maize or forages can have an impact on weed seed viability with reductions for some species, ranging from 70% to 100% (Table 1.4) (Zahnley and Fitch, 1941; Blackshaw and Rode, 1991). The reduction in seed viability differed by weed species, with the viability of grass species being reduced more than that of broadleaved species. Tildesley (1937) reported all seeds of 18 species were destroyed within 14 days in a silo, while one species, *Chenopodium album*, required 21 days. However, seeds at the top of the silo were not impacted by the ensiling process.

Livestock farmers need to recall that rotating pasture systems, where animals remain for a short time in each pasture and later go to an adjacent area or even to other properties, may also contribute to weed dispersion. In order to prevent weed seed dispersal, animals could be kept in a confined area for a period of time to allow for clearing their digestive tract of viable weed seeds prior to moving to uninfested areas. The period of time is dependent on seed shape, type of animal, and type of diet. Most studies have shown that at least 4 days is required to eliminate seeds from the digestive tract for a variety of livestock (Neto *et al.*, 1987; Gardener *et al.*, 1993; Willms *et al.*, 1995). However, this will not remove seeds that have adhered to the animals' fur or skin. Adhesion to animal fur or skin is also dependent on the characteristics of the seeds and animals (Couvreur *et al.*, 2005; Mouissie *et al.*, 2005).

Water and wind

As previously described in Section 1.2, seeds of some weeds possess specialized structures that facilitate their transport through water channels (Dastgheib, 1989; Lorenzi, 2000) and many species that do not have these special adaptations can still float temporarily and disperse through water (Wilson, 1980; Radosevich and Holt, 1984). When agricultural or food-processing plant wastes are deposited in rivers and irrigation canals (as previously described for sugarcane), weed seeds accompanying these wastes may be redistributed to nearby irrigated fields or dispersed to new habitats. Thus, irrigation of agricultural crops with surface waters may introduce new weed species to farm fields and/or deposit seeds of endemic species that grow along the water corridor or in nearby fields (Hope, 1927). Kelley and Bruns (1975) recorded seeds of 137 species in sources of irrigation water over the course of a season. Furthermore, they calculated that 10,000–94,000 seeds/ha would be distributed over the course of a season due to the contamination of irrigation water. Wilson (1980) monitored seeds in irrigation canals as well as a major river (Platte River) in Nebraska, USA, and found that weed seed content was higher in the irrigation canals. During one season, an irrigated field received 48,400 seeds/ha from 34 different species.

Seed of many species remain viable for months in water, sometimes for periods of 5 years. Seed viability can decrease with submersion time, depending on the species and duration of submersion (Kelley and Bruns, 1975; Comes *et al.*, 1978). In a limited number of situations, maintaining seeds in water can break dormancy and increase germination rates (Comes *et al.*, 1978).

Farm managers should consider the role that irrigation water may play in replenishing soil weed seed banks. In cases of severe contamination where the species number and seed density is high or it is likely that water contains problematic or invasive species, the water source might be changed or filters or decanting tanks installed. The most cost-effective method for farmers within an irrigation district may be to collectively maintain a zone adjacent to the irrigation canals where weed seed production is prevented in order to reduce the number of seeds in the irrigation water. Growers may be able to avoid the use of suspected irrigation water during times of peak seed shed. In the case of irrigation canals that are only used seasonally, it is wise to avoid the use of irrigation water early in the season, before the canals have been 'flushed' of deposited weed seeds.

Dispersal by wind was discussed in Section 1.2, and while many weed species are disseminated by wind, the special case of 'tumbleweeds' is worthy of closer examination. *Salsola kali* and *Kochia scoparia* are dispersed as whole plants that shed seed as they move across the landscape. Erecting fences has a limited impact on reducing seed dispersal of tumbleweeds, as strong winds can cause the weeds to be blown over the fences.

1.4 When Prevention is not Successful

Muenscher (1955) wrote that the aim of eradication is the elimination of a weed after it has become established in an area. That requires stopping the production of further propagules and the depletion of propagule reservoirs in the soil. As illustrated throughout this chapter, this needs to occur within the infested field as well as within distances that the species is capable of disseminating. Control methods will receive comprehensive discussion in other chapters within this book.

Control prior to seed development

Weed seed development is a complex phenomenon influenced by many variables. Farmers recognize that weeds are most detrimental to the crop during its establishment. This observation has been confirmed experimentally and has led to development of the 'critical period of competition' concept. One of the consequences of this approach is that farmers generally neglect weed control late in the season and/or do not focus on eliminating seed production. Weed growth late in the season replenishes the soil weed seed bank and perpetuates weed infestations. Some farmers in California, USA, who have practised scrupulous weed control for many decades, have succeeded in largely depleting the soil weed seed reservoir and, in so doing, have achieved extremely low annual weed control costs. Factors that influence the seed return include crop competition (with the level of seed production dependent upon the crop's competitiveness), time of weed emergence in relation to the crop's emergence, and time of the planting season.

Seed bank depletion

Complete depletion of the soil weed seed bank is not realistic under most circumstances. However, reducing the number of seeds in the soil ultimately reduces weed density and the opportunity presented by this approach should not be overlooked. Eliminating seeds entering into the seed bank is based on obtaining excellent (or complete) weed control. Weed species differ in the number of seeds they can produce as well as longevity of the seed in the soil seed banks (Davis *et al.*, 2005). The time required to reduce seed viability by 50% ranged from <1 to 12 years for a select number of species. Yet the time to reduce seed viability by 99% ranged from 2 to 78 years for this same group of species.

Destroying perennial root systems and other vegetative propagules

Perennial weeds are a unique challenge to prevention. In addition to preventing seed production, prevention also implies eliminating the production and storage of food reserves into underground organs. Many species develop perennial characteristics as early as 2–3 weeks after emerging (McWhorter, 1989; Bhowmik, 1994). Tillage is the most effective non-chemical means to eliminate vegetative propagules from the soil. Frequent removal of above-ground biomass over an extended period of time is necessary to effectively deplete the food reserves in the underground organs. For example, multiple cultivations at 2–3 week intervals for 3 or more years have been effective in controlling many perennial weed species (Anderson, 1999).

1.5 Integrating Prevention into Weed Management

Agricultural systems need to optimize weed management. The critical, but often overlooked, first step is preventing weed infestations. At every step of production (such as seed selection, field preparation, planting, fertilization, irrigation, weed control, harvest and transport), prevention can be implemented and can impact the crop and cropping patterns in future years. Prevention is not, and cannot be considered as an isolated activity. Prevention is awareness and as such, it should be a daily activity which needs

to be incorporated into the routines of all the people involved in agricultural production, at farm, state and national levels.

In this chapter, we have enumerated several agricultural practices that contribute to the long- and short-distance weed dispersal if complete weed control is not achieved. For the most part, managers can make relatively simple, cost-effective modifications to their practices to eliminate or greatly reduce the risk of introducing new weed seeds to the field; however, these opportunities require awareness and vigilance. Key considerations include:

- diligent monitoring for sources and vectors of new weed introductions to the ecosystem and to the individual farm property;
- proactive government laws and regulations that control the introduction and movement of plants and plant materials;
- preventing problems caused by perennial weeds by eradicating vegetative propagules;
- depleting the soil weed seed bank whenever possible;
- considering the probability of a devastating weed problem resulting from the introduction of a non-indigenous plant or transport of plant materials or soil from one location to another;
- producing and planting seed, seedling transplants and nursery stock that are free of weed propagules;
- preventing weed seed production in crop fields;
- preventing spread of weeds by farm machinery and transport and processing of agricultural commodities;
- adopting practices that minimize the presence of weed seed in livestock feed, manures and composts;
- preventing weed seed introduction into rivers and irrigation canals.

The focus of a prevention programme is twofold: to eliminate the introduction of new species as well as reducing the number of seeds in the soil seed bank. Once a species is introduced and is allowed to emerge, become established and produce seed, there is the potential to become a significant component of the weed seed bank in a relatively short period of time. For preventive approaches to be integrated into on-farm weed management practices we believe that agricultural educators, especially those who advise farmers, must reconsider the great importance of preventative weed control. As educators lead the way, prevention can be re-integrated into all aspects of agricultural production so that it is always the first line of defence.

1.6 Implications at the Farm Level

Manure application, irrigation water, use of plant material as organic matter (although not quantified in the literature), and use of weed-seed-contaminated crop seed all contribute thousands to hundreds of thousands of seeds per hectare (Table 1.1). Mt. Pleasant and Schlather (1994) calculated that >20 million seeds per hectare would be applied with cow manure in the worst-case scenario. In order to put this into perspective, in terms of contributing to the soil seed bank, we need a point of reference on the expected size of the seed bank. A study examining weed seed density in agricultural fields ranged from 6 million to 1.6 billion seeds per hectare at eight sites in the north-central region of the USA (Forcella *et al.*, 1992); 255 million seeds/ha was the second-highest seed density. A sample of 58 fields

throughout England, mostly used for vegetable production, had a range of 16 million to 861 million seeds/ha (Roberts and Stokes, 1966), with half of the fields having a density of 62 to 222 million seeds/ha. Roberts (1983) summarized six additional studies from Europe, representing 310 fields, averaging 225 million seeds/ha. The extremes were 2.5 million to 5 billion seeds/ha. Across all the soil seed bank surveys, a relatively small number of species comprised >70% of the seed bank.

The size of the seed bank can substantially increase under the combination of a low weed seed bank density and a high number of seeds in the manure. Likewise, a number of activities (i.e. composting, feed to animals) can reduce the viability of weed seeds and lessen the impact the seeds have on the weed seed bank. Activities to reduce weed seed viability seldom result in 100% loss of viability. In some cases, multiple practices complemented each other, such as ensiling and rumen digestion (Table 1.4) (Blackshaw and Rode, 1991). Due diligence is important in order to prevent the introduction of weed seeds into a farming system. As noted previously, the number of seeds introduced may not be as important as the introduction of a new species or weed biotype.

1.7 References

Anderson, W.P. (1999) *Perennial Weeds: Characteristics and Identification of Selected Herbaceous Species.* Iowa State University Press, Ames, IA, USA.

Anonymous (2005) *Manure Composting Manual.* Livestock Engineering Unit and Environmental Practices Unit, Technical Services Division, Alberta Agriculture, Food, and Rural Development, Edmonton, Alberta, Canada.

Baker, H.G. (1974) The evolution of weeds. *Annual Review of Ecology and Systematics* 5, 1–24.

Bhowmik, P.C. (1994) Biology and control of common milkweed (*Asclepias syriaca*). *Reviews of Weed Science* 6, 227–250.

Bischoff, A. (2005) Analysis of weed dispersal to predict chances of re-colonization. *Agriculture, Ecosystems and Environment* 106, 377–387.

Blackshaw, R.E. and Rode, L.M. (1991) Effect of ensiling and rumen digestion by cattle on weed seed viability. *Weed Science* 39, 104–108.

Blanco-Moreno, J.M., Chamorro, L., Masalles, R.M., Recasens, J. and Sans, F.X. (2004) Spatial distribution of *Lolium rigidum* seedlings following seed dispersal by combine harvesters. *Weed Research* 44, 375–387.

Bough, M., Colosi, J.C. and Cavers, P.B. (1986) The major weed biotypes of proso millet (*Panicum miliaceum*) in Canada. *Canadian Journal of Botany* 64, 1188–1198.

Cash, S.D., Zamora, D.L. and Lenssen, A.W. (1998) Viability of weed seeds in feed pellet processing. *Journal of Range Management* 51, 181–185.

Clark, G.H. and Fletcher, J. (1906) *Farm Weeds of Canada.* Fisher, Ministry of Agriculture, Ottawa, Canada.

Comes, R.D., Bruns, V.F. and Kelley, A.D. (1978) Longevity of certain weeds and crop seeds in fresh water. *Weed Science* 26, 336–344.

Cooperband, L. (2002) *The Art and Science of Composting: A Resource for Farmers and Compost Producers.* Center for Integrated Agricultural Systems, University of Wisconsin-Madison, WI, USA.

Couvreur, M., Verheyen, K. and Hermy, M. (2005) Experimental assessment of plant seed retention times in fur of cattle and horse. *Flora* 200, 136–147.

Cudney, D.W., Wright, S.D., Shultz, T.A. and Reints, J.S. (1992) Weed seed in dairy manure depends on collection site. *California Agriculture* 46, 31–32.

Dastgheib, F. (1989) Relative importance of crop seed, manure, and irrigation water as sources of weed infestation. *Weed Research* 29, 113–116.

Davis, A., Renner, K., Sprague, C., Dyer, L. and Mutch, D. (2005) *Integrated Weed Management: 'One Year's Seeding …'.* Michigan State University Extension Bulletin E-2931, East Lansing, MI, USA.

Derscheid, L.A. and Schultz, R.E. (1960) Achene development of Canada thistle and perennial sowthistle. *Weeds* 8, 55–62.

Donald, W.W. and Ogg, A.G. (1991) Biology and control of jointed goatgrass (*Aegilops cylindrica*): a review. *Weed Technology* 5, 3–17.

Eghball, B. and Lesoing, G.W. (2000) Viability of weed seeds following manure windrow composting. *Compost Science and Utilization* 8, 46–53.

Forcella, F., Wilson, R.G., Renner, K.A., Dekker, J., Harvey, R.G., Alm, D., Buhler, D.D. and Cardina, J. (1992) Weed seed banks of the U.S. corn belt: magnitude, variation, emergence, and application. *Weed Science* 40, 636–644.

Gardener, C.J., McIvor, J.G. and Jansen, A. (1993) Passage of legume and grass seeds through the digestive tract of cattle and their survival in faeces. *Journal of Applied Ecology* 30, 63–74.

Ghersa, C.M., Martinez-Ghersa, M.A., Satorre, E.H., Van Esso, M.L. and Chichotky, G. (1993) Seed dispersal, distribution and recruitment of seedlings of *Sorghum halepense* (L.) Pers. *Weed Research* 33, 79–88.

Gill, N.T. (1938) The viability of weed seeds at various stages of maturity. *Annals of Applied Biology* 25, 447–456.

Grundy, A.C., Green, J.M. and Lennartsson, M. (1998) The effect of temperature on the viability of weed seeds in compost. *Compost Science and Utilization* 6, 26–33.

Guglielmini, A.C. and Satorre, E.H. (2004) The effect of non-inversion tillage and light availability on dispersal and spatial growth of *Cynodon dactylon*. *Weed Research* 44, 366–374.

Harmon, G.W. and Keim, F.D. (1934) The percentage and viability of weed seeds recovered in the feces of farm animals and their longevity when buried in manure. *Journal of American Society of Agronomy* 26, 762–767.

Harper, J.L. (1977) *Population Biology of Plants.* Academic Press, London, UK.

Hope, A. (1927) The dissemination of weed seeds by irrigation water in Alberta. *Scientific Agriculture* 7, 268–276.

Katovich, J., Becker, R. and Doll, J. (2006) *Weed Seed Survival in Livestock Systems.* University of Minnesota, Extension Service and University of Wisconsin Extension [see http://www.manure.umn.edu/assets/WeedSeedSurvival.pdf].

Kelley, A.D. and Bruns, V.F. (1975) Dissemination of weed seeds by irrigation water. *Weed Science* 23, 486–493.

Kissmann, K.G. (1997) *Plantas infestantes e nocivas*, 2nd edn. v.1. BASF, São Paulo, Brazil.

Kissmann, K.G. and Groth, D. (1999) *Plantas Infestantes e Nocivas*, 2nd edn. v.2. BASF, São Paulo, Brazil.

Kissmann, K.G. and Groth, D. (2000) *Plantas Infestantes e Nocivas*, 2nd edn. v.3. BASF, São Paulo, Brazil.

Larney, F.J. and Blackshaw, R.E. (2003) Weed seed viability in composted beef cattle feedlot manure. *Journal of Environmental Quality* 32, 1105–1113.

LeBlanc, L. and Doohan, D. (1992) The introduction and spread of velvetleaf (*Abutilon theophrasti* L) in Nova Scotia. *WSSA Abstracts* 32, 111.

Lorenzi, H. (2000) *Plantas daninhas do Brasil: terrestres, aquáticas, parasitas e tóxicas.* Instituto Plantarum, Nova Odessa, Brazil.

Matthews, J.M., Llewellyn, R., Geeves, T.G., Jaeschke, R. and Powles, S.B. (1996) Catching weed seeds at harvest: a method to reduce annual weed populations. In: *8th Australian Agronomy Conference*, Australian Society of Agronomy, pp. 684–685.

Mayer, F., Albrecht, H. and Pfadenhauer, J. (2002) Secondary dispersal of seeds in the soil seed bank by cultivation. *Journal of Plant Diseases and Protection* 18, 551–560.

McWhorter, C.G. (1989) History, biology, and control of johnsongrass. *Reviews of Weed Science* 4, 85–121.

Medeiros, D. (2001) Efeitos da palha de cana-de-açúcar (*Saccharum* spp.) sobre o manejo de plantas daninhas e dinâmica do banco de sementes. MSc thesis, Escola Superior de Agricultura 'Luiz de Queiroz', University of São Paulo, Piracicaba, Brazil.

Mouissie, A.M., Lengkeek, W. and Van Diggelen, R. (2005) Estimating adhesive seed-dispersal distances: field experiments and correlated random walks. *Functional Ecology* 19, 478–486.

Mt. Pleasant, J.M. and Schlather, K.J. (1994) Incidence of weed seed in cow (*Bos* sp.) manure and its importance as a weed source for cropland. *Weed Technology* 8, 304–310.

Mueller, J.P., Lewis, W.M., Green, J.T. and Burns, J.C. (1993) Yield and quality of silage corn as altered by johnsongrass infestation. *Agronomy Journal* 85, 49–52.

Muenscher, W.C. (1955) *Weeds*, 2nd edn. Cornell University Press, Ithaca, NY, USA.

Neto, M.S., Jones, R.M. and Ratcliff, D. (1987) Recovery of pasture seed ingested by ruminants. I. Seed of six tropical pasture species fed to cattle, sheep and goats. *Australian Journal of Experimental Agriculture* 27, 239–246.

Norris, R.F., Caswell-Chen, E.P. and Kogan, M. (2003) *Concepts in integrated pest management.* Prentice Hall, Upper Saddle River, NJ, USA, 586 pp.

Pidgeon, N.F. and Beattie, J. (1998) The psychology of risk and uncertainty. In: Calow, P. (ed.) *Handbook of Environmental Risk Assessment and Management.* Blackwell Science, Oxford, UK, pp. 289–318.

Radosevich, S.R. and Holt, J.S. (1984) Dispersal. In: Radosevich, S.R. and Holt, J.S. (eds) *Weed Ecology.* John Wiley and Sons, New York, USA, pp. 53–68.

Rew, L.J. and Cussans, G.W. (1997) Horizontal movement of seed following tine and plough cultivation: implications for spatial dynamics of weed infestations. *Weed Research* 37, 247–256.

Rizzardi, M.A., Vargas, L., Roman, E.S. and Kissmann, K.G. (2004) Aspectos gerais do manejo e controle de plantas daninhas. In: Vargas, L. and Roman, E.S. (eds) *Manual de manejo e controle de plantas daninhas.* Embrapa Uva e Vinho, Bento Gonçalves, Brazil, pp. 105–144.

Roberts, H.A. (1983) Weed seeds in horticultural soils. *Scientific Horticulture* 34, 1–11.

Roberts, H.A. and Stokes, F.G. (1966) Studies on the weeds of vegetable crops. VI. Seed populations of soil under commercial cropping *Journal of Applied Ecology* 3, 181–190.

Rupende, E., Chivinge, O.A. and Mariga, I.K. (1998) Effect of storage time on weed seedling emergence and nutrient release in cattle manure. *Experimental Agriculture* 34, 277–285.

Sebuka, A. and Deressa, A. (2000) Validating farmer's indigenous social networks for local seed supply in central Rift Valley of Ethiopia. *Journal of Agricultural Education and Extension* 6, 245–254.

Shields, E.J., Dauer, J.T., VanGessel, M.J. and Neumann, G. (2006) Horseweed (*Conyza canadensis*) seed collected in the planetary boundary layer. *Weed Science* 55, 185 185.

Shirtliffe, S.J. and Entz, M.H. (2005) Chaff collection reduces seed dispersal of wild oat (*Avena fatua*) by combine harvester. *Weed Science* 53, 465–470.

Slovic, P. (1987) Perception of risk. *Science* 36, 280–285.

Stanford, K., Wallins, G.L., Smart, W.G., Stanton, R., Piltz, J., Pratley, J., Kaiser, A., Hudson, D. and McAllister, T.A. (2000) Effects of feeding canola screenings on apparent digestibility, growth performance and carcass characteristics of feedlot lambs. *Canadian Journal of Animal Science* 80, 355–362.

Stanton, R., Piltz, J., Pratley, J., Kaiser, A., Hudson, D. and Dill, G. (2002) Annual ryegrass (*Lolium rigidum*) seed survival and digestibility in cattle and sheep. *Australian Journal of Experimental Agriculture* 42, 111–115.

Thill, D.C. and Mallory-Smith, C.A. (1997) The nature and consequence of weed spread in cropping systems. *Weed Science* 45, 337–342.

Thill, D.C., Zamora, D.L. and Kambitsch, D.L. (1986) The germination and viability of excreted common crupina (*Crupina vulgaris*) achenes. *Weed Science* 34, 273–241.

Tildesley, W.T. (1937) A study of some ingredients found in ensilage juice and its effect on the vitality of certain weed seeds. *Scientific Agriculture* 17, 492–501.

Tompkins, D.K., Chaw, D. and Abiola, A.T. (1998) Effect of windrow composting on weed seed germination and viability. *Compost Science and Utilization* 6, 30–34.

Wallander, R.T., Olson, B.E. and Lacey, J.R. (1995) Spotted knapweed seed viability after passing through sheep and mule deer. *Journal of Range Management* 48, 145–149.

Willms, W.D., Acharya, S.N. and Rode, L.M. (1995) Feasibility of using cattle to disperse cicer milkvetch (*Astragalus cicer* L.) seed in pastures. *Canadian Journal of Animal Science* 75, 173–175.

Wilson, R. (1980) Dissemination of weed seeds by surface irrigation water in western Nebraska. *Weed Science* 28, 87–92.

Woolcock, J.L. and Cousens, R. (2000) A mathematical analysis of factors affecting the rate of spread of patches of annual weeds in an arable field. *Weed Science* 48, 27–34.

Zahnley, J.W. and Fitch, J.B. (1941) Effect of ensiling on the viability of weed seeds. *Agronomy Journal* 33, 816–822.

Zamora, D.L. and Olivarez, J.P. (1994) The viability of seeds in feed pellets. *Weed Technology* 8, 148–153.

Zimdahl, R.L. (1999) *Fundamentals of Weed Science.* Academic Press, Fort Collins, CO, USA.

2 Understanding Weed–Crop Interactions to Manage Weed Problems

B.D. Maxwell[1] and J.T. O'Donovan[2]

[1]*Land Resources and Environmental Science Department, 334 Leon Johnson Hall, Montana State University, Bozeman, MT 59717, USA*; [2]*Grains and Oilseeds Team, Agriculture and Agri-Food Canada, Lacombe, AB, Canada T4L 1W1*

2.1 Introduction

Successful non-chemical weed management requires a keen knowledge of first principles of the ecology and biology of weeds. The first principles of a science are generalizations based on many empirical observations. Aristotle argued that first principles are a point in the development of knowledge when there is no need for proof because the first principles are self-evident. That is, that a particular consistent outcome (first principle) has a logical explanation. Thus, first principles become the basic tenets that form the foundation of a science discipline. In this chapter we seek to identify the first principles of weed ecology and biology that relate to weed–crop interactions and demonstrate how they may be used to assess weed management alternatives, including non-chemical approaches.

Imagine an agricultural system based on perfect knowledge of the impact of weeds on crop yield. Perfect knowledge would be an understanding of which weed species, under what conditions, have a quantifiable impact. For example, if one observed a patch of a particular weed species in a field, there would be a known amount of impact on crop yield. The weed-caused reduction in yield would be based on the distribution of density and spatial extent of the weed patch. One could then assign an economic cost of the weed based on crop yield reduction and compare that cost with the cost of weed management to make a decision on whether or not to impose management. Knowledge of the processes determining population dynamics would also be required in order to determine whether a decision to not manage the weed infestation would cause an unmanageable or costly problem in the future. These components of knowledge, first principles of weed ecology, would allow all forms of management, chemical or non-chemical, to be compared on the basis of economic outcomes.

What components of this perfect knowledge do we know and what do we not know? We have a relatively precise knowledge of the economic cost of different weed management practices. Although we know the general principles or factors that may determine weed occurrence, we have an imperfect or imprecise knowledge of when and where weeds may occur. We also have an imprecise knowledge of the impact of weeds on the crop, but we do know the best general quantitative methods to characterize the relationship between the weed infestation and crop yield. We know about the general processes that govern weed population dynamics, but we have an imprecise knowledge of all the factors that influence those processes. These generalizations about the relationships between weed infestation and crop yield loss and the factors that determine occurrence and population dynamics can be thought of as first principles.

Assessing the merits of any set of alternative weed management strategies (e.g. physical or cultural control methods) is dependent upon application of first principles in understanding the impact of a particular weed in a given crop production system. We use the first principles to guide us through gaining knowledge about the biology, ecology and agronomics of weed populations in a crop or rotations of crops. Non-chemical weed management is often reliant on a mix of tactics, some causing mortality and some that suppress populations. The complexity of these mixed tactics requires principles based on knowledge of plant population dynamics to predict the outcomes of management (Jordon, 1993). More broadly, one may begin to develop knowledge to predict outcomes from an assessment of fundamental ecological theories that form the foundation for how plant species interact.

The relative abundance of species in a weed–crop community provides one line of evidence for which common traits weed species possess for persistence and relative success, given the environment and history of management. Two prominent theories account for how species interact to determine species assemblages in plant communities. The first is based on Gause's competitive exclusion principle (Gause, 1932) and has evolved into *niche assembly theory*. This theory is based on the idea that plant species fitness and dominance is most fundamentally determined by the accumulation of traits associated with competition for resources (Tilman, 1982; Booth and Swanton, 2002). The second has its root in island biogeography (MacArthur and Wilson, 1967) and is based on the belief that all species are more or less neutral in their ability to capture resources. The membership and dominance of each species in a plant community are most fundamentally determined based on chance availability to the community and traits associated with dispersal. The latter has become known as the *unified neutral theory* (Hubbell, 2001). The importance of these theories to understanding weed–crop interactions is the formation of a platform of theory from which we may build the first principles of weed ecology (Maxwell and Luschei, 2004).

Farmers, even prior to biblical times, observed that the presence of weeds reduced crop yield, so it is logical to assume that competition for resources (niche assembly theory) may best explain the presence and persistence of any given weed species. In addition, it becomes clear that assessment of the weed–crop interaction is a simple task of pitting one against the other and measuring the outcome based on how well each accumulates biomass in the presence of the other. We will address the details of this approach in the next section. However, one should keep in mind that other processes, such as dispersal, should not be ignored for their contribution in determining the density and spatial extent of any given weed species and ultimately its economic importance. One can imagine that a species, although not very competitive, may still have a significant impact on a crop if it is widely distributed and present in high densities. Alternatively, it may interfere with harvesting or planting, or simply degrade the market value of the crop even when it is present at low densities. Regardless of the impact, consideration of all possible mechanisms of weed population regulation is crucial to developing the widest set of alternative management strategies (Maxwell and Luschei, 2004).

2.2 Experimental Designs and Regression Techniques to Study and Interpret Weed–Crop Interactions

The weed–crop interaction can be explored experimentally with a range of experimental designs depending on the specific objectives (Weigelt and Jolliffe, 2003). These include replacement series, addition series, neighbourhood and additive designs. The statistical analyses and modelling approaches used to describe the nature of the responses with data will be strongly influenced by the experimental design (Roush *et al.*, 1989; Cousens, 1991).

Replacement series design

If the objective is to account for the reciprocal impacts of the weed population on the crop, and the crop on the weed population, then a replacement series (DeWit, 1960) or an addition series design (Radosevich, 1988) would be appropriate. Replacement series and addition series experimental designs are used to quantify the impact of the weed on the crop yield and the crop on the weed seed production and/or vegetative reproduction. The replacement series design holds the total density of plants constant and varies the proportion of each species in a two-species mixture (Fig. 2.1) (DeWit, 1960).

A wide range of quantitative characterizations have been developed for replacement series data and these were thoroughly reviewed by Williams and McCarthy (2001). This experimental design has been widely applied to identify candidate intercrop systems, which has application to the first principle of filling available niches with desired species so that weeds are less likely to compete for resources. The primary limitation of this design is that the relationship between the two species can vary significantly by changing the total density of individuals (Firbank and Watkinson, 1985).

Addition series design

The addition series varies both the density and the proportion of each species in a full-factorial design to create a wide range of possible combinations of the two species (Fig. 2.2) (Connolly, 1986; Roush *et al.*, 1989).

The quantification of impacts using the data from the addition series experiment is based on the average response of individual plants (w) of each species (i) by dividing the amount of yield (biomass or seed produced) per unit area by the number of reproductive shoots. The average response per plant is then a non-linear function of the density of each species in the mixture (here shown for a two-species, i and j, mixture):

$$w_i = w_{max\,i}\left[a_i\left(N_i + \alpha_{ji} \cdot N_j\right)\right]^{-b_i} \quad (1)$$

$$w_j = w_{max\,j}\left[a_j\left(N_j + \alpha_{ij} \cdot N_i\right)\right]^{-b_j} \quad (2)$$

where N_i is density of species i; N_j is the density of species j; a, b and α are species-specific

parameters estimated with non-linear regression and have specific biological interpretations (Firbank and Watkinson, 1985). For example, the parameter α_{ji} is the equivalence of a single plant of species j relative to a plant of species i in terms of impact on the mean size of species i (w_i). More species can be added to the function by adding more α parameters with the corresponding density of each additional species. This approach to quantifying the impact of weed on crop and crop on weed can be regarded as a first principle based the reciprocal yield law. That is, mean individual plant weight will decline following a negative hyperbola with increasing population density so that the response can be made linear by taking the reciprocal of the mean plant weight (i.e. $b = -1$ in Eqns 1 and 2). Quantifying this response is limited by having to create a full range of mostly high densities of both species. It is very useful for quantifying the relative competitive abilities of different species in a crop–weed community, but its application to most agricultural situations, where weeds have been held to relatively low densities, may be limited. In addition, selecting plots within a production field, rather than experimentally creating the community conditions, should be avoided due to the bias that is created by lack of knowledge of the factors that may have been acting to create a particular community (e.g. mix or densities of different species).

The use of crop competition to discourage weeds is an important non-chemical weed management approach. Replacement series and addition series designs can be very useful in selecting intercrops, cover crops, and green manure crops that maximize crop production and at the same time maximize impact on the weeds. The addition series design can be modified to incorporate more than two species (Radosevich, 1988) and could be employed

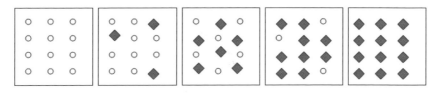

Fig. 2.1. Replacement series experimental design; empty circles represent weeds and filled diamonds represent crop plants.

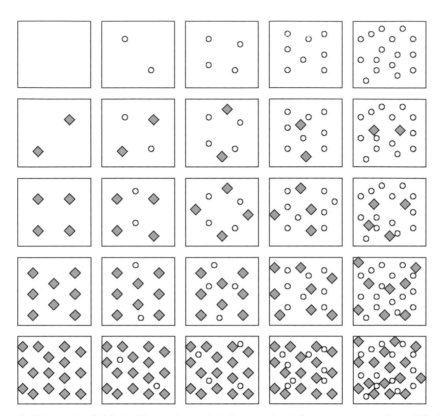

Fig. 2.2. Addition series full-factorial experimental design; empty circles represent weeds and filled diamonds represent crop plants.

more often to determine optimum crop mixtures (polycultures) to maximize impact on weeds.

Neighbourhood designs

Understanding the interactions among plant populations in a crop ecosystem when there are more than two species is increasingly difficult and inefficient as species are added into the experimental designs that have been discussed so far. One approach to overcome the limitations of the previously described designs is to change the focus from the population to individual plant responses to neighbouring plants (Stoll and Weiner, 2000). Experiments can be designed to quantify the impact of neighbouring plants on a target plant where the target can be a crop plant or a weed plant.

Individual-plant-centred experimental designs may provide a more fundamental understanding of the interactions between a target plant and its neighbouring plants as well as the influence of abiotic conditions on the interaction. Typically, target plants are selected and repeatedly measured through a growing season and the number, identity, distance to, spatial arrangement, relative time of emergence and size of neighbours is recorded and used to predict the growth rate, size, or reproductive output of the target plant (Weiner, 1982; Pacala and Silander, 1990; Bussler et al., 1995; Wagner and Radosevich, 1998). The target can be a crop or weed plant. The response variables are typically empirically assessed to identify the neighbourhood factors having an impact on the target. Thus, this method is useful for developing an understanding of how the spatial arrangement of the plants in a crop–weed community determines outcomes. This understanding can be applied to weed management as a first principle. For example, Lindquist et al. (1994) found several optimal spatial arrangements to maximize the yield of target plants in a competitive environment. This approach could be used to create crop planting arrangements that minimize intraspecific competition and maximize impact on the weeds.

Creating neighbourhood conditions experimentally has considerably greater flexibility compared with the previously described designs. Neighbourhood models may be more predictive of interplant interactions because they do not use average individual response from the population ($w = B/N$, where B is total plot biomass). Replication (i.e. selecting target plants) that represent the range of neighbourhood conditions (crop–weed communities) in a field is crucial for scaling up results to the field. Individual-plant-based models are increasingly applied in studies looking for sensitive measures to detect positive or negative interactions among plants, and represent a fruitful approach for discovering first principles associated with weed–crop interactions (Ellison *et al.*, 1994; Stoll and Weiner, 2000).

Additive design

Experiments that intentionally create a range of weed densities (plants/m^2) in plots where a crop is planted under standard agronomic conditions (seeding rate, distance between rows, fertilization,

etc) have been historically used to quantify weed impact on crop yield. The additive design is directly applicable to assessing weed–crop interactions in a way that can directly relate to weed management. The design is called an additive experiment (Fig. 2.3).

The proportional (or percentage) yield loss (*YL*) with increasing density of the weed (*Nw*) is calculated by dividing the yield at each density by the weed-free yield (Cousens, 1985a). The data are then fitted to a non-linear regression model (rectangular hyperbola):

$$YL = \frac{i \cdot Nw}{1 + i \cdot Nw / a} \qquad (3)$$

where *i* is the initial slope (proportional yield loss as weed density approaches zero) and *a* is the asymptote (maximum yield loss as weed density approaches infinity). There is a tendency to assume that the proportional yield decrease in a particular crop by a particular weed species will be consistent over time and space, allowing generalized weed management recommendations across regions where a particular crop–weed association occurs. Although the form of the non-linear response is usually consistent across different weed–crop relationships, there are two problems with its application. First, the weed-free

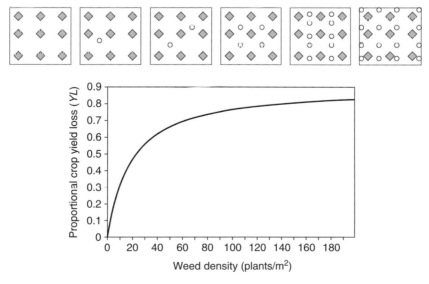

Fig. 2.3. The additive experimental design and the typical resultant hyperbolic crop yield loss (*YL*) relationship to increasing weed density; empty circles represent weeds and filled diamonds represent crop plants.

crop yield is often not the highest yield in the experiment. This can result in the calculation of negative values of YL at low densities. Second, the parameter values i and a are not always consistent over time and over regions where a particular crop–weed association occurs (Lindquist *et al.*, 1996; Jasieniuk *et al.*, 1999; Maxwell and Luschei, 2005; O'Donovan *et al.*, 2005). However, the consistency of the response can still be applied as a first principle, but it must be recognized that it may require local parameterization and experiments conducted over a number of years to estimate variation in the parameters over time. Then the quantification of the weed impact on the crop can effectively be applied to management decisions with an estimate of confidence in the recommendation.

A modification of the functional weed–crop relationship can be made to apply it under the assumption of local parameterization, where a third parameter, the maximum yield (Y_{max}) is fitted, and yield (Y) calculated directly (Fig. 2.4). Alternatively, Y_{max} can be set to the highest yield recorded in the experiment or field for a given year (Jasieniuk *et al.*, 2001).

$$Y = Y_{max} \left[1 - \frac{i \cdot Nw}{1 + i \cdot Nw / a} \right] \qquad (4)$$

Another limitation associated with the implementation of these relatively simple yield loss functions is the assumption that weed density is the paramount factor that should be considered when estimating crop yield loss due to weeds. This is not always the case. Other factors, including relative time of emergence of the crop and weed (Cousens *et al.*, 1987; Bosnic and

Swanton, 1997; O'Donovan and McClay, 2002), crop density (Carlson and Hill, 1985; Weaver and Ivany, 1998; O'Donovan *et al.*, 1999; Jasieniuk *et al.*, 2001), and even crop seed size (Stougaard and Xue, 2005), can influence YL or Y.

It makes intuitive sense that, regardless of weed density, the species that emerges first will have a competitive advantage over the other species. Thus, differences in relative times of emergence of weeds and crops may account, at least partly, for variability in Y_{max}, i and a across regions and years (O'Donovan *et al.*, 2005). This hypothesis was tested by re-parameterizing Eqn 2 to include a relative time of emergence parameter (Cousens *et al.*, 1987). The assumption is that if the weed emerges, for example, 2 days before the crop ($T = 2$) then it will have a significantly greater impact than if it emerges 2 days after the crop ($T = -2$) (Fig. 2.5). Relative emergence time is included in the negative hyperbola equation as follows:

$$Y = Y_{max} \left[1 - \frac{i \cdot Nw}{e^{cT} + i \cdot Nw / a} \right] \qquad (5)$$

where c is a new parameter influencing the impact of relative time of emergence, T is the number of days between emergence of the crop and weed, and e is the base of the natural log. Typically, T is calculated based on the difference in time (days) between when 50% of the weed population versus 50% of the crop population has emerged. Thus, at a given weed density (Nw), crop yield (Y) declines exponentially for every day the weed emerges ahead of the crop ($T < 0$) (Fig. 2.5). T can also be estimated based on the phenological stages of the crop and weed (Dew, 1980).

The response of Y or YL to relative time of emergence identifies an important first principle, pertinent to non-chemical weed management strategies. Improving crop competitiveness by adopting agronomic practices that ensure that crops emerge as early as possible ahead of weeds can maximize crop yield and minimize economic losses as well as reduce weed seed production. Early crop emergence, as a first principle, can be promoted by planting vigorous crop seed at relatively shallow depths as soon as possible after a weed-control operation. Otherwise, weed seed present in the soil may begin germinating even before the crop is planted, resulting in weeds

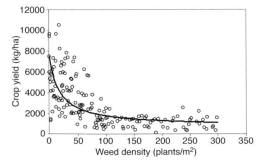

Fig. 2.4. Typical data scatter and Eqn 4 fit line for crop (barley) yield plotted against weed (wild oat) density.

emerging ahead of the crop with negative consequences for the crop. The critical period of weed control, defined by Nieto et al. (1968) as the time period during which weeds must be controlled to prevent yield losses, can be influenced considerably by relative time of emergence of the weed and crop. For example, delaying weed emergence by 3–5 weeks prevented significant yield losses in maize (Hall et al., 1992) and soybean (Van Acker et al., 1993).

Crop density is another important variable relevant to non-chemical weed management that should be included in weed–crop competition models. Crop density can vary within farmers' fields as well as across years and locations. For example, wheat plant density was found to differ by an average of 25 plants/m^2 in adjacent quadrats in farmers' fields, and wheat yield loss estimates due to weeds were highly distorted if crop density was not taken into account (Hume, 1985). Thus some of the reported instability associated with weed-density-dependent crop yield loss models (e.g. Lindquist et al., 1996; Jasieniuk et al., 1999; Maxwell and Luschei, 2005) may have occurred because of inconsistent crop density.

Several regression models have been proposed to describe the relationship between crop yield and both weed and crop density

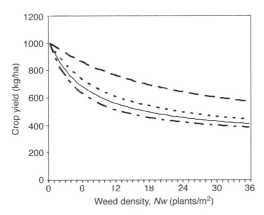

Fig. 2.5. The influence of weed emergence time, relative to the crop, on crop yield over a range of weed densities. If $T = 21$ (dotted and dashed line) the weed emerged 1 day ahead of the crop, if $T = 0$ (solid line) the weed emerged on the same day as the crop, if $T = 1$ (dotted line) the weed emerged 1 day after the crop, and if $T = 4$ (dashed line) the weed emerged 4 days after the crop.

(Carlson and Hill, 1985; Cousens, 1985b; Martin et al., 1987; Weaver and Ivany, 1998; O'Donovan et al., 1999; Jasieniuk et al., 2001). These generally involve modifications of the crop yield functions (Eqn 4) and can provide a way to quantify the weed–crop interaction that can be more directly applied to assessment of management alternatives under the typical weed and crop densities found in agronomic systems. The addition series design (described above) is an effective approach for creating data sets to apply these new functions, but care should be taken to ensure a set of densities of the crop and weed to cover the range from very low to very high.

The functional form for crop yield can be modified to represent a double hyperbola. Thus, a positive hyperbola is substituted for Y_{max} in Eqn 4 to account for increasing crop yields with increasing densities or seeding rates of the crop (Nc) (Jasieniuk et al., 2001).

$$Y = \frac{j \cdot Nc}{1 + j \cdot Nc / Y_{max}} \left[1 - \frac{i \cdot Nw}{1 + i \cdot Nw / a} \right] \quad (6)$$

This model was evaluated against two other models and found to be superior for the weed damage function in that it was less likely to cause bias in yield predictions based on data sets of minimum size (Jasieniuk et al., 2001). This is important, since experiments required to fully populate a data matrix that is appropriate for accurate estimates of many parameters (e.g. weed and crop density, and relative time of emergence) can be too large to be logistically manageable. Yield predictions from Eqn 6 and other published equations indicate that, at a given weed density, crop yield increases with increasing crop density. However, Eqn 6 does not adequately capture the influence of increasing crop density on weed density impact on yield, which can be a critical component of a non-chemical weed management approach. Therefore the equation is modified by including total plant density ($Nw + Nc$) into the denominator of the proportional yield loss part of Eqn (6):

$$Y = \frac{j \cdot Nc}{1 + j \cdot Nc / Y_{max}} \left[1 - \frac{i \cdot Nw}{1 + i(Nw + Nc) / a} \right] \quad (7)$$

The new equation (Eqn 7) accurately reflects the combined effect of crop and weed density on crop yield (Fig. 2.6). This is an example of another important first principle for non-chemical weed management. Increased crop density, achieved by increasing the seeding rate, can reduce the impact of the weed on the crop.

The additive experiment design allows for assessment of the biological interaction between a crop and a weed species. However, it is based on the assumption that the interaction is determined by competition for resources and that the result will be a negative impact of the weed on the crop. There may be situations where the crop could gain a net benefit from the presence of weeds because they are harbouring beneficial insects (crop pollinators, pest predators, etc) or providing other positive feedbacks (Andow, 1988).

The experimental approaches described above can provide data to quantify first principles for assessing weed–crop interactions, but have largely been applied to the case of a single-species weed community. This may seem like a restricted case with marginal relevance, because most crop fields have more than one weed species. However, one may argue that the prevalence of high aggregation (patchiness) in managed weed communities (Dieleman *et al.*, 2000) restricts most weed–crop interactions to a two-species competition for resources in a large proportion of production fields.

One must be appropriately sceptical when

applying the results of plot experiments to production fields because there is generally a greater range of crop yield responses for any given weed density under field conditions than from plots in an experiment (Jordon, 1993). This is especially true in systems or locations where resources are most limiting. Under extreme resource-limited conditions, plant species diversity will tend to be greater in areas where resources are present with the greatest availability (Stohlgren *et al.*, 2003). Thus, one may hypothesize that weeds growing with crops under highly limited resource conditions may also correlate with resource availability and that crop yields are also likely to be highest in the areas of the fields with greatest resources. Crop yields plotted against weed densities in fields under extreme resource limitations can be very noisy and it is often impossible to identify the negative hyperbola signal (Maxwell and Luschei, 2005). In summary, the interpretation of the first-principle negative hyperbola response from plot experiments will be most useful when one allows for the uncertainty associated with extrapolating the results across regions and years, especially when using the response to inform management decisions.

2.3 Making Weed Management Decisions Based on Weed Density Thresholds

The weed economic threshold or economic injury level (EIL) can be defined as the weed infestation (density) at which the weed management costs equal the value of the recovered crop yield. Thus, at densities below the EIL there is no economic incentive to manage the weed, and when the weed density exceeds the EIL the cost of the weed management is more than offset by the return in extra crop yield. Quantifying weed impacts is important because it allows managers to assess management alternatives. Although the assessments of management alternatives are generally based on economics, they could also be used to evaluate the management alternatives based on the environmental impacts of the weed or the practices used to manage the weed. For example, if tillage is used to control weeds and this causes increased silting of adjacent streams, the cost of

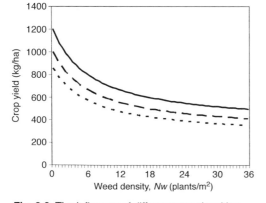

Fig. 2.6. The influence of different crop densities (Nc; plants/m^2) on crop yield (Y; kg/ha) response to weed density (Nw; plants/m^2), where Nc = 150 (dotted line), 200 (dashed line) and 300 plants/m^2 (solid line).

mitigating the siltation could be set against any increased crop yield value from weed removal to determine the break-even point (EIL).

Single-season application of first principles

The decision to manage a weed population can be based on whether or not the population exceeds some economic or environmental impact threshold in a single growing season. The concept is usually applied to conventional cropping systems to help rationalize herbicide application for weed control, but could just as well be applied to non-chemical weed management that may include cultural practices such as tillage and flaming, or agronomic strategies such as increased crop seeding rates or early versus late planting times. For example, one can calculate the density of a weed population that causes enough impact so that the crop value gained by removing the weed is equal to the cost of the management using the net return equation:

$$NR = (Y \cdot P) - (H + W) \qquad (8)$$

where NR ($/ha) is the net return per unit area to the producer, Y is the crop yield (e.g. kg/ha), P ($/kg) is the price received per unit of crop yield, W ($/ha) is the cost of weed control per unit area, and H ($/ha) represents all other costs required to produce the crop. The EIL is calculated by determining when net return with weed control (NR_W) is equal to net return with no weed control (NR_0) and can be graphically identified as the intersection of NR_W and NR_0 when plotted against weed density (Fig. 2.7).

In the case presented above (see Fig. 2.7), the EIL was 12 plants/m^2, so that if the density of this weed was below the EIL (<12 plants/m^2) then the producer would be losing money by applying the selected management approach. If the density of the weed was greater than the EIL, then the producer would increase net returns using the selected management approach. Therefore, the calculation of the EIL allows assessment of different management alternatives based on the weed density found in a given field and the decrease in weed density that results from the selected form of management. The EIL is therefore a good example of a first principle that extends the mathematical

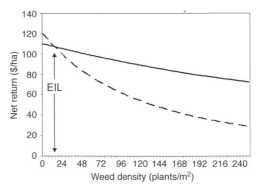

Fig. 2.7. Net return ($/ha) over a range of weed densities when weed management is (solid line) or is not imposed (dashed line). The density corresponding to the point where the two lines intersect is the economic injury level (EIL).

characterization of the weed–crop interaction to weed management decisions. Several computerized weed management decision support systems based on the EIL concept are now available, some of which are accessible through the world-wide-web (Wilkerson *et al.*, 2002).

In conventional (herbicide-dependent) cropping systems, the EIL has not been widely adopted as a tool for aiding management decisions (Wilkerson *et al.*, 2002). This is partly because the EIL does not take into account future weed population if there is a decision not to manage (i.e. the weed population is below the EIL). Many farmers are reluctant to intentionally leave weeds in a field because of their perceived future threat. In addition, the dominance of chemical methods of weed control has created expectations of complete removal of weeds. Non-chemical methods are not typically as thorough at weed removal, but to be realistically regarded as an alternative, they need to be quantitatively assessed for their effect on current and future weed populations. Weed reproduction is important for determining future population impacts if all of the weeds are not removed by a management practice or if there was a decision to not manage the weeds because the population was below the EIL.

Long-term application of first principles

The single-season EIL concept has been described as an unsound management practice because the long-term implications of seed produced by weed populations below the EIL is not considered (Norris, 1992; Sattin et al., 1992). Thus there have been attempts to develop models that account for seed production by uncontrolled weeds and estimate an EIL that optimizes returns over a number of years (Cousens et al., 1986; Doyle et al., 1986; Baur and Mortensen, 1992). This concept has been referred to as the 'economic optimum threshold' and can be as much as four times lower than the single-season threshold (Cousens et al., 1985). There is a need to incorporate the future weed population that can result from any particular management practice in order to choose an optimum approach (Holst et al., 2007), whether the optimum is based on economics or other selected outcomes. Thus, weed population dynamics in response to management decisions, particularly the decision not to manage, is important for determining density thresholds to optimize net return over a given time horizon.

To demonstrate the use of predicted population dynamics to calculate profit-maximizing management, a highly simplified weed population projection model (Freckleton and Watkinson, 1998) can be used:

$$Nw_{t+1} = p \cdot \left(Nw_t + SP_t\right) \cdot p_{wm} \qquad (9)$$

where Nw_t is the density of the weed population in the seed bank at the beginning of the growing season, p is the average probability of an individual weed seed surviving from one growing season to the next, and p_{wm} is the average probability of a plant surviving to reproduction following weed management. Weed seed production per unit area (SP) in the current growing season (t) can be estimated as follows:

$$SP = \frac{k \cdot Nw}{1 + k \cdot Nw / SP_{max}} \left[1 - \frac{q \cdot Nc}{1 + q \cdot (Nc + Nw) / z}\right] \qquad (10)$$

where SP_{max}, k, q and z are all estimated with non-linear regression analysis. For the purpose of demonstration, one can use the simplifying assumptions that the crop is always planted to achieve density Nc and the crop and weed always emerge at the same time, then SP and Y will strictly be a function of weed density (Nw_t) in the current generation. One can then calculate a long-term economic threshold or a

maximum net return for a management approach that causes $1 - p_{wm}$ mortality in the weed population over a specified time period. For example, solving Eqns 9, 7, 8 and 10 in sequence will determine the initial weed density (N_0) below which the population will not require management for the specified time in order to maximize net returns.

To calculate net return without management (NR_{no}):

$$Nw_t = p \cdot \left(Nw_{t-1} + SP_{t-1}\right)$$

$$Y = \frac{j \cdot Nc}{1 + j \cdot Nc / Y_{max}} \left[1 - \frac{i \cdot Nw}{1 + i(Nw_t + Nc) / a}\right]$$

$$NR_{no} = (Y \cdot P) - (H)$$

$$SP_t = \frac{k \cdot Nw_t}{1 + k \cdot Nw_t / SP_{max}} \left[1 - \frac{q \cdot Nc}{1 + q(Nc + Nw) / z}\right]$$

To calculate net return with management (NR_{wm}):

$$Nw_t = p \cdot \left(Nw_{t-1} + SP_{t-1}\right) \cdot p_{wm}$$

$$Y = \frac{j \cdot Nc}{1 + j \cdot Nc / Y_{max}} \left[1 - \frac{i \cdot Nw}{1 + i(Nw_t + Nc) / a}\right]$$

$$NR_{wm} = (Y \cdot P) - (H + W)$$

$$SP_t = \frac{k \cdot Nw_t}{1 + k \cdot Nw_t / SP_{max}} \left[1 - \frac{q \cdot Nc}{1 + q(Nc + Nw) / z}\right]$$

Therefore, N_0 can be identified by calculating all possible combinations of years with, and years without, weed management for a specified time (number of years) to determine the highest N_0 that maximizes net return when no weed management is applied for each year of the specified time period. This first-principle model produces reasonable projections of weed population dynamics (Fig. 2.8) and weed impact (Fig. 2.9) and thus allows the comparison of different weed management practices based on their potential to augment the weed mortality rate ($1 - p_{wm}$) and decrease the weed seed production caused by management.

Several weed management generalizations result from this combination of first principles included in the long-term threshold model. First, the single-year weed density threshold for the decision to manage or not to manage based on economic net return (EIL) is lower when multiple years are included in the calculation. Second, the weed density threshold for maximizing net return

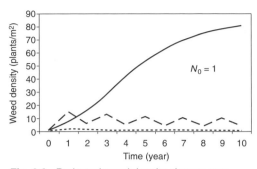

Fig. 2.8. Projected weed density change over time for a range of weed management strategies starting from a population density of 1 plant/m². No weed management (solid line), weed management alternate years (dashed line), weed management every year (dotted line).

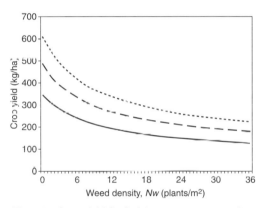

Fig. 2.9. Crop yield (kg/ha) in response to weed density (Nw; plants/m²) at three crop densities (Nc): 100 (solid line), 150 (dashed line) and 200 plants/m² (dotted line).

over several cropping years increases with increasing crop density. Third, weed management practices that reduce the current year weed density (e.g. tillage, herbicide, flaming, etc), combined with increased crop seeding rates, will maximize profits when applied the first few years, but are dropped near the end of the specified time horizon for calculating accumulated net returns. Fourth, decreasing weed reproduction at low weed densities is critical for maximizing net returns. Non-chemical management that best accomplishes reduced weed reproduction includes rotation to highly competitive crops that compete for resources, thereby minimizing weed seed production and increasing mortality rates

through the disruption of weed life cycles. We arrived at these outcomes by systematically varying the parameter values in the first-principle models associated with each process.

These generalized results emphasize the important first principles related to non-chemical weed management. For example, increasing crop density, achieved by increasing seeding rates, decreases the impact of the weed on the crop. The dynamic interactions involved in determining the long-term thresholds and incremental production advantage from increasing crop density are complex. Four processes, characterized as first principles, interact to determine the outcome of increased crop density:

1. Increasing crop density increases the crop yield per unit area, but the increase becomes small as constant final yield is approached (see left side of Eqn 7)
2. Increasing crop density decreases the proportional impact of a fixed density of weeds on the crop yield (see right side of Eqn 7)
3. Increased crop density decreases weed seed production directly through competition (see right side of Eqn 10)
4. Decreased weed seed production is further decreased by density-independent mortality from natural factors (p) and management (p_{wm}) (Eqn 9).

First-principle model outcomes have generally been supported with empirical studies. Blackshaw (1993) provided an excellent demonstration of the negative influence of crop density on weed abundance. Mohler (2001) provided an exhaustive list of cases where weed responses to crop density were assessed. In addition, crop density has a strong impact on future weed densities by decreasing weed reproduction (Mohler, 1996). Empirical studies, although critical for developing first principles, are logistically limited for assessing the complex interactions that can be studied more effectively with models that integrate the first principles.

Limitations to implementation of the threshold concept

The weed density threshold concept is an excellent example of a first principle that directly relates to weed management decisions. However, limited knowledge about the true risk of leaving weeds unmanaged when the density is below the threshold, how multiple weed species may interact to determine the threshold for any of the species, and how to determine on what spatial scale it should be implemented, all serve to limit its current utility. Research that identifies the first principles that drive these limitations will improve utilization of the threshold concept.

2.4 Single Versus Multiple Weed Species Populations

Virtually all fields that are seeded to crops contain more than one weed species. Weed species diversity has been shown to be greater in non-chemical than in conventional cropping systems (Menalled *et al.*, 2001; Hyvonen *et al.*, 2003; Davis *et al.*, 2005). However, most weed interference experiments have tended to investigate the effects of a single weed species on crop yield. The standard additive design experiments (Fig. 2.3) make it difficult to include multiple species into calculations of the EIL or other thresholds. Individual-plant-based neighbourhood assessment of impacts under high, locally diverse, weed communities may offer improved assessments of weed–crop interactions. Scaling responses from individual plants in a neighbourhood to a scale where management decisions are made will require further research to determine adequate sampling to best estimate the true distribution of neighbourhood types.

2.5 Making Estimates of Weed and Crop Density to Insert into Models

Weed density is a very important component of weed–crop interference models (Weaver, 1996; Swanton *et al.*, 1999). However, assessing weed density over large areas can be difficult and time-consuming (O'Donovan, 1996). The suggested number of sampling units required to achieve accurate estimates of weed density vary considerably. For example, in one study conducted in Germany it was suggested that as few as 20–30 random quadrats in a 3–5 ha field would be sufficient (Gerowitt and Heitefuss, 1990), while another conducted in the UK indicated that precise estimates of grass weed density could be

achieved only at a high sampling intensity of at least 18 locations per hectare (Marshall, 1988). It is unlikely that this high sampling intensity would be cost-effective, especially in non-chemical systems where weed densities and species diversity would probably be high.

Sampling can also be confounded by the fact that weeds in a field are rarely distributed randomly, but tend to be patchy or aggregated (Wiles *et al.*, 1993). In the UK, crop yield loss estimates based on random and aggregated weed distributions were compared (Brain and Cousens, 1990) and it was concluded that, under most farm conditions, models that assume a random distribution would overestimate crop yield loss predictions. A soybean decision model (HERB) was modified to incorporate uncertainty due to weed patchiness (Wiles *et al.*, 1993). This resulted in improved decision making in most, but not all, cases. Identifying the appropriate scale to assess weed populations and apply management in fields will be an important aspect of developing useful management decision aids.

There is an obvious need for increased research effort to determine realistic weed monitoring procedures for crop production systems if the EIL concept is to be accepted more readily by growers. Research on weed sampling methods based on digital image analysis is showing promise for identifying and locating some species (Gerhards and Oebel, 2006). Calculation of weed density thresholds in systems with non-chemical management is generally more constrained by a less aggregated distribution of weed populations and weed species diversity is generally higher than in conventional cropping systems (Davis *et al.*, 2005). These constraints should not eliminate the use of thresholds in non-chemical weed management systems; it simply means that greater knowledge is required in order to implement them with a level of certainty that will be acceptable to farmers.

2.6 Empirical Versus Mechanistic Modelling to Understand First Principles

Thus far, we have mostly described empirical models and their application as first principles. A more complete review of empirical models and their advantages and disadvantages can be found in the literature (Cousens, 1991; Cousens and Mortimer, 1995; Holst *et al.*, 2007). In addition, we feel compelled to introduce more mechanistic models that have been used in order to develop a more fundamental understanding of weed–crop interactions. These models attempt to capture physiological processes and morphological characteristics associated with acquiring and allocating resources for plant growth as first principles. Models that include the critical processes associated with competition logically should account for the variation in the interaction between crops and weeds (Graf *et al.*, 1990; Ryel *et al.*, 1990; Wilkerson *et al.*, 1990; Kiniry *et al.*, 1992; Kropff *et al.*, 1994; Caton *et al.*, 1999). Kropff and Spitters (1991) offered a more mechanistic form of a standard weed–crop empirical model that accounted for time of emergence based on the relative development of leaf area of the crop and weed. Kropff and van Laar (1993) went on to develop a physiologically based weed–crop model that explained the empirical results. This was an excellent example of the power of developing more mechanistic approaches. Although not concentrating on weed–crop interactions, Tilman (1982) has developed mechanistic models that explain the behaviour of species interacting through a resource pool in a plant community. The models were successful under single limited resource conditions of the systems where they conducted their research, but have not been widely tested elsewhere. Lindquist (2001) tested the robustness of the INTERCOM model in maize fields infested with velvetleaf (*Abutilon theophrasti*) across the north-central region of the USA and found encouraging results for prediction of maize yield loss and subsequent EIL estimates for velvetleaf. It has not been demonstrated that the physiological growth simulation models perform any better than empirical models for making management decisions (Deen *et al.*, 2003), but their utility in gaining a more complete understanding of first-principle processes is clearly their most attractive attribute.

2.7 Application of First Principles to Make Weed Management Recommendations

Applying crop–weed interaction first principles to the management of weeds can take two different philosophical approaches. First, one can simply try to apply the first principle to field data and quantitatively identify management options and make a prescriptive management recommendation. Second, one can apply the first principles as trends or generalities to make recommendations. The first approach is effective when management specifically suppresses the variation in the crop–weed interactions so that the response is more predictable. Using a high-efficacy herbicide to impose a high level of mortality on seedling weeds results in a reduction in variation in the crop–weed interaction and the outcome is very predictable. In addition, the effects of the treatments are so overwhelming that results from small plot experiments can be extrapolated to fields near and far with relatively high certainty. However, with many non-chemical approaches, which more gently impose stress on the weeds, the variation in crop response often obscures the first principles, making it difficult to predict the outcomes of the crop–weed interaction or weed population dynamics. Maxwell and Luschei (2005) demonstrated, with simulation models, that predicting weed impact on crop yield with an abundance of small-scale experimental plots within a relatively short distance of a farm field would only predict crop response with high uncertainty. Uncertainty in predictions about the crop–weed interaction should not be interpreted as a failing of the first principles or the way in which they are included in the models. One should simply learn to incorporate uncertainty into the recommendations from the models. Recommendations should be made as probabilities of outcomes using the variation observed in experiments and devising ways to estimate the variability that may join with extrapolation in time (over years) and space (out to farms) to increase confidence in the probabilistic recommendation.

The second approach draws on the knowledge that variability is high and therefore certainty of outcomes is low. Non-chemical weed management recommendations that involve biotic interactions are likely to be risky if they are prescriptive (Maxwell and Luschei,

2004). Therefore, the first-principle relationships may be best used to recommend general rules of management rather than prescription. For example, if an experiment indicates that the crop is more competitive than the weed, one could make the recommendation that increased crop seeding rates may decrease future weed pressure. However, it would not be wise to use specific seeding rates and subsequent quantified crop yield returns from an experiment as specific recommendations for farms, even if they were in close proximity to the experiment. A first principle of non-chemical weed management could be that the higher the uncertainty in a crop–weed interaction, the more management emphasis should be placed on weed prevention and less on causing weed mortality in the crop.

Farmers, managers, crop consultants and agriculture industry representatives are accustomed to receiving prescriptive recommendations; therefore there must be a shift in understanding that recommendations of non-chemical weed management approaches are inherently different. The recommendations must shift away from the tools of management and focus more on how to most effectively and efficiently influence the biology of the target species (Jordon, 1993). The application of the tools and reacting to the weeds must become secondary to understanding the biology of the crop–weed interaction. Understanding the biology has a first-principle component (generalities about the behaviour and impacts of the weed species) and a local component. The local component is the farmer or manager developing a keen knowledge of the site-specific biology of their system through observations and measurements. Weed scientists must develop methods for discovering the first principles and methods to enable the practitioners to efficiently gain an understanding of their site-specific weed biology in order to allow effective decisions about the application of management tools.

2.8 References

Andow, D.A. (1988) Management of weeds for insect manipulation in agroecosystems. In: Altieri, M.A. and Liebman, M. (eds) *Weed Management in Agroecosystems: Ecological Approaches.* CRC, Boca Raton, FL, USA, pp. 265–301.

Baur, T.A. and Mortensen, D.A. (1992) A comparison of economic and economic optimum thresholds for two annual weeds in soybeans. *Weed Technology* 6, 225–235.

Blackshaw, R.E. (1993) Safflower (*Carthamus tinctorius*) density and row spacing effects on competition with green foxtail (*Setaria viridis*). *Weed Science* 41, 403–408.

Booth, B.D. and Swanton, C.J. (2002) Assembly theory applied to weed communities. *Weed Science* 50, 2–13.

Bosnic A.C. and Swanton, C.J. (1997) Influence of barnyardgrass (*Echinochloa crus-galli*) time of emergence and density on corn (*Zea mays*). *Weed Science* 45, 276–282.

Brain, P. and Cousens, R. (1990) The effect of weed distribution on predictions of yield loss. *Journal of Applied Ecology* 27, 735–742.

Bussler, B.H., Maxwell, B.D. and Puettmann, K.J. (1995) Using plant volume to quantify interference in corn (*Zea mays*) neighborhoods. *Weed Science* 43, 586–594.

Carlson, H.L. and Hill, J.E. (1985) Wild oat (*Avena fatua*) competition in spring wheat: plant density effects. *Weed Science* 33, 176–181

Caton, B.P., Foin, T.C. and Hill, J.E. (1999) A plant growth model for integrated weed management in direct-seeded rice. I. Development and sensitivity analyses of monoculture growth. *Field Crops Research* 62, 129–143.

Connolly, J. (1986) On difficulties with replacement-series methodology. *Journal of Applied Ecology* 23, 125–137.

Cousens, R. (1985a) A simple model relating yield loss to weed density. *Annals of Applied Biology* 107, 239–252.

Cousens, R. (1985b) An empirical model relating crop yield to weed and crop density and a statistical comparison with other models. *Journal of Agricultural Science* 105, 513–521.

Cousens, R. (1991) Design and interpretation of competition experiments. *Weed Technology* 5, 664–673.

Cousens, R. and Mortimer, M. (1995) *Dynamics of Weed Populations.* Cambridge University Press, Cambridge, UK, 332 pp.

Cousens, R., Wilson, B.J. and Cussans, G. (1985) To spray or not to spray: the theory behind the practice. In: *Proceedings of the British Crop Protection Conference, Weeds* 3, pp. 671–678.

Cousens, R., Doyle, C.J., Wilson, B.J. and Cousens, G.W. (1986) Modelling the economics of controlling *Avena fatua* in winter wheat. *Pesticide Science* 17, 1–12.

Cousens, R., Brain, P., O'Donovan, J.T. and O'Sullivan, P.A. (1987) The use of biologically realistic equations to describe the effect of weed density and relative time of emergence on crop yield. *Weed Science* 35, 720–725.

Davis, A.S., Renner, K.A. and Gross, K.L. (2005) Weed seedbank and community shifts in a long-term cropping systems experiment. *Weed Science* 53, 296–306.

Deen, W., Cousens, R., Warringa, J., Bastiaans, L., Carberry, P., Rebel, K., Riha, S., Murphy, C., Benjamin, L.R., Cloughley, C., Cussans, J., Forcella, F., Hunt, T., Jamieson, P., Lindquist, J. and Wang, E. (2003) An evaluation of four crop : weed competition models using a common data set. *Weed Research* 43, 116–129

Dew, D.A. (1980) Relationship between leaf stages of cereal crops on grassy weeds and days from seeding. *Canadian Journal of Plant Science* 60, 1263–1267.

DeWit, C.T. (1960) On competition. *Verslagen van Landbouwkundige Onderzoekingen* 66, 1–82.

Dieleman, J.A., Mortensen, D.A., Buhler, D.D. and Ferguson, R.B. (2000) Identifying associations among site properties and weed species abundance. II. Hypothesis generation. *Weed Science* 48, 576–587.

Doyle, C.J., Cousens, R. and Moss, S.R. (1986) A model of the economics of controlling *Alopecurus myosuroides* Huds. in winter wheat. *Crop Protection* 5, 143–150.

Ellison, A.M., Dixon, P.M. and Ngai, J. (1994) A null model for neighborhood models of plant competitive interactions. *Oikos* 71, 225–238.

Firbank, L.G. and Watkinson, A.R. (1985) On the analysis of competition within two-species mixtures of plants. *Journal of Applied Ecology* 22, 503–517.

Freckleton, R.P. and Watkinson, A.R. (1998) Predicting the determinants of weed abundance: a model for the population dynamics of *Chenopodium album* in sugar beet. *Journal of Applied Ecology* 35, 904–920.

Gause, G.F. (1932) Experimental studies on the struggle for existence. *Journal of Experimental Biology* 9, 389–402.

Gerhards, R. and Oebel, H. (2006) Practical experiences with a system for site-specific weed control in arable crops using real-time image analysis and GPS-controlled patch spraying. *Weed Research* 46, 185–193.

Gerowitt, B. and Heitefuss, R. (1990) Weed economic thresholds in cereals in the Federal Republic of Germany. *Crop Protection* 9, 323–331.

Graf, B., Gutierrez, A.P., Rakotobe, O., Zahner, P. and Delucchi, W. (1990) A simulation model for the dynamics of rice growth and development. Part II. The competition with weeds for nitrogen and light. *Agricultural Systems* 32, 367–392.

Hall, M.R., Swanton, C.J. and Anderson, G.W. (1992) The critical period of weed control in grain corn (*Zea mays*). *Weed Science* 40, 441–447.

Holst, N., Rasmussen, I.A. and Bastiaans, L. (2007) Field weed population dynamics: a review of model approaches and applications. *Weed Research* 47, 1–14.

Hubbell, S.P. (2001) *The Unified Neutral Theory of Biodiversity and Biogeography*. Monographs in Population Biology. Princeton Press, Princeton, NJ, USA.

Hume, L. (1985) Crop losses in wheat (*Triticum aestivum*) as determined using weeded and non-weeded quadrats. *Weed Science* 33, 734–740.

Hyvonen, T., Ketoja, E., Salonen, J., Jalli, H. and Tiainen, J. (2003) Weed species diversity and community composition in organic and conventional cropping of spring cereals. *Agriculture, Ecosystems, and Environment* 97, 131–149.

Jasieniuk, M., Maxwell, B.D., Anderson, R.L., Evans, J.O., Lyon, D.J., Miller, S.D., Morishita, D.W., Ogg, A.G., Jr, Seefeldt, S.S., Stahlman, P.W., Northam, F.E., Westra, P., Kebede, Z. and Wicks, G.A. (1999) Site-to-site and year-to-year variation in *Triticum aestivum–Aegilops cylindrica* interference relationships. *Weed Science* 47, 529–537.

Jasieniuk, M., Maxwell, B.D., Anderson, R.L., Evans, J.O., Lyon, D.J., Miller, S.D., Morishita, D.W., Ogg, A.G., Jr, Seefeldt, S.S., Stahlman, P.W., Northam, F.E., Westra, P., Kebede, Z. and Wicks, G.A. (2001) Evaluation of models predicting winter wheat yield as a function of winter wheat and jointed goatgrass densities. *Weed Science* 49, 48–60.

Jordon, N. (1993) Prospects for weed control through crop interference. *Ecological Applications* 3, 84–91.

Kiniry, J.R., Williams, J.R., Gassman, P.W. and Debaeke, P. (1992) A general, process oriented model for two competing plant species. *Transactions of the American Society of Agricultural Engineers* 35, 801–810.

Kropff, M.J. and Spitters, C.J.T. (1991) A simple model of crop loss by weed competition from early observations on relative leaf area of the weeds. *Weed Research* 31, 97–105.

Kropff, M.J. and van Laar, H.H. (1993) *Modelling Crop–Weed Interactions*. CAB International, Wallingford, UK.

Kropff, M.J., Moody, K., Lindquist, J.L., Migo, T.R. and Fajardo, F.F. (1994) Models to predict yield loss due to weeds in rice ecosystems. *Philippines Journal of Weed Science* (Special Issue), 29–44.

Lindquist, J.L. (2001) Performance of INTERCOM for predicting corn–velvetleaf interference across north-central United States. *Weed Science* 49, 195–201.

Lindquist, J.L., Rhode, D., Puettmann, K.J. and Maxwell, B.D. (1994) The influence of plant population spatial arrangement on individual plant yield. *Ecological Applications* 4, 518–524.

Lindquist, J.L., Mortensen, D.A., Clay, S.A., Schmenk, R., Kells, J.J., Howatt, K. and Westra, P. (1996) Stability of corn (*Zea mays*)–velvetleaf (*Abutilon theophrasti*) interference relationships. *Weed Science* 44, 309–313.

MacArthur, R.H. and Wilson, E.O. (1967) *The Theory of Island Biogeography*. Monographs in Population Biology. Princeton University Press, Princeton, NJ, USA.

Marshall, E.J.P. (1988) Field estimates of grass weed populations in arable land. *Weed Research* 28, 191–198.

Martin, R.J., Cullis, B.R. and McNamara, D.W. (1987) Prediction of wheat yield loss due to competition by wild oats (*Avena* spp.). *Australian Journal of Agricultural Research* 38, 487–499.

Maxwell, B.D. and Luschei, E. (2004) The ecology of crop–weed interactions: toward a more complete model of weed communities in agroecosystems. *Journal of Crop Improvement* 11, 137–154.

Maxwell, B.D. and Luschei, L.C. (2005) Ecological justification for site-specific weed management. *Weed Science* 53, 221–227.

Menalled, F.D., Gross, K.L. and Hammond, M. (2001) Weed aboveground and seedbank community responses to agricultural management systems. *Ecological Applications* 11, 1586–1601.

Mohler, C.L. (1996) Ecological bases for the cultural control of annual weeds. *Journal of Production Agriculture* 9, 468–474.

Mohler, C.L. (2001) Enhancing the competitive ability of crops. In: Liebman, M., Mohler, C.L. and Staver, C.P. (eds) *Ecological Management of Agricultural Weeds*. Cambridge University Press, Cambridge, UK, pp. 269–321.

Nieto, J.H., Brondo, M.A. and Gonzales, J.T. (1968) Critical periods of the crop growth cycle for competition with weeds. *PANS* (C) 14, 159–166.

Norris, R.F. (1992) Case history of weed competition/population ecology: barnyardgrass (*Echinochloa crus-galli*) in sugarbeets (*Beta vulgaris*). *Weed Technology* 6, 220–227.

O'Donovan, J.T. (1996) Weed economic thresholds: useful agronomic tool or pipe dream? *Phytoprotection* 77, 13–28.

O'Donovan, J.T. and McClay, A.S. (2002) Relationship between relative time of emergence of Tartary buckwheat (*Fagopyrum tataricum*) and yield loss of barley. *Canadian Journal of Plant Science* 82, 861–863.

O'Donovan, J.T., Newman, J.C., Harker, K.N., Blackshaw, R.E. and McAndrew, D.W. (1999) Effect of barley plant density on wild oat interference, shoot biomass and seed yield under zero tillage. *Canadian Journal of Plant Science* 79, 655–662.

O'Donovan, J.T., Blackshaw, R.E., Harker, K.N., Clayton, G.W. and Maurice, D.C. (2005) Field evaluation of regression equations to estimate crop yield losses due to weeds. *Canadian Journal of Plant Science* 85, 955–962.

Pacala, S. and Silander, J. (1990) Field tests of neighborhood population dynamic models of two annual weed species. *Ecological Monographs* 60, 113–134.

Radosevich, S.R. (1988) Methods to study crop and weed interactions. In: Altieri, M.A. and Liebman, M. (eds) *Weed Management in Agroecosystems: Ecological Approaches*. CRC Press, Boca Raton, FL, USA.

Roush, M.L., Radosevich, S.R., Wagner, R.G., Maxwell, B.D. and Petersen, T.D. (1989) A comparison of methods for measuring effects of density and proportion in plant competition experiments. *Weed Science* 37, 268–275.

Ryel, R., Barnes, P.W., Beyschlag, W., Caldwell, M.M. and Flint, S.D. (1990) Plant competition for light analyzed with a multispecies canopy model. I. Model development and influence of enhanced UV-B conditions on photosynthesis in mixed wheat and wild oat canopies. *Oecologia* 82, 304–310.

Sattin, M., Zanin, G. and Berti, A. (1992) Case history for weed competition/population ecology: velvetleaf (*Abutilon theophrasti*) in corn (*Zea mays*). *Weed Technology* 6, 213–219.

Stohlgren, T.J., Barnett, D.T. and Kartesz, J.T. (2003) The rich get richer: patterns of plant invasions in the United States. *Frontiers in Ecology* 1, 11–14.

Stoll, P. and Weiner, J. (2000) A neighborhood view of interactions among individual plants. In: Dieckmann, U., Law, R. and Metz, J.A.J. (eds) *The Geometry of Ecological Interactions: Simplifying Spatial Complexity*. Cambridge University Press, New York, pp. 11–27.

Stougaard, R.N. and Xue, Q. (2005) Quality versus quantity: spring wheat seed size and seeding rate effects on *Avena fatua* interference, economic returns and economic thresholds. *Weed Research* 45, 351–360.

Swanton, C.J., Weaver, S., Cowan, P., Van Acker, R., Deen, W. and Shresta, A. (1999) Weed thresholds: theory and applicability. *Journal of Crop Production* 2, 9–29.

Tilman, D. (1982) *Resource Competition and Community Structure.* Princeton University Press, Princeton, NJ, USA.

Van Acker, R.C., Swanton, C.J. and Wiese, S.F. (1993) The critical period of weed control in soybean (*Glycine max*). *Weed Science* 41, 194–220.

Wagner, R.G. and Radosevich, S.R. (1998) Neighborhood approach for quantifying interspecific competition in coastal Oregon forests. *Ecological Applications* 8, 779–794.

Weaver, S.E. (1996) Simulation of crop–weed competition: models and their applications. *Phytoprotection* 77, 3–11.

Weaver, S.E. and Ivany, J.A. (1998) Economic thresholds for wild radish, wild oat, hemp-nettle and corn spurry in spring barley. *Canadian Journal of Plant Science* 78, 357–361.

Weigelt, A. and Jolliffe, P. (2003) Indices of plant competition. *Journal of Ecology* 91, 707–720.

Weiner, J. (1982) A neighborhood model of annual-plant interference. *Ecology* 63, 1237–1241.

Wiles, L.J., Gold, H.J. and Wilkerson, G.G. (1993) Modelling the uncertainty of weed density estimates to improve post-emergence herbicide control decisions. *Weed Research* 33, 241–252.

Wilkerson, G.G., Jones, J.W., Coble, H.D. and Gunsolus, J.L. (1990) SOYWEED: a simulation model of soybean and common cocklebur growth and competition. *Agronomy Journal* 82, 1003–1010.

Wilkerson, G.G., Wiles, L.J. and Bennet, A.C. (2002) Weed management decision models: pitfalls, perceptions, and possibilities of the economic threshold approach. *Weed Science* 50, 411–424.

Williams, A.C. and McCarthy, B.C. (2001) A new index of interspecific competition for replacement and additive designs. *Ecological Research* 16, 29–40.

3 Cultural Weed Management

R.E. Blackshaw,[1] R.L. Anderson[2] and D. Lemerle[3]

[1]*Agriculture and Agri-Food Canada, Lethbridge, AB, Canada, T1J 4B1;*
[2]*USDA-ARS, Brookings, SD 57006, USA;* [3]*EH Graham Centre for Agricultural Innovation, New South Wales Department of Primary Industries and Charles Sturt University, Wagga Wagga, NSW 2678, Australia*

3.1 Introduction

Successful long-term weed management requires a shift away from simply controlling problem weeds to systems that reduce weed establishment and minimize weed competition with crops. Research indicates that numerous 'cultural' practices can be utilized to effectively control weeds. This chapter outlines various cultural weed management practices and discusses the benefits of combining practices to develop more effective and sustainable weed management systems.

3.2 Crop Rotation

Diverse crop rotations are integral components of improved weed management systems. Weeds tend to associate with crops that have similar life cycles. For example, the winter annual weed downy brome (*Bromus tectorum*) proliferates in winter wheat because seedling emergence and flowering periods coincide, thus enabling downy brome to produce viable seed during the crop season (Moyer *et al.*, 1994). Another common association is wild oat (*Avena fatua*) with spring cereals.

Rotating crops with different life cycles can disrupt the development of weed–crop associations (Karlen *et al.*, 1994; Derksen *et al.*, 2002). Liebman and Ohno (1998) summarized the results of 25 crops by rotation combinations and found that weed densities in rotation were less than in monoculture in 19 of 25 cases. Different planting and harvest dates of diverse crops provide opportunities for producers to prevent either weed establishment or seed production. For example, downy brome emerges in autumn or early spring; thus farmers have the opportunity to control it before spring-planted crops such as maize (*Zea mays*) or sunflower (*Helianthus annuus*).

Diverse crop rotations can aid in reducing the weed seedbank. Seeds in soil can germinate, die of natural causes, or be consumed by fauna or microorganisms; consequently, the number of live seeds in soil declines with time. With downy brome, approximately 20% of seeds are alive 1 year after seed shed, whereas less than 3% are alive after 2 years (Anderson, 2003). This rapid decline in seed viability with time is typical of many annual weed species (Roberts, 1981).

On the Canadian prairies, rotating canola (*Brassica rapa*) with winter wheat (*Triticum aestivum*) reduced downy brome density to less than 50 plants/m^2 compared with 740 plants/m^2 with 6 years of continuous winter wheat (Blackshaw, 1994). Downy brome and jointed goatgrass (*Aegilops cylindrica*) populations were almost eliminated when maize or sunflower was inserted into a winter wheat–fallow rotation in the USA (Daugovish *et al.*, 1999). The 2-year interval, where weeds with different life cycles from those of the crops are well controlled, significantly reduced densities of these weeds in subsequent crops.

In semiarid regions of North America, no-till systems and crop residue management have improved water relations such that continuous cropping is now more commonly practised (Derksen *et al.*, 2002; Anderson, 2005a). Long-term rotation studies in this region show that arranging crops in a 4-year cycle, with two cool-season crops followed by two warm-season crops, results in the lowest weed community density (Anderson, 2003). Weed seedling emergence was eight times greater in two-crop rotations compared with four-crop rotations. However, the four-crop rotation was effective only if crops with different planting dates were grown within a life-cycle interval. For example, downy brome or jointed goat-grass densities rapidly increased when winter wheat was grown two years in a row, even if the rotation included two years of warm-season crops.

3.3 Crop Competition

Crop competition is an important and cost-effective tactic for enhancing weed suppression and optimizing crop yield. The balance between crop–weed competition can be manipulated to favour crop growth by careful planning and management of agronomic practices. During the last 15 years, as herbicide resistance has become more widespread, considerable research has examined the benefits of crop competition for integrated weed management (Lemerle *et al.*, 2001a; Mohler, 2001).

Competition is defined as neighbour effects due to the consumption of resources in limited supply, or when two or more organisms compete for a common resource whose supply falls below their combined demand (Harper, 1977). Weeds compete with crops for essential resources such as nutrients, water and light.

The competitive ability of a crop can be measured either as the suppression of weed growth and the weed seed bank, or the ability of the crop to maintain yield in the presence of weeds (i.e. 'tolerate' weed competition) (Goldberg, 1990). Ideally, a competitive crop will both suppress and exhibit tolerance to weeds, but as the mechanisms underpinning these responses may be different, this is not always the case (Jordon, 1993).

Early crop competitiveness is essential and depends on optimal crop establishment and early vigour. Poor seed quality and susceptibility to disease can reduce competitiveness. Seeding machinery as well as good agronomy can facilitate the natural competitive characteristics of crop plants, including rapid emergence, root development, height, canopy closure, high leaf area index, and profuse tillering or branching. These factors are also influenced by crop species, cultivar, seed characteristics and interactions with the prevailing soil and environmental conditions. Agronomic factors that can be manipulated to favour crop competitiveness include species, cultivar, seed quality, seed rate, row spacing, seed placement in soil, and fertilizer management.

3.4 Crop Species

Many studies have examined the competitiveness of various crop species (Table 3.1). Poorly competitive crops are generally short in stature with low early vigour, such as legumes like lentil (*Lens culinaris*). Generally, cereal crops are more competitive than grain legumes, and oilseed crops are intermediate.

Choosing a strongly competitive crop requires no additional cost to farmers apart from the extra planning required. Poorly competitive crops should be sown in fields where weed populations are low due to good management in previous crops in the rotation. Growing strongly competitive crops will reduce dependence on herbicides, slow the development of herbicide resistance, and reduce chemical inputs into the environment.

Crop species ranking in terms of competitiveness is influenced by environmental conditions and thus can vary with location and year. In the future, as resistance to herbicides increases, the benefits of choosing strongly competitive species will become increasingly important.

3.5 Cultivar Selection

Cultivars within crop species differ in competitiveness with weeds. Such variation is due to morphological and physiological differences between types and can also interact strongly with environmental factors. Older, taller crop cultivars are often more competitive than the modern,

Table 3.1 Relative competitive ability of crop species with weeds.

Order of decreasing crop competitiveness	Country	Reference
Barley > rye > wheat > linseed	Canada	Pavlychenko and Harrington, 1934
Barley > oats = canola = field peas	UK	Lutman *et al.*, 1993
Rye > oilseed rape = field peas	Denmark	Melander, 1993
Oats = rye = triticale > canola > spring wheat = spring barley > field pea = lupin	Australia	Lemerle *et al.*, 1995
Oats > barley > wheat	UK	Seavers and Wright, 1999
Spring wheat > canola > sunflower	USA	Holman *et al.*, 2004

semi-dwarf types (Lindquist *et al.*, 1998; Gibson and Fischer, 2004). In Australia, rankings of current wheat cultivars for competitiveness with weeds varied considerably with environmental conditions and thus were generally too unreliable to make recommendations to farmers (Cousens and Mokhtari, 1998; Lemerle *et al.*, 2001b). Greater benefits would probably be gained from selecting or breeding for competitiveness.

The need for incorporating selection for competitive ability into a wheat breeding programme has been previously proposed (Pester *et al.*, 1999). Successful selection will depend on strong genetic control of the morphological and physiological traits linked with competitiveness (Mokhtari *et al.*, 2002), with no associated penalties such as loss of disease resistance or reduced yield or quality. Morphological traits associated with wheat competitiveness are tiller number, height and early vigour, but no one set of characteristics indicates strongly competitive wheat plants in all situations.

3.6 Early and Uniform Seedling Emergence and Fast Seedling Growth

Early and uniform crop establishment is essential for crops to successfully compete with weeds. Healthy crop seed is required to optimize seed germination and emergence. Optimal seeding depth for the soil and crop type and uniform seeding depth result in synchronous crop emergence. Sowing too shallow can result in uneven germination if the soil is dry, while sowing too deep depletes seed reserves for emergence. Sowing crops into unsuitable soils will reduce crop emergence. Tillage can help achieve a suitable seedbed for crop growth but

excellent crop emergence can also be attained with zero tillage.

Osmopriming or osmoconditioning has recently been examined to increase crop seed germinability, water uptake, emergence and competitive ability with weeds. In wheat and barley (*Hordeum vulgare*), pre-sowing osmotic seed treatments of polyethylene glycol (PEG) 8000 or KCl increased germination percentage (Al-Karaki, 1998). The potential of osmoconditioning to enhance crop competitiveness with weeds has also been demonstrated with soybean (*Glycine max*) (Nunes *et al.*, 2003). Further research is needed to substantiate the benefits of this technology as a practical way to improve crop competitive ability.

3.7 Planting Pattern, Row Spacing and Crop Density

Many crops are grown in wide rows to reduce seed costs, enable better stubble management, provide less soil disturbance, and facilitate weed control between rows. The disadvantages of wide rows include reduced competition with weeds and reduced yield in some situations.

Increasing crop density and reducing row spacing increases the competitive ability of crops with weeds (Jordon, 1993; Lemerle *et al.*, 2001a; Mohler, 2001). Generally, closer row spacing will improve crop competition for light, soil moisture and soil-borne nutrients as postulated mathematically by Fischer and Miles (1973). Time to canopy closure is an inverse function of row spacing and plant population and is also influenced by crop species and cultivar.

Weiner *et al.* (2001) demonstrated 30% less weed biomass and 9% higher grain yield in wheat sown in a grid pattern with 4 cm between rows and 2.5 cm between plants in the row compared with 12.8 cm row spacing. They concluded that a 'more crowded, uniform, distribution of some crops could contribute to a strategy to reduce the use of herbicides and energy-intensive forms of weed control'. More recently, Olsen *et al.* (2005) confirmed the benefits of increased wheat density and spatial uniformity to increase weed suppression and grain yield. However, dense crops can also increase the risks of lodging and disease development. Furthermore, there is a large gap between the theoretical benefits of high crop densities and actual practice because farmers are reluctant to adopt higher seeding rates due to concerns about reduced yield or quality. However, recent studies have demonstrated no such penalties at wheat seeding rates required for weed suppression over a wide range of environments in Australia (Lemerle *et al.*, 2004).

3.8 Transplanting

Transplanting crops by hand or machine can provide a competitive advantage against weeds. However, direct seeding of rice (*Oryza sativa*) is becoming more widespread due to higher labour costs and increased water efficiency. In India, similar yields were recorded in direct-seeded compared with transplanted rice, but a number of weed species were more abundant with direct seeding (Singh *et al.*, 2003).

In Africa, transplanting compared with seeding maize highly infested with *Striga* spp. more than doubled crop yield (Oswald *et al.*, 2001). An incentive to using this method by small-scale farmers would be that the main input at risk is their own labour. However, the establishment of nurseries and the timing of the transplanting operation require a certain level of farm management that could constrain adoption of this technique.

3.9 Delayed Seeding

Weeds display a characteristic emergence pattern during the growing season, which offers farmers a control opportunity. For example, wild oat emerges early in the spring; therefore if

producers delay seeding of spring-planted crops they can control the first flush of wild oat seedlings. However, this practice is seldom effective with weed community management because of the diversity of emergence patterns among weed species. When seeding of spring wheat is delayed to reduce wild oat density, a later-germinating species such as green foxtail (*Setaria viridis*) often proliferates (Donald and Nalewaja, 1990).

A second limitation of delayed seeding is its inconsistency, especially in semiarid regions. Delayed seeding of winter wheat in the western USA reduced downy brome density in only one of six years (Anderson, 1996). Delayed seeding is of little value if dry soils prevent weed germination. Nevertheless, delayed seeding for weed management has merits and is widely practised by organic farmers (Smith *et al.*, 2004).

3.10 Flooding

Flooding suppresses weeds by imposing anaerobic conditions in the soil, which is lethal to many weed seeds, seedlings and perennial storage organs. This practice is commonly used to control weeds in rice because rice tolerates anaerobic conditions. Flooding can also control weeds in other crops. McWhorter (1972) found that flooding fields in the southern USA for 2–4 weeks before planting reduced johnsongrass (*Sorghum halepense*) density without negatively affecting soybean yield. With adapted crops, an adequate water source and appropriate soils, flooding can be a component of integrated weed management systems.

3.11 Crop Fertilization

Fertilizer use in cropping systems alters soil nutrient levels that can affect weed demographic processes and crop–weed competitive interactions. Nitrogen fertilizer has been documented in breaking dormancy of certain weed species (Agenbag and Villiers, 1989) and thus may directly affect weed infestation densities. Many weeds are high consumers of nitrogen (Qasem, 1992), and are therefore capable of reducing available nitrogen for crop growth. Not only can weeds reduce the amount of nitrogen available to crops but also the growth of many weed species is enhanced by higher soil nitrogen levels (Supasilapa *et al.*, 1992). Indeed, shoot and root growth of many agricultural weeds was found to be more responsive than crops to higher soil nitrogen and phosphorus levels (Blackshaw *et al.*, 2003, 2004a). This can lead to a worst-case scenario where fertilization increases the competitive ability of weeds more than that of the crop, and crop yield remains unchanged or decreases (Supasilapa *et al.*, 1992; Dhima and Eleftherohorinos, 2001).

There is good potential to manipulate fertilizer timing, dose and placement to reduce weed interference in crops (DiTomaso, 1995). Forcella (1984) documented that rigid ryegrass (*Lolium rigidum*) was less competitive when N was applied before the three-leaf stage of wheat compared with later applications. Spring compared with autumn-applied fertilizer often reduced weed biomass and increased yields of spring-planted wheat, barley, canola (*Brassica napus*) and peas (*Pisum sativum*) (Blackshaw *et al.*, 2004b, 2005a,b).

Nitrogen fertilizer placed as narrow in-soil bands, rather than surface broadcast, has been found to reduce the competitive ability of several weed species (Rasmussen *et al.*, 1996; Blackshaw *et al.*, 2004b). Indeed, banding of nitrogen fertilizer for four consecutive years of zero-till barley production reduced green foxtail levels below those causing economic loss (O'Donovan *et al.*, 1997). A field study utilizing ^{15}N-enriched liquid nitrogen fertilizer clearly demonstrated greater nitrogen uptake by wheat, and often lower nitrogen uptake by weeds, when nitrogen was placed 10 cm below the soil surface (away from surface germinating weeds) compared to when it was surface broadcast (Blackshaw *et al.*, 2002).

Weed seed-bank data indicate that nitrogen fertilizer timing and application method are important components of long-term weed management (Figs 3.1 and 3.2). Clearly, manipulation of crop fertilization has the potential not only to protect crop yield but also to contribute to long-term weed management.

3.12 Silage, Green Manure and Cover Crops

Inclusion of silage crops in rotations can markedly reduce weed populations over time. Silage crops are often harvested before weeds produce mature seed, thus limiting seed return to the soil seed bank. Harker *et al.* (2003) reported that barley silage effectively reduced wild oat densities in subsequent years.

Green manure and cover crops have been used for weed management for centuries (Caamal-Maldonado *et al.*, 2001). Living cover crops suppress weeds by competing for resources, and their decaying residues inhibit weeds through physical, biotic and allelopathic interactions (Weston, 1996; Hartwig and Ammon, 2002). Cover crops that establish quickly and have high biomass production are well suited for weed management (Teasdale,

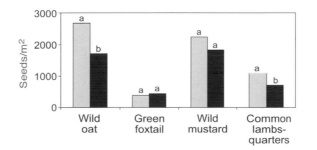

Fig. 3.1. Effect of nitrogen fertilizer applied either in autumn (grey columns) or spring (black columns) in four consecutive years on the weed seed bank (seeds/m[2]) of four weed species at the conclusion of the 4-year experiment. Bars on the graph within a weed species with the same letter are not significantly different according to Fisher's protected LSD test at the 5% probability level. Adapted from Blackshaw *et al.* (2004b).

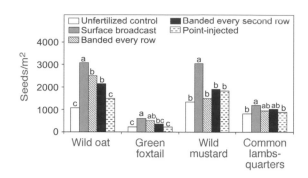

Fig. 3.2. Effect of nitrogen fertilizer application method in four consecutive years on the weed seed bank at the conclusion of the 4-year experiment. Bars on the graph within a weed species with the same letter are not significantly different according to Fisher's protected LSD test at the 5% probability level. Adapted from Blackshaw *et al.* (2004b).

1996; Ekeleme *et al.*, 2003). Weed control is improved with dense cover crop plantings and when they are allowed to grow for the longest time possible. Living cover crops have greater weed suppression capacity than dead ones but can be quite competitive with crops and thus are probably best used in orchards, vineyards and some transplanted horticultural crops (Sarrantonio and Gallandt, 2003).

Weed germination and emergence can be markedly reduced by living cover crops and their residues. Many weed seeds require light or fluctuating temperature and moisture conditions to trigger germination, and cover crops may effectively negate these environmental cues (Teasdale, 1993; Sarrantonio and Gallandt, 2003). Cover crop residues on the soil surface can be strong physical barriers to weed emergence and establishment (Teasdale, 1996).

Cover crops such as rye (*Secale cereale*), oats (*Avena sativa*), barley, mustard (*Brassica* spp.) and sweetclover (*Melilotus officinalis*) contain allelopathic compounds that inhibit weed germination and growth (Weston, 1996; Masiunas, 1998). These allelopathic chemicals may be secreted by living plants but are more commonly released from decaying cover-crop residues and may require conversion from inactive to phytotoxic compounds by soil microbes. Allelopathic compounds usually degrade rapidly in the environment and would be expected to inhibit weeds for only a few weeks. However, the physical suppression effects of cover crops may be present for several months

depending on the species, amount of biomass, killing method, whether the residues are incorporated or left on the soil surface, and environmental conditions (Teasdale, 1996; Masiunas, 1998).

In summary, cover crops can aid in weed management through diverse mechanisms such as resource competition, reducing weed establishment and growth through physical and chemical means, and promoting establishment of vesicular-arbuscular mychorrhizae that may benefit crops over non-mycorrhizal weed species (Teasdale, 1996; Hartwig and Ammon, 2002).

3.13 Intercropping

Intercropping refers to growing two or more crops simultaneously (Vandermeer, 1989). It may involve mixtures of annual crops with other annuals, annuals with perennials, or perennials with perennials. Increasing crop productivity (while simultaneously reducing the risk of total crop failure) and managing weeds are the major objectives of intercropping systems (Liebman and Dyck, 1993). Indeed, the second crop in some intercropping systems is grown for the sole purpose of weed management. Chikoye *et al.* (2001) documented the many benefits of velvetbean (*Mucuna cochinchinensis*), lablab (*Lablab purpureus*), and tropical kudzu (*Pueraria phaseoloides*) grown as intercrops with maize or cassava

(*Manihot esculenta*) for the express purpose of controlling cogongrass (*Imperata cylindrica*).

The use of intercropping to manage weeds is more commonly practised in Africa, Asia and Latin America than in Europe or North America due to the heavy reliance on herbicides in those latter regions. Carsky *et al.* (1994) found that a sorghum (*Sorghum halepense*)–cowpea (*Vigna unguiculata*) intercrop in Cameroon more effectively reduced growth and seed production of *Striga* spp. compared with either crop grown alone. Kale (*Brassica oleracea*)–bean (*Phaseolus vulgaris*) intercrops in Kenya effectively suppressed redroot pigweed (*Amaranthus retroflexus*) growth and increased crop yield (Itulya and Aguyoh, 1998).

Weed management in non-competitive vegetable crops may be improved with intercropping. For example, a leek (*Allium porrum*)–celery (*Apium graveolens*) intercrop in the Netherlands inhibited both weed emergence and growth compared with leek grown alone (Baumann *et al.*, 2000). There is increasing interest in intercropping of field crops in North America. Wheat–lentil, wheat–canola, wheat–canola–pea and barley–medic (*Medicago* spp.) intercrops have shown potential to reduce herbicide use while maintaining adequate levels of weed management (Carr *et al.*, 1995; Szumigalski and Van Acker, 2006).

Intercrops may inhibit weeds by limiting resource capture by weeds or through allelopathic interactions (Liebman and Dyck, 1993). Intercropping results in spatial diversification of crops that may aid in competitive interactions with weeds. Studies have reported that intercrops often shade weeds to a greater extent compared with sole crops (Liebman and Dyck, 1993; Itulya and Aguyoh, 1998). Crop competitive ability for nutrients and water can also be greater in intercrop than in monoculture systems (Hauggaard-Nielsen *et al.*, 2001). Liebman and Dyck (1993) reviewed the literature and found that weed biomass was reduced in 90% of the cases when a main crop was intercropped with a 'smother' crop. When two or more main crops were intercropped, weed biomass was lower than in all component individual crops in 50% of the cases, intermediate between component individual crops in 42% of the cases, and higher than all individual crops in 8% of the cases.

Utilization of intercrops must be economically feasible for widespread adoption by farmers (Liebman and Dyck, 1993). Unamma *et al.* (1986) found that the ranking of economic returns in maize and cassava systems was intercrops > herbicides > hand-weeding. Self-regenerating intercrops reduce establishment costs and can provide weed suppression over years (Martin, 1996).

3.14 Timing of Weed Control

Crops can tolerate some weed interference, and thus a key question is knowing when to control weeds. O'Donovan *et al.* (1985) documented that 50 wild oats/m^2 reduced spring wheat yield by 35% if they emerged with the crop, but only 15% if they emerged 6 days later. Similarly, Stahlman and Miller (1990) found that 40 downy brome/m^2 reduced winter wheat yield by 15% if both species emerged together but yield was unaffected if downy brome emerged 21 days after wheat. With all crops, the greater the delay in weed emergence relative to the crop, the more tolerant the crop is to weed interference.

Crops also can tolerate weed interference early in the growing season if weeds are eventually controlled. Zimdahl (1988) reported that maize can tolerate 2–6 weeks of early-season weed growth without losing yield. This 2–6 week tolerance range reflects the varying impacts of weed species, time of weed and maize emergence, weed density, crop management and environmental conditions.

The concept of critical period of control, the interval during the crop season when weeds need to be controlled to avoid yield loss, has been developed to guide weed management decisions. This concept reflects crop tolerance to both early-season and late-season weed interference. Quantifying this critical period for a specific crop has been difficult due to the numerous factors affecting crop–weed interactions. However, Zimdahl (1988) suggested the guideline that crops need a weed-free period of at least one-third of their growing period. A key principle is that controlling weeds early in the crop season is most favourable for preserving crop yield.

3.15 Site-specific Weed Management Utilizing Cultural Methods

Numerous studies have documented that weeds often occur in patches of varying size and density, with some areas of a field containing few or no weeds (Clay et al., 1999; Wiles, 2005). The concept of site-specific weed management (SSWM) is to identify, analyse and manage site-specific spatial and/or temporal variability of weed populations in order to optimize economic returns, sustainability of cropping systems, and environmental protection (Shaw, 2005). In recent years, technological developments such as remote sensing, geographic information systems (GIS), and the global positioning system (GPS) have markedly increased the potential for SSWM.

Much research has been conducted on using SSWM technologies to increase the efficiency of herbicide use in agricultural systems (Gerhards et al., 2002; Shaw, 2005). However, precise knowledge of when and where weeds occur in a field will also facilitate increased efficiencies of cultural techniques. Crop seed rate could be increased or planting pattern altered in dense weed patches to reduce weed competition. Timing or application method of fertilizers could possibly be manipulated according to weed spatial data to reduce weed establishment and competitive ability with the crop.

3.16 Merits of Combining Cropping Practices

Individual cultural practices for weed suppression are variable due to complex interactions between crops and weeds that are often strongly influenced by environment. For example, planting winter wheat in narrower rows reduced cheatgrass (Bromus secalinus) biomass only 60% of the time (Koscelny et al., 1991). Kappler et al. (2002) reported that increased winter wheat seeding rate reduced jointed goatgrass biomass in only about 50% of the cases.

Consistency of weed management can be greatly improved by combining several cultural practices. Anderson (2003) examined three cultural tactics for weed management in sunflower: a plant population of 18,200 plants/ha, 50-cm row spacing and a 2-week delay in planting was compared to the conventional system with 16,000 plants/ha, 78-cm row spacing and normal planting date. Weed biomass was reduced only 5–10% by any single cultural practice compared with the conventional system (Fig. 3.3), whereas combining two cultural practices suppressed weed biomass 20–25%. A surprising trend, however, occurred when the three practices were combined; weed biomass was reduced almost 90%, a fourfold increase compared to treatments with two cultural practices. A similar trend occurred with maize when narrow row spacing, higher plant population and N fertilizer placement were combined; weed suppression was six times greater than with any single cultural practice (Fig. 3.3).

In addition to controlling weeds in any given crop, growers also seek to manage weed populations over years. Cultural practices can disrupt weed population growth during several stages of their life cycle. Prominent weeds infesting crops in the Great Plains of the USA are annuals, where seed is the key component of population dynamics. Producers are using a multi-tactical approach based on cultural practices to reduce the seed bank, reduce weed seedling establishment, and minimize seed production by individual plants (Anderson, 2005b). Cultural practices related to rotation design, crop sequencing, no-till, crop residue management, and competitive crop canopies (Fig. 3.4) are integrated to reduce weed densities over years.

This population management approach has reduced weed community density such that weeds can be controlled with lower herbicide doses (Anderson, 2003). Indeed, herbicides are not needed in some competitive crops. However, farmers have observed that this approach requires integration of tactics from all components (Fig. 3.4); whereas systems comprising only two or three components are often ineffective.

Farmers in Canada are also exploring this approach with traditional small-grain rotations (Derksen et al., 2002). Rotations have been expanded to include canola, pulses and perennial forages, which increases the range of planting and harvesting dates, diversity in canopy development, and timing of weed control operations. Producers have more opportunities

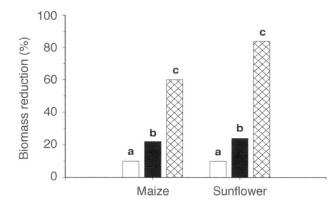

Fig. 3.3. Benefits of combining different numbers of cultural practices (one tactic, white columns; two tactics, black columns; three tactics, cross-hatched columns) on suppression of weed biomass in maize and sunflower. Cultural practices included higher seeding rates, narrower row spacing, banded fertilizer, and delayed planting date, with treatments compared with the conventional system used by farmers. Bars with an identical letter within a crop are not significantly different according to Fisher's protected LSD test at the 5% probability level. Means for single cultural practice treatments did not differ from the conventional system. Adapted from Anderson (2003).

to disrupt weed population growth and weed densities gradually decrease with time.

3.17 Economics of Cultural Weed Control Practices

Economics concerns the optimal allocation of scarce resources with the goal of maximizing profits within a sustainable system. Economic

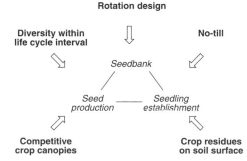

Fig. 3.4. Five components of an ecologically based weed management system for cropping systems in the central Great Plains of the USA. Cultural strategies in each component disrupt weed population dynamics by minimizing weed seed survival in the soil (seed bank), seedling establishment, or seed production. Adapted from Anderson (2005b).

analyses can be used to determine whether weeds should be controlled. A number of economic analyses have been developed specifically dealing with weed management (Auld *et al.*, 1987; Jones and Medd, 1997).

One of the simplest management techniques to decide whether to control weeds in a field is called the 'economic threshold'. This can be a critical weed density threshold above which the financial benefits from controlling the weed exceed the costs. The need for integrated weed management utilizing many control inputs has led to more complex economic frameworks in recent years to determine longer-term optimal benefits, including the weed seed bank. The future impact from weed control is accounted for by introducing the population dynamics of a weed, which results in more intensive control being optimal than if the benefits were calculated from the current period alone (Jones and Medd, 1997).

The integration of a range of weed control tactics, such as cultivar and seed rate (Korres and Froud-Williams, 2002), will lead to long-term reductions in weed populations and greater profits for farmers (Smith *et al.*, 2006). Increasingly, weed management will need to consider longer-term, socio-economic, political and environmental considerations as well as short-term profit scenarios.

3.18 Summary

Cultural weed control has been practised by farmers for centuries. However, recent research has improved our knowledge of such techniques and has facilitated their inclusion in modern agricultural systems. Cultural weed control practices can be effective when utilized individually, but consistency and level of weed management are much greater when they are combined and when they are utilized within a multi-year approach. Clearly, cultural weed management practices have merit and show good potential for greater utilization in the future.

3.19 References

Agenbag, G.A. and Villiers, O.T. (1989) The effect of nitrogen fertilizers on the germination and seedling emergence of wild oat (*Avena fatua* L.) seed in different soil types. *Weed Research* 29, 239–245.

Al-Karaki, G.N. (1998) Response of wheat and barley during germination to seed osmopriming at different water potential. *Journal of Agronomy and Crop Science* 181, 229–235.

Anderson, R.L. (1996) Downy brome (*Bromus tectorum*) emergence variability in a semiarid region. *Weed Technology* 10, 750–753.

Anderson, R.L. (2003) An ecological approach to strengthen weed management in the semiarid Great Plains. *Advances in Agronomy* 80, 33–62.

Anderson, R.L. (2005a) Improving sustainability of cropping systems in the Central Great Plains. *Journal of Sustainable Agriculture* 26, 97–114.

Anderson, R.L. (2005b) A multi-tactic approach to manage weed population dynamics in crop rotations. *Agronomy Journal* 97, 1579–1583.

Auld, B.A., Menz, K.M. and Tisdell, C.A. (1987) *Weed Control Economics*. Academic Press, London, UK.

Baumann, D.T., Kropff, M.J. and Bastiaans, L. (2000) Intercropping leeks to suppress weeds. *Weed Research* 40, 359–374.

Blackshaw, R.E. (1994) Rotation affects downy brome (*Bromus tectorum*) in winter wheat (*Triticum aestivum*). *Weed Technology* 8, 728–732.

Blackshaw, R.E., Semach, G. and Janzen, H.H. (2002) Fertilizer application method affects nitrogen uptake in weeds and wheat. *Weed Science* 50, 634–641.

Blackshaw, R.E., Brandt, R.N., Janzen, H.H., Entz, T., Grant, C.A. and Derksen, D.A. (2003) Differential response of weed species to added nitrogen. *Weed Science* 51, 532–539.

Blackshaw, R.E., Brandt, R.N., Janzen, H.H. and Entz, T. (2004a) Weed species response to phosphorus fertilization. *Weed Science* 52, 406–412.

Blackshaw, R.E., Molnar, L.J. and Janzen, H.H. (2004b) Nitrogen fertilizer timing and application method affect weed growth and competition with spring wheat. *Weed Science* 52, 614–622.

Blackshaw, R.E., Beckie, H.J., Molnar, L.J. and Moyer, J.R. (2005a) Combining agronomic practices and herbicides improves weed management in wheat–canola rotations within zero-tillage production systems. *Weed Science* 53, 528–535.

Blackshaw, R.E., Moyer, J.R., Harker, K.N. and Clayton, G.W. (2005b) Integration of agronomic practices and herbicides for sustainable weed management in a zero-till barley field pea rotation. *Weed Technology* 19, 190–196.

Caamal-Maldonado, J.A., Jimenez-Osornio, J.J., Torres-Barragan, A. and Anaya, A.L. (2001) The use of allelopathic legume cover and mulch species for weed control in cropping systems. *Agronomy Journal* 93, 27–36.

Carr, P.M., Gardner, J.C., Schatz, B.G., Zwinger, S.W. and Guldan, S.J. (1995) Grain yield and weed biomass of a wheat–lentil intercrop. *Agronomy Journal* 87, 574–579.

Carsky, R.J., Singh, L. and Ndikawa, R. (1994) Suppression of *Striga hermonthica* on sorghum using a cowpea intercrop. *Experimental Agriculture* 30, 349–358.

Chikoye, D., Ekeleme, F. and Udensi, U.E. (2001) Cogongrass suppression by intercropping cover crops in corn/cassava systems. *Weed Science* 49, 658–667.

Clay, S.A., Lems, G.J., Clay, D.E., Forcella, F., Ellsbury, M.E. and Carlson, C.G. (1999) Sampling weed spatial variability on a field-wise scale. *Weed Science* 47, 674–681.

Cousens, R.D. and Mokhtari, S. (1998) Seasonal and site variability in the tolerance of wheat varieties to interference from *Lolium rigidum*. *Weed Research* 38, 301–307.

Daugovish, O., Lyon, D.J. and Baltensperger, D.D. (1999) Cropping systems to control winter annual grasses in winter wheat (*Triticum aestivum*). *Weed Technology* 13, 120–126.

Derksen, D.A., Anderson, R.L., Blackshaw, R.E. and Maxwell, B. (2002) Weed dynamics and management strategies for cropping systems in the northern Great Plains. *Agronomy Journal* 94, 174–185.

Dhima, K.V. and Eleftherohorinos, L.G. (2001) Influence of nitrogen on competition between winter cereals and sterile oat. *Weed Science* 49, 77–82.

DiTomaso, J.M. (1995) Approaches for improving crop competitiveness through the manipulation of fertilization strategies. *Weed Science* 43, 491–497.

Donald, W.W. and Nalewaja, J.D. (1990) Northern Great Plains. In: Donald, W.W. (ed.) *Systems of Weed Control in Wheat in North America*. Weed Science Society of America, Lawrence, KS, USA, pp. 90–126.

Ekeleme, F., Akobundu, I.O., Fadayomi, R.O., Chikoye, D. and Abayomi, Y.A. (2003) Characterization of legume cover crops for weed suppression in the moist savanna of Nigeria. *Weed Technology* 17, 1–13.

Fischer, R.A. and Miles, R.E. (1973) The role of spatial pattern in competition between crop plants and weeds: a theoretical analysis. *Mathematical Biosciences* 18, 335–350.

Forcella, F. (1984) Wheat and ryegrass competition for pulses of mineral nitrogen. *Australian Journal of Experimental Agriculture and Animal Husbandry* 24, 421–425.

Gerhards, R., Sökefeld, M., Timmermann, C. and Kühbauch, W. (2002) Site-specific weed control in maize, sugar beet, winter wheat, and winter barley. *Precision Agriculture* 3, 25–35.

Gibson, K.D. and Fischer, A.J. (2004) Competitiveness of rice cultivars as a tool for crop-based weed management. *Weed Biology and Management* 4, 517–537.

Goldberg, D.E. (1990) Components of resource competition in plant communities. In: Grace, J.B. and Tilman, D. (eds) *Perspectives on Plant Competition*. Academic Press, New York, pp. 27–49.

Harker, K.N., Kirkland, K.J., Baron, V.S. and Clayton, G.W. (2003) Early harvest barley (*Hordeum vulgare*) silage reduces wild oat (*Avena fatua*) densities under zero tillage. *Weed Technology* 17, 102–110.

Harper, J.L. (1977) *Population Biology of Plants*. Academic Press, London.

Hartwig, N.L. and Ammon, H.U. (2002) Cover crops and living mulches. *Weed Science* 50, 688–699.

Hauggaard-Nielsen, H., Ambus, P. and Jensen, E.S. (2001) Interspecific competition, N use and interference with weeds in pea–barley intercropping. *Field Crops Research* 70, 101–109.

Holman, J.D., Bussan, A.J., Maxwell, B.D., Miller, P.R. and Mickelson, J. (2004) Spring wheat, canola, and sunflower response to Persian darnel (*Lolium persicum*) interference. *Weed Technology* 18, 509–520.

Itulya, F.M. and Aguyoh, J.N. (1998) The effects of intercropping kale with beans on yield and suppression of redroot pigweed under high altitude conditions in Kenya. *Experimental Agriculture* 34, 171–176.

Jones, R. and Medd, R.W. (1997) Economic analysis of integrated management of wild oats involving fallow, herbicide and crop rotational options. *Australian Journal of Experimental Agriculture* 37, 683–691.

Jordon, N. (1993) Prospects for weed control through weed suppression. *Ecological Applications* 3, 84–91.

Kappler, B.F., Lyon, D.J., Stahlman, P.W., Miller, S.D. and Eskridge, K.M. (2002) Wheat plant density influences jointed goatgrass (*Aegilops cylindrica*) competitiveness. *Weed Technology* 16, 102–108.

Karlen, D.L., Varvel, G.E., Bullock, D.G. and Cruse, R.M. (1994) Crop rotations for the 21st century. *Advances in Agronomy* 53, 1–45.

Korres, N.E. and Froud-Williams, R.J. (2002) Effects of winter wheat cultivars and seed rate on the biological characteristics of naturally occurring weed flora. *Weed Research* 42, 417–428.

Koscelny, J.A., Peeper, T.F., Solie, J.B. and Solomon, S.G., Jr (1991) Effect of wheat (*Triticum aestivum*) row spacing, seeding rate, and cultivar on yield loss from cheat (*Bromus secalinus*). *Weed Technology* 5, 707–712.

Lemerle, D., Verbeek, B. and Coombes, N. (1995) Losses in grain yield of winter crops from *Lolium rigidum* competition depends on crop species, cultivar and season. *Weed Research* 35, 503–509.

Lemerle, D., Gill, G.S., Murphy, C.E., Walker, S.R., Cousens, R.D., Mokhtari, S., Peltzer, S.J., Coleman, R. and Luckett, D.J. (2001a) Genetic improvement and agronomy for enhanced wheat competitiveness with weeds. *Australian Journal of Agricultural Research* 52, 527–548.

Lemerle, D., Verbeek, B. and Orchard, B. (2001b) Ranking the ability of wheat varieties to compete with *Lolium rigidum*. *Weed Research* 41, 197–209.

Lemerle, D., Cousens, R.D., Gill, G.S., Peltzer, S.J., Moerkerk, M., Murphy, C.E., Collins, D. and Cullis, B.R. (2004) Reliability of higher seeding rates of wheat for increased competitiveness with weeds in low rainfall environments. *Journal of Agricultural Science* 142, 395–409.

Liebman, M. and Dyck, E. (1993) Crop rotation and intercropping strategies for weed management. *Ecological Applications* 3, 92–122.

Liebman, M. and Ohno, T. (1998) Crop rotation and legume residue effects on weed emergence and growth: applications for weed management. In: Hatfield, J.L., Buhler, D.D. and Stewart, B.A. (eds) *Integrated Weed and Soil Management*. Ann Arbor Press, Chelsea, MI, USA, pp. 181–221.

Lindquist, J.L., Mortensen, D.A. and Johnson, B.E. (1998) Mechanisms for crop tolerance and velvetleaf suppressive ability. *Agronomy Journal* 90, 787–794.

Lutman, P.J.W., Dixon, F.L. and Risiott, R. (1993) The response of four spring-sown combinable arable crops to weed competition. *Weed Research* 34, 137–146.

Martin, C.C. (1996) Weed control in tropical ley farming systems: a review. *Australian Journal of Experimental Agriculture* 36, 1013–1023.

Masiunas, J.B. (1998) Production of vegetables using cover crop and living mulches: a review. *Journal of Vegetable Crop Production* 4, 11–31.

McWhorter, C.G. (1972) Flooding for johnsongrass control. *Weed Science* 20, 238–241.

Melander, B. (1993) Modelling the effects of *Elymus repens* (L.) Gould competition on yield of cereals, peas and oilseed rape. *Weed Research* 33, 99–108.

Mohler, C.L. (2001) Enhancing the competitive ability of crops. In: Liebman, M., Mohler, C.L. and Staver, C.P. (eds) *Ecological Management of Agricultural Weeds*. Cambridge University Press, Cambridge, UK, pp. 269–321.

Mokhtari, S., Galwey, N.W., Cousens, R.D. and Thurling, N. (2002) The genetic basis of variation among wheat F3 genotypes in tolerance to competition by ryegrass (*Lolium rigidum*). *Euphytica* 124, 355–364.

Moyer, J.R., Roman, E.S., Lindwall, C.W. and Blackshaw, R.E. (1994) Weed management in conservation tillage systems for wheat production in North and South America. *Crop Protection* 13, 243–258.

Nunes, U.R., Silva, A.A., Reis, M.S., Sediyama, C.S. and Sediyama, T. (2003) Soybean seed osmoconditioning effect on the crop competitive ability against weeds. *Planta Daninha* 21, 27–35.

O'Donovan, J.T., de St. Remy, E.A., O'Sullivan, P.A., Dew, D.A. and Sharma, A.K. (1985) Influence of the relative time of emergence of wild oat (*Avena fatua*) on yield loss of barley (*Hordeum vulgare*) and wheat (*Triticum aestivum*). *Weed Science* 33, 498–503.

O'Donovan, J.T., McAndrew, D.A. and Thomas, A.G. (1997) Tillage and nitrogen influence weed population dynamics in barley (*Hordeum vulgare*). *Weed Technology* 11, 502–509.

Olsen, J., Kristensen, L., Weiner, J. and Griepentrog, H.W. (2005) Increasing density and spatial uniformity increase weed suppression by spring wheat. *Weed Research* 45, 316–21.

Oswald, A., Ransom, J.K., Kroschel, J. and Sauerborn, J. (2001) Transplanting maize and sorghum reduces *Striga hermonthica* damage. *Weed Science* 49, 346–353.

Pavlychenko, T.K. and Harrington, J.B. (1934) Competitive efficiency of weeds and cereal crops. *Canadian Journal of Research* 10, 77–94.

Pester, T.A., Burnside, O.C. and Orf, J.H. (1999) Increasing crop competitiveness to weeds through crop breeding. *Journal of Crop Production* 2, 59–76.

Qasem, J.R. (1992) Nutrient accumulation by weeds and their associated vegetable crops. *Journal of Horticultural Science* 67, 189–195.

Rasmussen, K., Rasmussen, J. and Petersen, J. (1996) Effects of fertilizer placement on weeds in weed harrowed spring barley. *Acta Agriculturae Scandinavica B, Soil and Plant Science* 45, 1–5.

Roberts, H.A. (1981) Seed banks in soils. *Advances in Applied Biology* 6, 1–55.

Sarrantonio, M. and Gallandt, E.R. (2003) The role of cover crops in North American cropping systems. *Journal of Crop Production* 8, 53–73.

Seavers, G.P. and Wright, K.J. (1999) Crop canopy development and structure influence weed suppression. *Weed Research* 39, 319–328.

Shaw, D.R. (2005) Translation of remote sensing data into weed management decisions. *Weed Science* 53, 264–273.

Singh, G., Singh, Y., Singh, V.P., Singh, R.K., Singh, P., Johnson, D.E., Mortimer, M. and Orr, A. (2003) Direct seeding as an alternative to transplanting rice for the rice–wheat systems of the Indo-Gangetic plains: sustainability issues related to weed management. In: *Proceedings of the BCPC International Congress: Crop Science and Technology*. British Crop Protection Council, Alton, UK, pp. 1035–1040.

Smith, E.G., Clapperton, M.J. and Blackshaw, R.E. (2004) Profitability and risk of organic production systems in the northern Great Plains. *Renewable Agriculture and Food Systems* 19, 152–158.

Smith, E.G., Upadhyay, B.M., Blackshaw, R.E., Beckie, H.J., Harker, K.N. and Clayton, G.W. (2006) Economic benefits of integrated weed management systems for field crops in the Dark Brown and Black soil zones of western Canada. *Canadian Journal of Plant Science* 86, 1273–1279.

Stahlman, P.W. and Miller, S.D. (1990) Downy brome (*Bromus tectorum*) interference and economic thresholds in winter wheat (*Triticum aestivum*). *Weed Science* 38, 224–228.

Supasilapa, S., Steer, B.T. and Milroy, S.P. (1992) Competition between lupin (*Lupinus angustifolia* L.) and great brome (*Bromus diandrus* Roth.): development of leaf area, light interception and yields. *Australian Journal of Experimental Agriculture* 32, 71–81.

Szumigalski, A. and Van Acker, R. (2006) Weed suppression and crop production in annual intercrops. *Weed Science* 53, 813–825.

Teasdale, J.R. (1993) Interaction of light, soil moisture, and temperature with weed suppression by hairy vetch residue. *Weed Science* 41, 46–52.

Teasdale, J.R. (1996) Contribution of cover crops to weed management in sustainable agricultural systems. *Journal of Production Agriculture* 9, 475–479.

Unamma, R.P.A., Ene, L.S.O., Odurukwe, S.O. and Enyinnia, T. (1986) Integrated weed management for cassava intercropped with maize. *Weed Research* 26, 9–17.

Vandermeer, J. (1989) *The Ecology of Intercropping*. Cambridge University Press, Cambridge, UK.

Weiner, J., Griepentrog, H.W. and Kristenses, L. (2001) Suppression of weeds by spring wheat (*Triticum aestivum*) increases with crop density and spatial uniformity. *Journal of Applied Ecology* 313, 31–51.

Weston, L.A. (1996) Utilization of allelopathy for weed management in agroecosystems. *Agronomy Journal* 88, 860–866.

Wiles, L.J. (2005) Sampling to make maps for site-specific weed management. *Weed Science* 53, 228–235.

Zimdahl, R.L. (1988) The concept and application of the critical weed-free period. In: Altieri, M.A. and Liebman, M. (eds) *Weed Management in Agroecosystems: Ecological Approaches*. CRC Press, Boca Raton, FL, USA, pp. 145–155.

4 Cover Crops and Weed Management

J.R. Teasdale,[1] L.O. Brandsæter,[2] A. Calegari[3] and F. Skora Neto[4]

[1]United States Department of Agriculture, Agricultural Research Service, Beltsville, MD 20705, USA; [2]Norwegian Institute for Agricultural and Environmental Research, Ås, Norway; [3]Instituto Agronômico do Paraná, Londrina, PR, Brazil; [4]Instituto Agronômico do Paraná, Ponta Grossa, PR, Brazil

4.1 Introduction

The term 'cover crops' will be used in this chapter as a general term to encompass a wide range of plants that are grown for various ecological benefits other than as a cash crop. They may be grown in rotations during periods when cash crops are not grown, or they may grow simultaneously during part or all of a cash-cropping season. Various terms such as 'green manure', 'smother crop', 'living mulch' and 'catch crop' refer to specific uses of cover crops.

Cover crops have multiple influences on the agroecosystem (Sustainable Agriculture Network, 1998; Sarrantonio and Gallandt, 2003). They intercept incoming radiation, thereby affecting the temperature environment and biological activity at various trophic levels in the leaf canopy and underlying soils. They fix carbon and capture nutrients, thereby changing the dynamics and availability of nutrients. They reduce rain droplet energy and influence the overall distribution of moisture in the soil profile. They influence the movement of soils, nutrients and agrochemicals into and away from agricultural fields. They can change the dynamics of weeds, pests and pathogens as well as of beneficial organisms. Thus, the introduction of cover crops into the agroecosystem offers opportunities for managing many aspects of the system simultaneously. However, cover cropping also adds a higher level of complexity and potential interactions that may be more difficult to predict and manage.

This chapter addresses the complexity of managing cover crops in selected growing regions of the world. It focuses on the contributions that cover crops can make to weed management and the trade-offs that may be required between achieving weed management, crop production, and environmental benefits. Since cover crops can play a significant role in mitigating environmental impacts worldwide, interactions between weed management and management to enhance environmental protection will be emphasized.

4.2 Impact of Cover Crops on Weeds and Crops

Cover crop impact on weeds

Cover crops can influence weeds either in the form of living plants or as plant residue remaining after the cover crop is killed. Different weed life stages will be affected by different mechanisms depending on whether the cover crop is acting during its living phase or as post-mortem residue. Management of the cover crop may also be influenced by whether the goal is to suppress weeds during the living or post-mortem phases.

There is wide agreement in the literature that a vigorous living cover crop will suppress weeds growing at the same time as the cover crop

(Stivers-Young, 1998; Akobundu *et al.*, 2000; Creamer and Baldwin, 2000; Blackshaw *et al.* 2001; Favero *et al.*, 2001; Grimmer and Masiunas, 2004; Peachey *et al.*, 2004; Brennan and Smith, 2005). There is often a negative correlation between cover crop and weed biomass (Akemo *et al.*, 2000; Ross *et al.*, 2001; Sheaffer *et al.*, 2002). Table 4.1 lists the degree of weed suppression by live cover crops grown in different areas of the world. Generally, vigorous cover crop species such as velvetbean (*Mucuna* spp.), jack bean (*Canavalia ensiformis* (L.) DC.), cowpea (*Vigna unguiculata* (L.) Walp.), and sorghum-sudangrass (*Sorghum bicolor* (L.) Moench × *S. sudanense* (Piper) Stapf), which are well adapted to growth in hot climates, are effective smother crops in warm-season environments. Yellow sweetclover (*Melilotus officinalis* (L.) Lam.), a biennial cover crop, was effective at suppressing weeds during a 20-month fallow on the Canadian Great Plains (Blackshaw *et al.*, 2001). Annual cover crops more adapted to cool conditions such as rye (*Secale cereale* L.), hairy vetch (*Vicia villosa* Roth) and various clovers (*Trifolium* spp.) are less effective as summer smother crops (Table 4.1). Many of these same cool-season species are more effective as winter annual cover crops (Peachey *et al.*, 2004); in fact, it is probably because of their effectiveness that there is so little literature documenting the suppression of winter weeds by these species. In Mediterranean climates with relatively mild winters, suppression of winter weeds may be more difficult, particularly with cover crops that often do not provide complete ground cover, such as subterranean clover (*Trifolium subterraneum* L.) and crimson clover (*Trifolium incarnatum* L.) in Italy (Barbari and Mazzoncini, 2001) or legume/oat (*Avena sativa* L.) mixes in central California (Brennan and Smith, 2005). Winter-killed cover crops such as mustard species (*Brassica* spp.) can establish quickly and suppress weeds during the autumn months but may allow spring weed establishment unless used preceding early spring cash crops (Grimmer and Masiunas, 2004).

Dead cover crop residue does not suppress weeds as consistently as live cover crops do (Teasdale and Daughtry, 1993; Reddy and Koger, 2004). The magnitude of weed suppression by residue is usually higher for weed emergence measured early in the season than for weed density or biomass measured later in the season. Table 4.2 outlines authors' estimates of the degree of weed suppression by living cover crops versus cover crop residue. Generally, living cover crops will suppress weeds more completely and at more phases of the weed life cycle than will cover crop residue. Some important mechanisms contrasting weed suppression by cover crop residue versus living cover crops are discussed below.

Cover crop residue can affect weed germination in soils through effects on the radiation and chemical environment of the seed. Cover crop residue on the soil surface can inhibit weed germination by creating conditions similar to those deeper in the soil, i.e. lower light and lower daily temperature amplitude (Teasdale and Mohler, 1993). Residue also can inhibit emergence by physically impeding the progress of seedlings from accessing light (Teasdale and Mohler, 2000) as well as by releasing phytotoxins that inhibit seedling growth (Yenish *et al.*, 1995; Blackshaw *et al.*, 2001). When fresh residue is incorporated into soils, decomposition processes can release pulses of phytotoxins or pathogens that inhibit germination and early growth of weeds (Dabney *et al.*, 1996; Blackshaw *et al.*, 2001; Davis and Liebman, 2003; Sarrantonio and Gallandt, 2003). Once seedlings become established, cover crop residue will usually have a negligible impact on weed growth and seed production or may even stimulate these processes through conservation of soil moisture and release of nutrients (Teasdale and Daughtry, 1993; Haramoto and Gallandt, 2005). Residue can provide a more favourable habitat for predators of weed seed on or near the surface of soils (Gallandt *et al.*, 2005); however, residue was found to have no effect on the survival of perennial structures or seeds in some experiments (Akobundu *et al.*, 2000; J.R. Teasdale *et al.*, unpublished data).

Live cover crops have a greater suppressive effect on all weed life cycle stages than cover crop residue (Table 4.2). A living cover crop absorbs red light and will reduce the red : far-red ratio sufficiently to inhibit phytochrome-mediated seed germination, whereas cover crop residue has a minimal affect on this ratio (Teasdale and Daughtry, 1993). A living cover crop competes with emerging and growing weeds for essential resources and inhibits

Table 4.1. Suppression of weeds that are growing at the same time as a live cover crop during summer or winter periods.

Period of growth	Location	Cover crop	Percentage weed biomass reduction[a]	Reference
Summer fallow	Nigeria	Velvetbean	85 (83–87)	Akobundu *et al.* (2000)
	Brazil savanna	Jack bean	72	Favero *et al.* (2001)
		Black mucuna	96	
		Lablab, pigeonpea	35 (22–48)	
	North Carolina, USA	Cowpea, sesbania, trailing soybean, buckwheat	85	Creamer and Baldwin (2000)
		Soybean, lablab	48	
		Sorghum-sudangrass, millet spp.	94	
	Maryland, USA	Hairy vetch	58 (52–70)	Teasdale and Daughtry (1993)
	Japan	Hairy vetch	66	Araki and Ito (1999)
		Wheat	39	
	Alberta, Canada	Yellow sweetclover	91 (77–99)	Blackshaw *et al.* (2001)
	Alberta, Canada	Berseem clover	58 (51–70)	Ross *et al.* (2001)
		Alsiko, balansa, crimson, Persian, red, white clover	35 (9–56)	
		Rye	64 (31–89)	
Summer intercrop	Brazil (southern)	Black mucuna, smooth rattlebox	97 (95–99)	Skora Neto (1993)
		Jack bean, pigeonpea	83 (71–90)	
		Cowpea	39 (29–48)	
	Mississippi, USA	Hairy vetch	79	Reddy and Koger (2004)
	New York, USA	Rye	61 (37–76)	Brainard and Bellinder (2004)
	Norway	Subterranean, white clover	48 (45–51)	Brandsaeter *et al.* (1998)
Winter-surviving annuals	Oregon, USA	Rye	97 (94–99)	Peachey *et al.* (2004)
		Oats	89 (81–96)	
		Barley	89 (78–99)	
	Italy	Rye	83 (54–99)	Barbari and Mazzoncini (2001)
		Subterranean, crimson clover	32 (0–67)	
Winter-killed annuals	New York, USA	Oilseed radish, mustard	94 (81–99)	Stivers-Young (1998)
		Oats	71 (19–95)	
	Michigan, USA	Annual medics, berseem clover	54 (18–88)	Fisk *et al.* (2001)
	Illinois, USA	Mustard	93	Grimmer and Masiunas (2004)
		Barley	94	
		Oats	76	

[a] Mean percentage reduction relative to a control without cover crop. Data that summarizes more than one year and/or location are presented with the range shown in parentheses. Where cover crop management treatments were included in the research, conditions that represented the optimum growth of the cover crop were chosen for this summary.

Table 4.2. Potential impact of typical cover crop residue or live cover crop on inhibition of weeds at various life cycle stages.

Weed life cycle stage	Cover crop residue	Live cover crop
Germination	Moderate	High
Emergence/establishment	Moderate	High
Growth	Low	High
Seed production	Low	Moderate
Seed survival	None?[a]	Moderate?[a]
Perennial structure survival	None?[a]	Low–moderate?[a,b]

[a] More research is needed to provide definitive estimates of cover crop influences on these processes.
[b] When cover crops are combined with other practices such as soil disturbance or mowing, perennial structure survival may be more effectively reduced, as discussed in Dock Gustavsson (1994) and Graglia *et al.* (2006).

emergence and growth more than cover crop residue does (Teasdale and Daughtry, 1993; Reddy and Koger, 2004). If growth suppression is sufficient, a live cover crop can also inhibit weed seed production (Brainard and Bellinder, 2004; Brennan and Smith, 2005). Weed seed predation at the soil surface was higher when living cover crop vegetation was present (Davis and Liebman, 2003; Gallandt *et al.*, 2005), suggesting a role for living cover crops in enhancing weed seed mortality.

Cover crop impact on perennial and parasitic weeds

Perennial weeds are often better competitors, and are more difficult to control with cover crops than annual weeds are, because of larger nutritional reserves and faster rates of establishment. However, several reports have shown the capability for suppressing perennial weeds with living cover crops during fallow periods. Blackshaw *et al.* (2001) found that yellow sweetclover controlled dandelion (*Taraxacum officinale* Weber ex Wiggers) and perennial sowthistle (*Sonchus arvensis* L.) as well as several annual weeds in Canada. Cultivation in combination with a competitive cover crop controlled important perennial weeds such as quackgrass (*Elytrigia repens* (L.) Nevski), perennial sowthistle and Canada thistle (*Cirsium arvense* (L.) Scop.) in cereal-dominated rotations in Scandinavia (Håkansson, 2003). Cover-cropping systems will probably be most effective if maximum disturbance of, or competition with, perennial weeds occurs at the compensation point which may be

defined as that time where the source–sink dynamic of carbohydrate reserves shifts from the underground organs as the source and the above-ground organs as the sink, to the reverse (Håkansson, 2003).

Many regions of Africa have heavy infestations of aggressive perennial weeds that multiply by seeds and rhizomes, such as cogongrass (*Imperata cylindrica* (L.) Beauv.), bermudagrass (*Cynodon dactylon* (L.) Pers.) and sedges (*Cyperus* spp.). Farmers cannot produce crops economically and have abandoned their fields when these weeds are not controlled. To overcome these constraints, various cover-crop species were evaluated under on-farm conditions, and up to 90% weed reduction was achieved (Taimo *et al.*, 2005). The results obtained in several districts in Sofala Province, Mozambique, with the use of black and grey mucuna (*Mucuna pruriens* (L.) DC.), calopo (*Calopogonium mucunoides* Desv.), sunn hemp (*Crotalaria juncea* L.), jack bean and Brazilian jack bean (*Canavalia brasiliensis* Mart. ex Benth.) are very encouraging and showed that they effectively suppressed bermudagrass, sedges and cogongrass. After cleaning the fields, farmers saved on labour time/costs and were able to grow soybean (*Glycine max* (L.) Merr.), beans and cereals successfully. Generally, live cover crops that establish an early leaf canopy cover are most competitive with weeds. Akobundu *et al.* (2000) found that development of early ground cover was more important than the quantity of dry matter produced for suppression of cogongrass by velvetbean accessions.

Some of Africa's worst agricultural pests are parasitic weeds, including witchweed (*Striga asiatica* (L.) Ktze.) Normally, the severity of these parasitic weeds is highly linked with continuous monocropping and also with soil fertility depletion. These weeds withdraw resources from the crop and, consequently, lead to very low crop yields. This means that measures to shift from the common practice of monocropping to crop rotation and enhanced soil organic matter and fertility must be implemented. Soil management that aims to increase soil fertility by crop rotation has included the use of the cover crops tropical kudzu (*Pueraria phaseoloides* (Roxb.) Benth.) and calopo (Table 4.3). Tropical kudzu was the best option to control witchweed in northern Ivory Coast. Velvetbean and some varieties of cowpea safely reduced the population of witchweed and eradicated it after two seasons (Calegari *et al.*, 2005a).

Cover crop impact on crops

Crops respond to cover crops in many of the same ways as weeds do. Numerous reports have documented that live cover crops that are competitive enough to suppress weeds will also suppress a cash intercrop. Brandsæter *et al.* (1998) showed that a white clover (*Trifolium repens* L.) or subterranean clover living mulch suppressed both weeds and cabbage (*Brassica oleracea* L. convar. *capitata* (L.) Alef.). Sheaffer *et al.* (2002) found that annual medic (*Medicago* spp.) living mulch and weed growth were inversely related, but they also found an inverse relationship between medic growth and soybean yield. Maize (*Zea mays* L.) grain yield

was reduced by several annual legumes intercropped with maize for autumn forage (Alford, 2003) or by a hairy vetch living mulch (Reddy and Koger, 2004). Regrowth from a rye cover crop that was not adequately killed before planting a cash crop also reduced crop growth (Brainard and Bellinder, 2004; De Bruin *et al.*, 2005; Westgate *et al.*, 2005). Generally, crop suppression by living cover crops is the result of competition for essential resources.

Cover crop residue can suppress cash crop growth for many of the same reasons as weeds are suppressed by residue. Residue can interfere with crop establishment by physically interfering with seed placement in the soil, by maintaining cool soils, by releasing phytotoxins, or by enhancing seedling diseases (Dabney *et al.*, 1996, Davis and Liebman, 2003; Gallagher *et al.*, 2003; Westgate *et al.*, 2005). Reduced growth of crops in cover crop residue, particularly small-grain cover crops, has been associated with reduced availability of nitrogen, release of phytotoxins, and cooler soils (Norsworthy, 2004; Westgate *et al.*, 2005). On the other hand, cover crop residue on the soil surface has the capability of stimulating crop growth because of retention of soil moisture by a surface mulch (Araki and Ito, 1999; Gallagher *et al.*, 2003) and maintenance of cooler soils in a hot mid-season environment (Araki and Ito, 1999; Hutchinson and McGiffen, 2000). Also, legume cover crops can stimulate crop growth by increased availability of nitrogen (Gallagher *et al.*, 2003; Sarrantonio and Gallandt, 2003; Calegari *et al.*, 2005b) and promotion of genes that delay senescence and enhance disease resistance (Kumar *et al.*, 2004).

Table 4.3. Effect of cover crops on witchweed infection and maize yield in Africa.

Cover crop species	Maize plants infested by witchweed (%)	Maize yield (kg/ha)
Pueraria phaseoloides	3	2540
Calopogonium mucunoides	4	2260
Cassia rotundifolia	18	2310
Macroptilium atropurpureum	98	1250
Centrosema pubescens	100	1120
Tephrosia pedicellata	100	910
Control	100	730

Source: Charpentier *et al.* (1999).

4.3 Cover Crop Uses in Selected Climatic Regions

Northern temperate regions

The climate within this region is characterized by freezing winters and relatively cool and short summers (e.g. most areas in northern Europe and Canada), but this may be modified by latitude and distance from the coast. The opportunity to grow cover crops other than during the cash-cropping season decreases in northern and inland directions. In southern and coastal areas of this region (e.g. Denmark and southern Sweden), cover crops can be established after a cash crop is harvested (Thorup-Kristensen *et al.*, 2003). Sowing cover crops after early-harvested cash crops, such as early cultivars of potatoes (*Solanum tuberosum* L.) and vegetables, is also possible at locations with a short growing season. Danish studies have focused on root growth dynamics where specific catch crops are coupled to specific subsequent cash crops for two purposes: (i) optimal transfer of plant nutrients from year to year; and (ii) plant nutrient release in the most advantageous soil layer for the subsequent cash crop during the following year (Thorup-Kristensen *et al.*, 2003). The main purpose of using cover crops in northern regions has traditionally been to prevent erosion and nutrient leaching or for green manuring, but the focus on weed control in cover crop systems is increasing as well. We will focus on two commonly used cover crop systems in the region – undersown green manure and catch crops in cereals, and annual green manure cover crops in rotation with cash crops.

The most common cover crop practice in the Scandinavian region is undersowing of clover or clover–grass as a green manure (organic farms), or grass as a catch crop (conventional and organic farms) in cereals. When management is optimized for: (i) cereal and cover crop species and cultivar; (ii) sowing time and seeding rates of the cover crop; and (iii) soil fertility, there are often small or insignificant negative cover crop impacts on crop yield in these systems (Breland, 1996; Olesen *et al.*, 2002; Molteberg *et al.*, 2004). However, cereal yield depression because of competition from undersown cover crops has been reported

(Korsaeth *et al.*, 2002). Experiments in Norway have shown that pure stands of cereals are often outyielded by cereals undersown with white clover by as much as 500–1000 kg/ha in stockless cereal-dominated organic farming rotations (Henriksen, 2005). Studies in Norway have shown that ryegrass (*Lolium* spp.) as a catch crop established through undersowing in cereals retains 25–35 kg N/ha in the autumn (Molteberg *et al.*, 2004). Several studies have demonstrated that undersown green manure or catch crops reduce weed biomass (Hartl, 1989; Breland, 1996). The significance of undersown cover crop impacts on weed growth depends on whether the results are compared with an untreated control or with different levels of other treatments such as weed harrowing. A Danish study indicated that undersown cover crops gave equivalent weed control to low-intensity weed harrowing in plots without undersown cover crops; however, high-intensity weed harrowing gave better weed control than did cover crops (Rasmussen *et al.*, 2006). Although it is expected that a living cover crop may inhibit weed seedlings emerging from seed more than shoots from perennial storage organs (see Table 4.2), Dyke and Barnard (1976) found that Italian ryegrass (*Lolium multiflorum* Lam.) and red clover (*Trifolium pratense* L.) undersown in barley (*Hordeum vulgare* L.) suppressed quackgrass by more than 50% compared with barley alone. However, the promising result of this study may have been influenced by the reduction in the competitive ability of quackgrass because rhizomes were transplanted at a depth of 20 cm, which is much deeper than these organs normally reside. Preliminary results from Norway (L.O. Brandsæter *et al.*, unpublished data) indicate that red clover undersown in oat reduces the biomass of established stands of perennial sowthistle (*Sonchus arvensis* L.), and to some degree quackgrass, but does not suppress established stands of Canada thistle (*Cirsium arvense* (L.) Scop.). Rasmussen *et al.* (2005) has hypothesized that, because undersown cover crops keep plant nutrients in the upper soil layer, their presence favours crops with shallow roots over Canada thistle, which has deeper roots. Thus, the use of cover crops undersown in cereals may both increase crop nutrient supply for the subsequent crop in the rotation and decrease the growth of weeds. However, the use of cover crops also jeopardizes the use of mechanical weed control because farmers cannot weed-harrow in the crop after sowing the cover crop, and a growing cover crop in the autumn obstructs stubble cultivation for quackgrass control (Rasmussen *et. al.*, 2005), which is otherwise a standard non-chemical method for controlling this weed.

In Nordic organic stockless farming, the use of one entire growing season for a green manure cover crop is a common practice for many purposes, the most important of which are adding nitrogen to the soil and controlling perennial weeds. Generally, a 1-year green manure cover crop can be introduced into a cropping system by undersowing clover–grass in cereals the previous year (as described above), or by sowing the cover crop in the spring. One advantage of undersowing in the previous year is that few weeds will emerge in the spring after the green manure is established. Studies have shown that the soil weed seed bank decreases when this method is used (Sjursen, 2005). On the other hand, sowing the green manure crop in spring or early summer provides an opportunity for a period of soil cultivation in the autumn and/or spring before sowing the green manure cover crop. Fragmentation of roots or rhizomes by soil cultivation, followed by deep ploughing, is a classical approach for controlling perennial weeds (Håkansson, 2003). Furthermore, in classic experiments (Fig. 4.1), soil cultivation and ploughing followed by a competitive cash crop or cover crop has shown promising effects on quackgrass (Håkansson, 1968), perennial sowthistle (Håkansson and Wallgren, 1972) and Canada thistle (Thomsen *et al.*, 2004). This may offer a good approach for non-chemical control of creeping perennial weeds in cereal-dominated rotations, but additional research is needed in order to optimize these methods.

Generally, cover crop competition for 1 year alone is not sufficient to satisfactorily suppress perennial weeds such as Canada thistle and perennial sowthistle. These weeds also have to be mowed frequently at specific stages of development. Mechanical disturbance for weakening a perennial weed plant is theoretically most effective when the plant has reached the stage with minimum reserves in underground storage structures, although more research is needed to

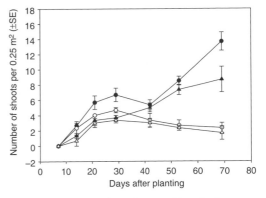

Fig. 4.1. Influence of root fragment length: 5-cm (triangles) vs 10-cm root pieces (circles); and competition: with (empty symbols) vs without cover crop (filled symbols) on the number of shoots of Canada thistle per 0.25 m². The root fragments were transplanted at a depth of 5 cm. Vertical bars represent standard error (SE). (*Source:* Thomsen *et al.*, 2004.)

more readily identify these stages (see overview in Håkansson, 2003). Factorial experiments are required to separate the effects of cover crop competition from the effects of mowing, and to determine potential interactions between mechanical and cover crop effects. In a field study conducted by Dock Gustavsson (1994) comparing times of 1 week and 5 weeks between mowing treatments, it was shown that Canada thistle growing in a red clover cover crop should preferably be mowed at intervals of 4 weeks to obtain the best suppression of the thistle without killing the red clover cover crop. More frequent mowing killed the clover plants or damaged them severely. The author also concluded that mowing in the spring and early summer suppressed weed growth more than mowing at later dates. In similar studies in Denmark with mixtures of white clover and grass as cover crops, Graglia *et al.* (2006) demonstrated an inverse linear relationship between the number of mowing passes up to six times between mid-May to late July and the above-ground biomass of Canada thistle in the subsequent year. The correlation between the weed control level and the yield of spring barley during the year following cover cropping was, however, not always positive. The reason was probably that mowing not only influenced weed growth but also the cover crop's ability to add nitrogen to the cropping system. Graglia et al. (2006) concluded that the presence of clover–grass cover crops strongly decreased the above-ground biomass of Canada thistle, presumably by suppressing the regrowth of shoots in the late-summer and autumn period that followed the ending of the mowing treatments. Hence, we can conclude that a continuous depletion of carbohydrates from the root system, resulting from a joint effect of mowing and competition by the cover crop, will decrease the regrowth capacity of Canada thistle as well as other perennials such as quackgrass and perennial sowthistle.

Warmer temperate regions

In this section we discuss the use of cover crops in regions characterized by cold, usually freezing, winters but with longer and warmer summers than in the northern regions discussed above (e.g. most areas of the USA, southern Europe, and Japan). In these regions there are suitable conditions for planting and growing winter annual cover crops after a cash crop is harvested in late summer/autumn and before the next cash crop is planted the following spring. Adapted species have the capability of: (i) reliably producing a uniform stand of established plants in autumn before the onset of cold weather; (ii) surviving freezing weather during winter; and (iii) rapidly growing during cool conditions in spring before planting a cash crop. Growth of a cover crop during this period has the advantage that summer annual weed species established before cover crop planting are destroyed by planting operations and those that become established after planting will be winterkilled. The only troublesome weed species that establish with winter annual cover crops are winter annuals and perennials that continue growth after the cover crop is terminated and the cash crop is planted. Typically, vigorous and well-adapted cover crops such as rye or hairy vetch will provide complete ground cover and be highly competitive, leaving relatively few weeds at the time of planting a spring crop.

Rye is a commonly used cover crop that is grown before summer annual cash crops and is representative of the use of small-grain cover crops in general. Rye can provide many bene-fits, including protecting the environment from loss of sediments, nutrients and agrochemicals (Sustainable Agriculture Network, 1998; Sarrantonio and Gallandt, 2003). It protects soils from water and wind erosion during the winter months and captures nutrients that may be leached during rainy periods when the soil is not frozen. If rye is terminated and the residue is left on the soil surface in conservation tillage systems, it can protect the environment from water and wind erosion during the period of crop establishment as well as from runoff losses of nutrients and agrochemicals. Other advantages include allowing earlier entry to fields in spring than would be possible with tilled soil. Long-term benefits include the sequestration of carbon and maintenance of soil organic matter, with related benefits for soil quality.

Residue of rye or other small-grain cover crops remaining on the soil surface after cover crop termination can suppress weed emergence and biomass in subsequent crops, particularly in the absence of herbicide (Kobayashi et al., 2004; Norsworthy, 2004). When rye termination was delayed, resulting in more residue biomass, greater weed suppression was achieved (Ashford and Reeves, 2003; Westgate et al., 2005). However, rye and small-grain cover crops also have been shown to reduce crop stands and yields because of interference with proper seed placement, cooler soils, and the release of phyto-toxins from decomposing residue (Reddy, 2001; Norsworthy, 2004; Westgate et al., 2005). Interseeding rye or small-grain cover crops tended to provide higher levels of weed suppression when interseeded at or near planting, but also tended to reduce crop yields under these conditions (Rajalahti et al., 1999; Brainard and Bellinder, 2004). Generally, management that increased weed suppression also tended to increase the risk of crop yield reductions. Weed suppression by rye or other small-grain cover crops without herbicide usually was not adequate on its own, and herbicide programmes were required in order to achieve maximum crop yield (Rajalahti et al., 1999; Reddy, 2001; Gallagher et al., 2003; Norsworthy, 2004; De Bruin et al., 2005). These results suggest that management of rye or small-grain cover crops should focus on optimization of the environmental rather than the weed-suppressive benefits of these cover crops.

Winter annual legume species represent another important group of cover crops in temperate climates. Hairy vetch is the most winter-hardy and reliable winter annual legume and is used primarily because of potential benefits to soil fertility (Sustainable Agriculture Network, 1998; Sarrantonio and Gallandt, 2003). The most important benefit is the release of nitrogen from killed vegetation to subsequent cash crops and the significant reduction in fertilizer nitrogen requirements. For this reason, it is often used preceding crops with a high nitrogen requirement such as maize or tomatoes. Hairy vetch mulch on the soil surface in no-tillage systems has increased soil moisture availability to crops during summer by increasing infiltration of rain and preventing evaporation on drought-prone soils. It also has been shown to trigger expression of genes that delay senescence and enhance disease resistance in tomatoes (Kumar *et. al.*, 2004) and to suppress certain pests, pathogens and weeds.

The impact of hairy vetch on weed emergence depends on many factors. Higher than naturally produced biomass levels on the soil surface (>5 t/ha of dry residue) can inhibit the emergence of many annual weed species. At naturally produced levels (usually 3–5 t/ha of dry residue), weed emergence may be suppressed, unaffected or stimulated, depending on species and conditions (Teasdale and Mohler, 2000). Araki and Ito (1999) showed a high level of weed suppression by hairy vetch residue in Japan. However, typically there is, at best, a temporary suppression of early emergence but little long-term control. The leguminous nature of this residue (low C:N ratio) results in more rapid degradation and less suppressive amounts of residue over time than rye residue (Mohler and Teasdale, 1993). Also, the release of inorganic nitrogenous compounds can trigger germination and stimulate emergence of selected weeds, e.g. *Amaranthus* spp. (Teasdale and Pillai, 2005). Attempts to allow hairy vetch to continue growth as a living mulch during early growth of cash crops have provided improved weed control but have also proved detrimental to crop populations, growth and yield (Czapar *et al.*, 2002; Reddy and Koger, 2004). Generally, as with a rye cover crop, most research shows that hairy vetch does not provide reliable full-season weed control and must be combined with additional weed management options, usually herbicides, in order to achieve acceptable control.

Many research projects have investigated the influence of cover crops in a factorial with other management practices. In most cases, management has a bigger influence on weed control than cover crops do. Barbari and Mazzoncini (2001) conducted a long-term factorial study of cover crops and management systems, including tillage and herbicide factors. They found that weed abundance was influenced most by management system rather than cover crop, although cover crop did influence weed community composition within a low-input, minimum-tillage management system. Swanton *et al.* (1999) determined that tillage was more important than nitrogen rate or a rye cover crop in having a long-term influence on weed density or species composition in a maize crop. Peachey *et al.* (2004) showed by variance partitioning that primary tillage was much more important than cover crop in regulating weed emergence in vegetable crops. Since minimum-tillage agriculture can make many important contributions to preserving and building soil quality and fertility, management of cover crops to enhance soil-building and environmental contributions to minimum-tillage systems appears to be more important than management for weed suppression in temperate cropping systems.

Organic production systems have become an increasingly important segment of agriculture in recent years. In the absence of herbicide and fertilizer products, cover crops play a more important role for weed management and fertility in organic than in conventional farming. Legumes are necessary cover crops, either alone or in mixtures, because of the need to produce nitrogen as part of the on-farm system. The use of living legume cover crops to suppress weeds during fallow periods can be successful (Blackshaw *et al.*, 2001; Fisk *et al.*, 2001; Ross *et al.*, 2001). This may be most important to organic weed management as a means to reduce weed seed production and accelerate weed seed predation within rotational programmes. The use of cover crops in minimum-tillage organic crop production is a worthy objective in order to realize the environmental benefits of both reduced tillage and organic farming, but it can be problematic on organic farms for several reasons. Mechanical implements must be used to terminate cover crops and the results can be inconsistent; however, cover crops mowed or rolled at flowering can be killed more effectively than when operations are performed while the cover crop remains vegetative (Ashford and Reeves, 2003; Teasdale and Rosecrance, 2003; De Bruin et al., 2005). As discussed earlier, residue on the soil surface will not consistently control weeds over a full season. Mechanical removal of weeds with a higher-residue cultivator has been shown to be less efficient in minimally tilled than in previously tilled soil (Teasdale and Rosecrance, 2003), thereby reducing the capacity for effective post-emergence weed control in minimum-tillage organic systems. The success of high-residue cover crop systems will depend on effective residue management to alleviate interference with crop production while maximizing interference with weed growth.

Subtropical/tropical South America

Agricultural conditions in warmer regions with potentially high rainfall make it difficult to maintain soil organic matter and to retain residue on the soil surface. Weed, nematode and pest populations can grow without interruption throughout the year. Bare soil is exposed to high levels of erosion from heavy rainfall, and soils can warm to temperatures that suppress productive root and biological activity. Cover crops can play an important role in alleviating all of these problems.

Concern over preserving soil and water in Brazil was not a priority until the 1970s. With the spread of annual crop production, monocultures, and tractor mechanization (which almost doubled in Paraná State in the 1970s), and with practically no conservation methods used, there was an acceleration of erosion and a decrease in organic matter and nutrients. This gave impetus to soil and water preservation efforts. The no-tillage system that has been developed includes the use of different species of cover crops and crop rotation as fundamentals in the structure of rational and sustainable management for annual crops. Almost all the advantages of the no-tillage system come from the permanent cover of the soil. Cover crops are planted primarily to protect the soil from the direct impact of raindrops. Protection is given

by the growing plants themselves as well as by their residues. A total cover of the soil with plant residues improves the infiltration of rainfall. At the same time, cover crops have the potential to improve soil fertility as green manure cover crops.

The use of cover crops and crop rotation, as well as permanent no-tillage, are the key factors for the unprecedented growth of no-tillage, especially in Brazil and Paraguay. Only those farmers who have understood the importance of these practices are obtaining the highest economic benefits from this system. The systematization of these practices through work in hydrological micro-basins has advanced to a point where these systems occupy more than 5.2 million ha in Paraná, and about 23 million ha in Brazil. Controlled studies conducted on the St Antonio farm in Floresta, North Paraná (500 ha), comparing both tillage systems on a cultivated area of 1.6 ha over a 6-year period, found that no-till systems yielded approximately 34% more soybeans and 14% more wheat (*Triticum aestivum* L.) than did conventional tillage systems. Growing these crops in rotation with cover crops rather than as a monoculture added 19% and 6%, respectively, to soybean and wheat yields (Calegari et al., 1998). A separate study on a 50 ha experimental site in North Paraná gave further evidence that a well-designed no-till system with soybeans in crop rotation can generate net income gains compared with conventional systems. Soybean production in a no-till system resulted in a US$3960 increase in revenue based on higher yields, and US$4942 in savings on machinery, fuel, labour and fertilizer compared with conventional tillage, resulting in a total benefit of US$8902 from 50 ha (Calegari et al., 1998). Thus, experimental results and farmers' practices in the tropics and temperate climates have shown the important effects of cover crop use, crop rotation and no-tillage production to improve soil properties, increase crop yields, and contribute to biodiversity and environmental equilibrium.

The most common cover crop species are black oats (*Avena strigosa* Schreb.) in subtropical areas and pearl millet (*Pennisetum americanum* (L.) Leeke) in tropical areas. The most frequent species used for mixtures with black oats are vetch (*Vicia* spp.), lupin (*Lupinus* spp.)

or radish (*Raphanus sativus* L.). Facility of seed production (and therefore lower price and greater availability on the market), good biomass production, and minimal input requirements are the reasons that farmers prefer black oats and pearl millet as cover crop species. They have good tolerance to pests and diseases and can grow in low-fertility conditions. Black oats are used on about 3.2 million ha in the states of Paraná and Rio Grande do Sul, Brazil, and on about 300,000 ha in Paraguay, mainly in mechanized farming systems.

One important characteristic of cover crops is their ability to suppress and smother weeds. Favero *et al.* (2001) found that cover crops modified the dynamics of weed species occurrence. Weed populations were reduced by different amounts depending on cover crop mass and species (Severino and Christoffoleti, 2004). Skora Neto and Campos (2004) demonstrated the effect of fallow period and the suppressive effect of cover crops on succeeding weed populations. In a period of 3 years, a weed population of 136 plants/m² was reduced to 9 plants/m² when cover crops were used during fallow periods (Fig. 4.2). One important aspect of a weed management programme is not leaving a niche between the harvesting of a crop and the sowing of the next, in which weeds are able to establish. The occupation of space during fallow is important not only during crop development, but also during the intervals between them. The use of cover crops in these intervals has a profound effect on weed populations; otherwise, fallow periods allow weeds to capture space and to replenish the seed bank.

Another option to maintain ground cover and produce more cash crops in the rotation is intercropping with cover crops. In small-scale farming, maize is one crop in which this operation is practised; cover crops suitable for intercropping are jack bean, dwarf pigeonpea (*Cajanus cajan* (L.) Millsp.) and showy rattlebox (*Crotalaria spectabilis* Roth). They are used primarily as a green manure; however, their smothering effect also provides good weed suppression at the harvest and postharvest stages of maize (Skora Neto, 1993).

Cover crop residues can also be effective for suppressing weeds through physical and allelopathic mechanisms. Mulch from cover crop species with high biomass production and with

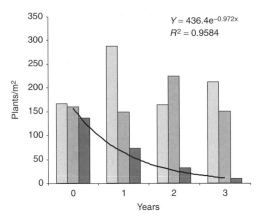

Fig. 4.2. Weed population change during three years with autumn and winter fallow (no cover crops; light grey), autumn fallow (cover crop in winter; dark grey), and without fallow (cover crop in autumn and winter; black). Solid line shows exponential decay for treatment without fallow. (*Source:* Skora Neto and Campos, 2004.)

slow decomposition (higher C:N ratio) are more effective for weed population reduction. Almeida and Rodrigues (1985) demonstrated a 2.5 t/ha weed biomass reduction for each 1.0 t/ha dry biomass of residues on the soil surface (Fig. 4.3). Soil tillage also affects weed density. Almeida (1991), verified, at 63 days after preparing the soil, that the weed infestation in conventional tillage (ploughing and harrowing) was 187% higher than with no tillage. The cumulative effect of absence of tillage and presence of mulching (physical and allelopathic effects) can be an important integrated strategy for reducing weed populations.

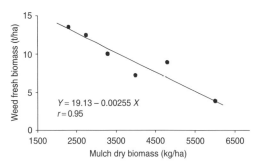

Fig. 4.3. Correlation between mulch dry biomass (kg/ha) and weed fresh biomass (t/ha) 85 days after mulch formation. (*Source:* Almeida and Rodrigues, 1985.)

Results demonstrating weed suppression by cover crops suggest the possibility of reducing the amounts of herbicides applied, and consequently reducing costs. Adegas (1998) describes a joint study by several institutes of an integrated weed management (IWM) programme on 58 farms in Parana State, Brazil, comparing the IWM approach including cover crops against the farmers' weed-control practices. The results after 3 years of evaluation were a 35% decrease in average costs and a herbicide reduction of 25%. Bianchi (1995) shows that, across 34 local areas at Rio Grande do Sul State, Brazil, the IWM programme reduced weed control costs by 42% compared with farmers' practices. These results demonstrate the agronomic, economic and ecological viability of IWM including cover crops.

Although a reduction in herbicide use has been observed on farms where good no-tillage practices are used, the total elimination of herbicides in crop production seems difficult, especially on large-scale farms. For small farms, where labour is available and the weed density is low, it is possible. To eliminate the herbicides before planting it is necessary to use cover crops that can be managed mechanically (knife-rolled). For example, oats, rye, radish, lupin and sunn hemp are some species that can be rolled down during the stage of seed formation without regrowing and which will form an effective mulch without herbicide. During the crop season, however, it is necessary to rely on manual labour and that can be time-consuming and full of drudgery. Skora Neto *et al.* (2003), in a study carried out at the farm level, verified the possibilities of no-tillage without herbicides; the constraint was the labour requirements for weed control. Areas with low weed populations were more suitable for no-tillage without herbicide. Skora Neto and Campos (2004) measured the hoeing time in two weed populations. With a high weed population (180 plants/m²), hoeing time was 231 h/ha, and at low weed density (9 plants/m²) it was 71 h/ha.

To overcome the constraints of labour requirements in no-till systems, Almeida (1991) recommends avoiding weed seed production as a way of reducing the weed seed bank and, as a consequence, the level of weed population and the inputs to control them. One way of reducing weed seed production is to occupy the area at all times with crops or cover crops (Almeida, 1991; Adegas, 1998). Kliewer (2003), in Paraguay, demonstrated the viability of practicing no-tillage without herbicides during successive years where the main strategy in a production system of soybean–wheat–soybean and maize–wheat–soybean was to use cover crops with fast growth and short cycle during the period between the summer crop and the wheat. Sunflower (*Helianthus annuus* L.) and sunn hemp eliminated the period of weed growth and reproduction. Therefore to reduce and eventually eliminate the use of herbicides, an appropriate rotation of ground covers and crops, a mulch effect on weed suppression, weed seed production control, and weed seed bank depletion are strategies to be pursued.

4.4 Conclusions

Cover crops can be used most reliably to suppress weeds during the vegetative growth phase of the cover crop (Table 4.1). Adapted warm-season cover crops can be used in rotations in subtropical and tropical areas to reduce populations of important weeds during fallow periods (Akobundu *et al.*, 2000; Skora Neto and Campos, 2004). Adapted cool-season cover crops can also suppress weeds during fallow years in northern temperate regions such as the semiarid Canadian Great Plains (Blackshaw *et al.*, 2001) or northern Europe (Thomsen *et al.*, 2004; Graglia *et al.*, 2006). It is noteworthy that live cover crops can be effective in suppressing important perennial weeds ranging from cogongrass in Africa to quackgrass and Canada thistle in Scandinavia. The maintenance of a vigorous ground cover during fallow periods in crop rotations represents an application of cover crops where the goals for weed management coincide well with other important environmental goals such as improving soil quality and fertility and reducing erosion.

Cover crops must be managed carefully to optimize environmental benefits and minimize potential liabilities for crop production. Cover crops that have been grown during any period unavailable for cash crops, whether a fallow period as discussed above or an off-season winter period in temperate production systems, will need to be managed before planting the next cash crop. The cover crop essentially becomes a weed that needs to be managed properly or it will become a liability rather than a benefit. These liabilities typically include consumption of soil moisture, interference with planting operations, negative effects from phytotoxins or nutrient sequestration, and direct competition with the cash crop for resources. Much of the cover-crop literature has focused on the determination of optimum management approaches for eliminating negative effects on cash crops and achieving maximum benefits.

Residue remaining after death of the cover crop is less reliable for suppressing weeds, particularly for the duration of a cash-crop season. This has led to many lines of research to enhance the inconsistent weed control achieved by cover crop residue. Attempts to increase residue biomass can enhance weed suppression but can also enhance the probability of crop suppression. Another strategy has been the use of the more effective weed-suppressive capabilities of live cover crops by developing various intercropping systems; however, research has shown that most live cover crops effective enough to suppress weeds will also suppress crops. Thus, the biggest trade-offs between optimizing weed control and enhancing environmental protection occur during the cash-cropping period that follows cover cropping. In this case, since the cover crop cannot be expected to adequately control weeds without interfering with the cash crop,

management of cover crops should focus on enhancing their environmental benefits to the agroecosystem rather than their contribution to weed management.

Cover crops may ultimately contribute most to weed management within subsequent cash crops by reducing weed populations during fallow periods. The agronomic goal would be to replace unmanageable weed populations with a more manageable cover crop population. As discussed above, live cover crops can significantly suppress weed biomass, seed production, and growth of perennial structures. In addition, research has suggested that live vegetation may be important for enhancing the activity of seed predators and the reduction of seed populations. More research is needed in order to understand the effects of cover crops on weed seed production and predation and on seed mortality in soil. More research is also needed on perennial weed responses to cover crops. Regulation of weed population dynamics and community structure could become an important objective for future weed management programmes using cover crops. Research in many areas of the world has shown that the suite of management practices deployed in association with cover-cropping rotations (e.g. tillage, herbicide, mowing, and the timing of these operations) often enhance weed management more than cover crops alone. Long-term cover-cropping strategies are needed that integrate cover-crop management, weed population regulation, and enhanced environmental services.

4.5 References

Adegas, F.S. (1998) Manejo integrado de plantas daninhas em plantio direto no Paraná. In: *Seminário Nacional sobre Manejo e Controle de Plantas Daninhas em Plantio Direto. 1: Passo Fundo*. Resumo de Palestras (ed.). Aldeia Norte, Passo Fundo, RS, Brazil, pp. 17–26.

Akemo, M.C., Regnier, E.E. and Bennett, M.A. (2000) Weed suppression in spring-sown rye–pea cover crop mixes. *Weed Technology* 14, 545–549.

Akobundu, I.O., Udensi, U.E. and Chikoye, D. (2000) Velvetbean suppresses speargrass and increases maize yield. *International Journal of Pest Management* 46, 103–108.

Alford, C.M. (2003) Intercropping irrigated corn with annual legumes for fall forage in the high plains. *Agronomy Journal* 95, 520–525.

Almeida, F.S. (1991) *Controle de plantas daninhas em plantio direto*. Circular 67, IAPAR, Londrina, PR, Brazil.

Almeida, F.S. and Rodrigues, B.N. (1985) Guia de herbicidas. In: *Contribuição para o uso adequado em plantio direto e convencional*. IAPAR, Londrina, PR, Brazil, 482 pp.

Araki, H. and Ito, M. (1999) Soil properties and vegetable production with organic mulch and no-tillage system. *Japanese Society of Farm Work Research* 34, 29–37.

Ashford, D.L. and Reeves, D.W. (2003) Use of a mechanical roller-crimper as an alternative kill method for cover crops. *American Journal of Alternative Agriculture* 18, 37–45.

Barbari, P. and Mazzoncini, M. (2001) Changes in weed community composition as influenced by cover crop and management system in continuous corn. *Weed Science* 49, 491–499.

Bianchi, M.A. (1995) *Programa de difusão de manejo integrado de plantas daninhas em soja no Rio Grande do Sul 1994/95.* Fundacep Fecotrigo, Cruz Alta, RS, Brazil, 31 pp.

Blackshaw, R.E., Moyer, J.R., Doram, R.C. and Boswell, A.L. (2001) Yellow sweetclover, green manure, and its residues effectively suppress weeds during fallow. *Weed Science* 49, 406–413.

Brainard, D.C. and Bellinder, R.R. (2004) Weed suppression in a broccoli–winter rye intercropping system. *Weed Science* 52, 281–290.

Brandsæter, L.O., Netland, J. and Meadow, R. (1998) Yields, weeds, pests and soil nitrogen in a white cabbage–living mulch system. *Biology, Agriculture and Horticulture* 16, 291–309.

Breland, T.A. (1996) Green manuring with clover and ryegrass catch crops undersown in small grains: crop development and yields. *Soil and Plant Science* 46, 30–40.

Brennan, E.B. and Smith, R.F. (2005) Winter cover crop growth and weed suppression on the central coast of California. *Weed Technology* 19, 1017–1024.

Calegari, A., Ferro, M. and Darolt, M. (1998) Towards sustainable agriculture with a no-tillage system. *Advances in GeoEcology* 31, 1205–1209.

Calegari, A., Ashburner, J. and Fowler, R. (2005a) *Conservation Agriculture in Africa.* FAO, Regional Office for Africa, Accra, Ghana, 91 pp.

Calegari, A., Ralisch, R. and Guimarães, M.F. (2005b) The effects of winter cover crops and no-tillage on soil chemical properties and maize yield in Brazil. *III World Congress on Conservation Agriculture.* CD ROM, Nairobi, Kenya, 8 pp.

Charpentier, H., Doumbia, S., Coulibaly, Z. and Zana, O. (1999) Fixation de l'agriculture au nord de la Côte d'Ivoire: quells nouveaux systèmes de culture? *Agriculture et Développement* 21, 4–70.

Creamer, N.G. and Baldwin, K.R. (2000) An evaluation of summer cover crops for use in vegetable production systems in north Carolina. *HortScience* 35, 600–603.

Czapar, G.F., Simmons, F.W. and Bullock, D.G. (2002) Delayed control of a hairy vetch cover crop in irrigated corn production. *Crop Protection* 21, 507–510.

Dabney, S.M., Schreiber, J.D., Rothrock, C.S. and Johnson, J.R. (1996) Cover crops affect sorghum seedling growth. *Agronomy Journal* 88, 961–970.

Davis, A.S. and Liebman, M. (2003) Cropping system effects on giant foxtail demography. I. Green manure and tillage timing. *Weed Science* 51, 919–929.

De Bruin, J.L., Porter, P.M. and Jordan, N.R. (2005) Use of a rye cover crop following corn in rotation with soybean in the upper Midwest. *Agronomy Journal* 97, 587–598.

Dock Gustavsson, A.-M. (1994) Åkertistelns reaktion på avslagning, omgrävning och konkurrens. Fakta Mark / växter 13, SLU, Uppsala.

Dyke, G.V. and Barnard, J. (1976) Suppression of couch grass by Italian ryegrass and broad red clover undersown in barley and field beans. *Journal of Agricultural Science* 87, 123–126.

Favero, C., Jucksch, I. and Alvarenga, R.C. (2001) Modifications in the population of spontaneous plants in the presence of green manure. *Pesquisa Agropecuária Brasileira* 36, 1355–1362.

Fisk, J.W., Hesterman, O.B., Shrestha, A., Kells, J.J., Harwood, R.R., Squire, J.M. and Sheaffer, C.C. (2001) Weed suppression by annual legume cover crops in no-tillage corn. *Agronomy Journal* 93, 319–325.

Gallagher, R.S., Cardina, J. and Loux, M. (2003) Integration of cover crops with postemergence herbicides in no-till corn and soybean. *Weed Science* 51, 995–1001.

Gallandt, E.R., Molloy, T., Lynch, R.P. and Drummond, F.A. (2005) Effect of cover-cropping systems on invertebrate seed predation. *Weed Science* 53, 69–76.

Graglia, E., Melander, B. and Jensen, R.K. (2006) Mechanical and cultural strategies to control *Cirsium arvense* in organic arable cropping systems. *Weed Research* 46, 304–312.

Grimmer, O.P. and Masiunas, J.B. (2004) Evaluation of winter-killed cover crops preceding snap pea. *HortTechnology* 14, 349–355.

Håkansson, S. (1968) Experiments with *Agropyron repens* (L.) Beauv. II. Production from rhizome pieces of different sizes and from seeds: various environmental conditions compared. *Annals of the Agricultural College of Sweden* 34, 3–29.

Håkansson, S. (2003) *Weeds and Weed Management on Arable Land: An Ecological Approach.* CABI Publishing, Wallingford, Oxon, UK, 274 pp.

Håkansson, S. and Wallgren, B. (1972) Experiments with *Sonchus arvensis* L. III. The development from reproductive roots cut into different lengths and planted at different depths, with and without competition from barley. *Swedish Journal of Agricultural Research* 2, 15–26.

Haramoto, E.R. and Gallandt, E.R. (2005) Brassica cover cropping. II. Effects on growth and interference of green bean and redroot pigweed. *Weed Science* 53, 702–708.

Hartl, W. (1989) Influence of undersown clovers on weeds and on the yield of winter wheat in organic farming. *Agricuture, Ecosystems and Environment* 27, 389–396.

Henriksen, T.M. (2005) Levende gjødsel. *Grønn Kunnskap* 9, 390–393.

Hutchinson, C.M. and McGiffen, M.E., Jr (2000) Cowpea cover crop mulch for weed control in desert pepper production. *HortScience* 35, 196–198.

Kliewer, I. (2003) Alternativas de controle de plantas daninhas sem herbicidas. In: *World Congress on Conservation Agriculture, 2.* 2003, Iguassu Falls. Lectures, v.I. Ponta Grossa: FEBRAPDP, pp. 107–110.

Kobayashi, H., Miura, S. and Oyanagi, A. (2004) Effects of winter barley as a cover crop on the weed vegetation in a no-tillage soybean. *Weed Biology and Management* 4, 195–205.

Korsaeth, A., Henriksen, T.M. and Bakken, L.R. (2002) Temporal changes in mineralization and immobilization of N during degradation of plant material: implications for the plant N supply and nitrogen losses. *Soil Biology and Biochemistry* 34, 789–799.

Kumar, V., Mills, D.J., Anderson, J.D. and Mattoo, A.K. (2004) An alternative agriculture system is defined by a distinct expression profile of select gene transcripts and proteins. *Proceedings of the National Academy of Sciences* 101, 10535–10540.

Mohler, C.L. and Teasdale, J.R. (1993) Response of weed emergence to rate of *Vicia villosa* Roth and *Secale cereale* L. residue. *Weed Research* 33, 487–499.

Molteberg, B., Henriksen, T.M. and Tangsveen, J. (2004) Use of catch crops in cereal production in Norway. *Grønn Kunnskap* 8(12), 57.

Norsworthy, J.K. (2004) Small-grain cover crop interaction with glyphosate-resistant corn. *Weed Technology* 18, 52–59.

Olesen, J.E., Rasmussen, I.A., Askegaard, M. and Kristensen, K. (2002) Whole-rotation dry matter and nitrogen grain yields from the first course of an organic farming crop rotation experiment. *Journal of Agricultural Science* 139, 361–370.

Peachey, R.E., William, R.D. and Mallory-Smith, C. (2004) Effect of no-till or conventional planting and cover crop residues on weed emergence in vegetable row crops. *Weed Technology* 18, 1023–1030.

Rajalahti, R.M., Bellinder, R.R. and Hoffman, M.P. (1999) Time of hilling and interseeding affects weed control and potato yield. *Weed Science* 47, 215–225.

Rasmussen, I.A., Askegaard, M. and Olesen, J.E. (2005) Long-term organic crop rotation experiments for cereal production: perennial weed control and nitrogen leaching. In: *NJF-Seminar 369: Organic Farming for a New Millennium – Status and Future Challenges.* Alnarp, Sweden, pp. 153–156.

Rasmussen, I.A., Askegaard, M., Olesen, J.E. and Kristiansen, K. (2006) Effects on weeds of management in newly converted organic crop rotations in Denmark. *Agriculture, Ecosystems and Environment* 113, 184–195.

Reddy, K.N. (2001) Effects of cereal and legume cover crop residues on weeds, yield, and net return in soybean. *Weed Technology* 15, 660–668.

Reddy, K.N. and Koger, C.H. (2004) Live and killed hairy vetch cover crop effects on weeds and yield in glyphosate-resistant corn. *Weed Technology* 18, 835–840.

Ross, S.M., King, J.R., Izaurralde, R.C. and O'Donovan, J.T. (2001) Weed suppression by seven clover species. *Agronomy Journal* 93, 820–827.

Sustainable Agriculture Network (1998) *Managing Cover Crops Profitably,* 2nd edn. Handbook Series Book 3. United States Department of Agriculture, Beltsville, MD, USA, 212 pp.

Sarrantonio, M. and Gallandt, E.R. (2003) The role of cover crops in North American cropping systems. *Journal of Crop Production* 8, 53–73.

Severino, F.J. and Christoffoleti, P.J. (2004) Weed suppression by smother crops and selective herbicides. *Scientia Agricola* 61, 21–26.

Sheaffer, C.C., Gunsolus, J.L., Grimsbo Jewett, J. and Lee, S.H. (2002) Annual *Medicago* as a smother crop in soybean. *Journal of Agronomy and Crop Science* 188, 408–416.

Sjursen, H. (2005) Sammenhengen mellom ugrasfrøbanken og -framspiring. *Grønn Kunnskap* 9(3), 1–36.

Skora Neto, F. (1993) Controle de plantas daninhas através de coberturas verdes consorciadas com milho. *Pesquisa Agropecuária Brasileira* 28, 1165–1171.

Skora Neto, F. and Campos, A.C. (2004) Alteração populacional da flora infestante pelo manejo pós-colheita e ocupação de curtos períodos de pousio com coberturas verdes. *Boletim Informativo da Sociedade Brasileira da Ciência das Plantas Daninhas (SBCPD)* 10 (Suppl.), 135.

Skora Neto, F., Milleo, R.D.S., Benassi, D., Ribeiro, M.F.S., Ahrens, D.C., Campos, A.C. and Gomes, E.P. (2003) No-tillage without herbicides: results with farmer-researchers at the central-southern region of the Parana State. In: *World Congress on Conservation Agriculture, 2, Iguassu Falls.* Extended Summary, vol. II. FEBRAPDP/CAAPAS. Ponta Grossa, PR, Brazil, pp. 543–545.

Stivers-Young, L. (1998) Growth, nitrogen accumulation, and weed suppression by fall cover crops following early harvest of vegetables. *HortScience* 33, 60–63.

Swanton, C.J., Shrestha, A., Roy, R.C., Ball-Coelho, B.R. and Knezevic, S.Z. (1999) Effect of tillage systems, N, and cover crop on the composition of weed flora. *Weed Science* 47, 454–461.

Taimo, J.P.C., Calegari, A. and Schug, M. (2005) Conservation agriculture approach for poverty reduction and food security in Sofala Province, Mozambique. *III World Congress on Conservation Agriculture: Linking Production, Livelihoods and Conservation.* 3–7 October, Nairobi, CD-ROM.

Teasdale, J.R. and Daughtry, C.S.T. (1993) Weed suppression by live and desiccated hairy vetch. *Weed Science* 41, 207–212.

Teasdale, J.R. and Mohler, C.L. (1993) Light transmittance, soil temperature, and soil moisture under residue of hairy vetch and rye. *Agronomy Journal* 85, 673–680.

Teasdale, J.R. and Mohler, C.L. (2000) The quantitative relationship between weed emergence and the physical properties of mulches. *Weed Science* 48, 385–392.

Teasdale, J.R. and Pillai, P. (2005) Contribution of ammonium to stimulation of smooth pigweed germination by extracts of hairy vetch residue. *Weed Biology and Management* 5, 19–25.

Teasdale, J.R. and Rosecrance, R.C. (2003) Mechanical versus herbicidal strategies for killing a hairy vetch cover crop and controlling weeds in minimum-tillage corn production. *American Journal of Alternative Agriculture* 18, 95–102.

Thomsen, M.G., Brandsæter, L.O. and Fykse, H. (2004) Temporal sensitivity of *Cirsium arvense* in relation to competition, and simulated premechanical treatment. In: *6th EWRS Workshop on Physical and Cultural Weed Control.* Lillehammer, Norway, p. 184.

Thorup-Kristensen, K., Magid, J. and Jensen, L.S. (2003) Catch crops and green manures as biological tools in nitrogen management in temperate zones. *Advances in Agronomy* 79, 227–302.

Westgate, L.R., Singer, J.W. and Kohler, K.A. (2005) Method and timing of rye control affects soybean development and resource utilization. *Agronomy Journal* 97, 806–816.

Yenish, J.P., Worsham, A.D. and Chilton, W.S. (1995) Disappearance of DIBOA-glucoside, DIBOA, and BOA from rye cover crop residue. *Weed Science* 43, 18–20.

5 Allelopathy: A Potential Tool in the Development of Strategies for Biorational Weed Management

L.A. Weston[1] and Inderjit[2]

[1]College of Agriculture and Life Sciences, Cornell University, Ithaca, NY 14853, USA; [2]Centre for Environmental Management of Degraded Ecosystems (CEMDE), University of Delhi, Delhi 110007, India

5.1 Introduction

Plants are known to release chemicals into the environment by several means which, depending upon edaphic and climatic factors, may influence the growth of neighbouring species (Inderjit and Weiner, 2001). The phenomenon is referred to as allelopathy, and could be exploited for the development of non-chemical weed management through the use of: (i) allelopathic cover crops; (ii) allelochemicals as natural herbicides; or (iii) allelopathic crop cultivars (Bhowmik and Inderjit, 2003; Weston and Duke, 2003). The extensive use of synthetic herbicides in landscapes and crop production systems is now receiving increased public scrutiny from the standpoint of both environmental and public health concerns (Macias, 1995). Although natural and organic products are often considered by the public to be safer and more environmentally sound management strategies, few offer the ability to achieve consistent and dependable weed control at present. In addition, natural products often exhibit reduced soil persistence and are often easily biodegradable, which can limit the longevity of weed management in many systems (Duke, 1988; Inderjit and Keating, 1999; Inderjit and Bhowmik, 2002). Although natural products are not widely utilized either as templates for herbicidal development or as products in their own right, allelopathic plants have potential for weed suppression in both agronomic and natural settings.

5.2 Allelopathic Interference in Landscape and Agronomic Settings

A systematic approach to weed management in agronomic and landscape settings is often needed in order to address both the economic and environmental consequences of invasive weeds and other pests (Lewis et al., 1997). Historically, certain cultivated crops or individual plants such as buckwheat, black mustard, sunflower and black walnut, and cereal crops such as sorghum, wheat, barley, oats and rye have been widely reported to suppress annual weed species (Weston, 1996, 2005). Residues of several cover crops including winter wheat (Triticum aestivum), barley (Hordeum vulgare), oats (Avena sativa), rye (Secale cereale), grain sorghum (Sorghum bicolor) and sudangrass (Sorghum sudanense) are widely utilized to suppress weeds in a variety of cropping systems (Einhellig and Leather, 1988; Rice, 1995; Miller, 1996; Weston, 1996, 2005). Velvetbean (Mucuna pruriens var. utilis), a legume cultivated as a green manure in Japan, has shown the ability to smother problematic perennials such as purple nutsedge and cogongrass (Fujii et al., 1992). Sorghum and its hybrid, sudex (S. bicolor × S. sudanense), are preferentially cultivated in the USA, Africa, Australia, India and South America as cover crops and green manures while assisting in weed management (Weston et al., 1999; Weston, 2005). While these cereals produce large amounts of biomass rapidly and have potential as weed-suppressive mulches, the foliage is a source of water-soluble phenolics and cyanogenic glucosides (Weston et al., 1999; Bertin et al., 2003). Macias et al. (2005a,b) have recently observed that residues of Secale cereale (winter rye) produce a number of bioactive compounds which are degraded to more toxic constituents over time in soil systems. These chemicals can persist for short periods in soil systems, and although they contribute to weed suppression and allelopathic interference, they are not thought to be a concern from a human health or environmental fate perspective, as they are readily degraded over time (Nair et al., 1990; Macias et al., 2005a,b). Interestingly, compounds that are generated in large quantities within plant roots and shoots, and released into the environment over time, have great potential to be altered or modified by environmental and chemical changes in microbially active soils. Although limited research has been performed in soil systems with respect to allelochemical degradation, it is certain that further scrutiny will show that a plethora of bioactive organic plant-derived constituents can be encountered in the soil environment, and subsequent environmental conditions will alter these fluxes (Blum et al., 2000).

Another system which has been studied in detail is that of black walnut (Juglans nigra), which produces the potent respiratory inhibitor, juglone (Jose, 2002; Weston and Duke, 2003; Weston, 2005). Juglone is produced by a variety of walnut plant parts, including bark, shoots, living roots and nuts, but is released in the greatest quantities over time by the living root system. For hundreds of years, the interference of black walnut and other related species such as the butternut have been described by landscapers and gardeners. Recently, studies performed in the USA and South America have shown that soil type, density and planting distances, as well as timing of walnut removal, all influence the

longevity of the phytotoxicity. Juglone has been shown to be a potent inhibitor of electron transport in respiration, and can remain active in soil systems for several months after the walnut harvest (Jose and Gillespie, 1998; Willis, 2000; Weston and Duke, 2003). Recent long-term observations on plant selectivity performed by L.A. Weston in the presence of established walnuts actively exuding juglone suggest that herbaceous species that are members of the Liliaceae, Malvaceae and Taxaceae are highly tolerant of black walnut, while members of other genera, including those of the Ericaceae and Aquifoliaceae, are very intolerant of the presence of walnut. The mechanism of tolerance or resistance to juglone is currently unknown, but one might suspect that the ability to prevent uptake and translocation of this compound by sequestration in unrelated species or the rapid metabolism of juglone into non-toxic forms may be responsible for enhanced tolerance in these landscape species (Weston and Duke, 2003; Weston, 2005).

A great deal of work has been performed to address the phenomenon of allelopathy in cultivated rice. Out of 111 rice cultivars, Olofsdotter and Navarez (1996) found that ten cultivars had allelopathic effects on the growth of barnyardgrass, and argued the need for further selection and breeding of weed-suppressive ability in rice. Olofsdotter and Navarez (1996) reported that one rice cultivar (Taichung native 1) potently suppressed the growth of barnyardgrass and several other common rice weeds. Dilday et al. (1991) examined 10,000 rice accessions for allelopathic activity upon ducksalad (Heteranthera limosa) and found that 412 rice accessions effectively suppressed ducksalad, 145 suppressed redstem (Ammannia coccinea), and 16 suppressed both species. Hassan et al. (1998) found that 30 rice accessions strongly suppressed barnyardgrass, 15 suppressed Cyperus difformis, and 5 suppressed both species.

Although the importance of allelopathic crop cultivars has been reviewed extensively in recent literature (Olofsdotter, 1998; Wu et al., 1999; Bhowmik and Inderjit, 2003; Weston, 2005), relatively little attention has been given by plant breeders to the concept of enhanced selection for allelopathic or weed-suppressive crop species. Although many researchers have investigated the allelopathic potential of rice, sorghum, winter rye, barley, wheat, sunflower, buckwheat, hairy vetch and cucumber, many agronomic, vegetable and landscape crops have not yet been evaluated for their inherent potential to suppress weeds (Weston, 2005). In studies performed by Bertholdsson (2004, 2005), older landraces of both wheat and barley were found to exhibit an enhanced ability to suppress weeds by potential production of bioactive root exudates, in comparison with newer, high-yielding cereal accessions. In addition, lodging potential was greater in the new cultivars than in older landraces. These traits will undoubtedly be of interest to future organic producers and those with an interest in minimizing herbicidal inputs over time (Bertholdsson, 2004, 2005).

Recently, our research group has initiated a long-term research project to evaluate the potential of both landscape ground covers and turfgrasses for their inherent ability to suppress weeds over time. In addition to allelopathic interference, we evaluated their potential to suppress weeds by biomass production, their ability to overwinter and resist pest attack, and their suppression of light at the base of the canopy to ultimately limit weed seed germination (Eom et al., 2006; Weston et al., 2006). In evaluations of over 100 herbaceous ornamental ground covers in multiple sites across New York state, it was found that certain ground covers performed exceptionally well when transplanted into full-sun conditions, and exhibited good drought and pest tolerance. In addition, these ground covers effectively suppress weeds with only limited intervention by hand-weeding or herbicide application. Successfully transplanted ground covers exhibited the ability to successfully overwinter in New York conditions, produce large quantities of biomass rapidly in late spring, withstand both droughty and salty soil conditions, and limit the amount of light received at the base of the canopy due to dense canopy formation. In addition, several ground covers also appeared to exhibit allelopathic ability to suppress weeds including *Nepeta* spp. (catmint) and *Solidago* spp. (ornamental goldenrod). Although allelopathy may not be the major or only mechanism of weed interference in these studies, it can contribute to weed suppression, and ultimately pest resistance, through the production of bioactive

compounds which limit the palatability of foliage for deer and other pests (Eom *et al.*, 2006).

We have also found that turfgrass species exhibit differential ability to suppress weeds by both allelopathic and competitive mechanisms related to crop establishment and subsequent growth. In trials conducted in low-maintenance field and roadside settings across New York state, several cultivars of fine fescue (primarily *Festuca rubra*) and perennial ryegrass (*Lolium perenne*) exhibited a consistent ability to suppress annual and perennial turf weeds over time. The perennial ryegrass cultivars Palmer and Prelude exhibited the ability to establish and effectively outcompete weeds. Perennial ryegrass has been noted for its ability to exhibit allelopathic potential in field and laboratory settings (Weston, 1990). Fine fescues have also been reported to be strongly weed-suppressive and, in our recent laboratory experiments, exhibited differential ability to inhibit weed growth due to the production of bioactive root exudates (Bertin *et al.*, 2003). The production of *m*-tyrosine and other potential inhibitors in bioactive root exudates is likely to be associated with the allelopathic interference observed by several fine fescue cultivars, including Intrigue, Wilma, Columbra and Sandpiper. *m*-Tyrosine is produced in large quantities by certain cultivars of chewings fescue and strong creeping fescue, which are very suppressive to weed growth in field and laboratory conditions. *m*-Tyrosine inhibits the root growth of sensitive weed seedlings by rapid death of the root tip or meristematic zone, although its specific mode of action is not clearly defined at present (Bertin *et al.*, 2007). Certain cultivars of chewings fescue, such as Intrigue, have the ability to be maintained without additional fertility or herbicide applications in nearly weed-free conditions, with timely mowing and rainfall. The study of allelopathic interference by perennial species which may be established for long periods of time at one site may have important economic and aesthetic implications for landscape and natural settings, where invasive weed management is of increasing concern. Increased attention to selection and evaluation of weed-suppressive landscape cultivars is likely, as most landscape managers are under increasing pressure to reduce time, labour and chemical inputs in roadside and landscape settings (Weston *et al.*, 2006).

5.3 Allelopathic Chemicals as Natural Herbicides

Although numerous crops, such as wheat, maize and rye, have the potential to produce novel allelochemicals such as hydroxamic acids, most plants produce mixtures of bioactive products with low specific activities when isolated (Weston and Duke, 2003; Weston, 2005). Often allelochemicals have been shown to strongly influence plant defence responses in the presence of pest complexes (Niemeyer and Perez, 1995). Although there are some natural herbicides of microbial origin which have been commercialized (Hoagland, 2001), natural herbicides from higher plants or microbial origins are generally not common. Products which have currently been successfully commercialized and which exhibit similarity to their natural-product counterparts include dicamba and cinnemethylin, synthetic derivatives of benzoic acid and 1,8-cineole, respectively.

More recently, mesotrione, a plant-derived benzoylhexane-1,3-dione compound, was discovered by Zeneca researchers in California. Mesotrione is now commercialized as a synthetic herbicide for the pre- and post-emergence control of broadleaved and grassy weeds in maize (Fig. 5.1) (Mitchell *et al.*, 2001). Syngenta has registered mesotrione under the trade name of Callisto, and it was initially developed from bioactive products isolated from *Callistemon citrinus* (Californian bottle brush). Mesotrione inhibits HPPD (hydroxyphenylpyruvate dioxygenase), a key enzyme involved in the carotenoid biosynthesis pathway. Inhibition of carotenoid biosynthesis results in bleaching symptoms in susceptible plant species (Fig. 5.1). After extensive field trials in Europe and the USA, mesotrione now effectively controls most economic weeds in maize, including *Xanthium strumarium*, *Abutilon theophrasti*, *Ambrosia trifida*, *Chenopodium* spp., *Amaranthus* spp., *Polygonum* spp., *Digitaria* spp. and *Echinochloa* spp. (Wichert *et al.*, 1999).

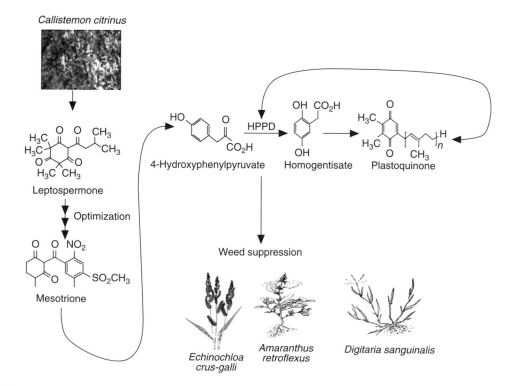

Fig. 5.1. Mesotrione, a commercially utilized pre-emergence herbicide which was developed from a natural plant product isolated from the Australian bottlebrush plant, *Callistemon citrinus*. Mesotrione is a potent inhibitor of plastoquinone biosynthesis in sensitive higher plants.

5.4 Production and Release of Allelochemicals by Various Plant Parts

Several allelochemicals have been reported to have potential for use as natural herbicides. These include artemisinin, ailanthone, momilactone, sorgoleone and L-DOPA (Fig. 5.2) (Bhowmik and Inderjit, 2003). Fujii *et al.* (1992) found that a non-protein amino-acid, L-DOPA (L-3,4-dihydroxyphenylalanine), from the leaves and seeds of velvetbean is mainly responsible for its allelopathic activity, and reported the seed germination response of different families to L-DOPA (Nishihara *et al.*, 2004). Recently, Nishihara *et al.* (2005) reported that L-DOPA was released in significant quantities from velvetbean roots.

Artemisinin

Artemisinin (Fig. 5.2), a non-volatile sesquiterpenoid lactone isolated from the glandular trichomes of annual wormwood (*Artemisia annua*), has been shown to have potential herbicidal activity in glasshouse and field experiments. Although artemisinin is distributed throughout the plant, its inflorescences could serve as a better option for the isolation of artemisinin on a commercial scale, as these have an artemisinin content which is more than 4–11 times higher (% dry weight) than that of the leaves (Ferreira *et al.*, 1995). Artemisinin has been shown to strongly inhibit the growth of *Amaranthus retroflexus*, *Ipomoea lacunose* and *Portulaca oleracea* (Duke *et al.*, 1987). The use of artemisinin directly or its development as a natural herbicide, has recently been discussed (Duke *et al.*, 2002; Inderjit and Bhowmik, 2002; Bhowmik and Inderjit, 2003).

Ailanthone

Ailanthone (Fig. 5.2) is a major phytotoxic compound isolated from *Ailanthus altissima* (tree of heaven) which is considered to be a reservoir for a number of quassinoid compounds including ailanthone, amarolide, acetyl amarolide, 2-dihydroailanthone, ailanthinone and chaparrin (Ishibashi *et al.*, 1981; Casinovi *et al.*, 1983; Polonsky, 1985) as well as alkaloids and other secondary products (Anderson *et al.*, 1983). The bark and root tissue of *A. altissima* is known to be most phytotoxic, followed by its leaves and then its wood (Heisey, 1990). Ailanthone possesses both pre- and post-emergence herbicidal activity, and seedlings treated with ailanthone show injury symptoms within 24–48 h of application, and are often dead after 48 h. Ailanthone reduced the biomass of both monocot and dicot weeds including *Chenopodium album*, *Amaranthus retroflexus*, *Sorghum halepense* and *Convolvulus arvensis* (Heisey and Heisey, 2003), *Lepidium sativum* (Heisey, 1990) and *Medicago sativa* (Tsao *et al.*, 2002).

Momilactone

Living roots of rice plants are known to exude numerous phenolic acids along with momilactone A and B. The involvement of phenolic acids in rice allelopathy has generally been ruled out because of their insufficient rate of release and their inability to accumulate at phytotoxic levels under field conditions so as to cause weed growth suppression (Olofsdotter *et al.*, 2002). However, momilactone B (Fig. 5.2) has been suggested as a potential natural herbicide since it appears to be released in sufficiently high concentrations to inhibit the growth

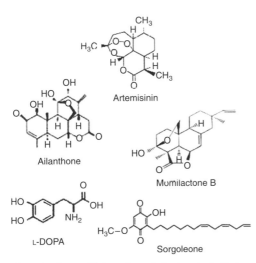

Fig. 5.2. Several interesting allelochemicals that have exhibited potential for development as natural herbicides.

of neighbouring weeds (Kato-Noguchi and Ino, 2005a,b). Momilactone B has also been shown to inhibit the germination of several weed species including *Amaranthus lividus*, *Poa annua*, and also the root and shoot growth of *Digitaria sanguinalis* (Lee *et al.*, 1999).

L-DOPA

L-DOPA (L-3,4-dihydroxyphenylalanine) (Fig. 5.2), is an amino-acid derivative produced in several plants. It is also a potent allelochemical when exuded from a tropical legume, *Mucuna pruriens*. It is present in very high concentrations in the leaves and roots of *M. pruriens* (10 g/kg) and also in the seeds (50–100 g/kg). Its high specific and total activity further suggests its use as a natural herbicide (Fujii and Hiradate, 2005). L-DOPA is reported to reduce the root growth of several species including *Miscanthus sinensis*, *Solidago altissima*, *Amaranthus lividus* and *Cucumis sativus* (Fujii *et al.*, 1992).

Matsumoto *et al.* (2004) observed that L-DOPA (at 0.01 and 0.1 mM) caused more than 50% inhibition of root elongation in 7 and 19 of 32 weed species, respectively. Interestingly, it was found to be more effective on broadleaved compared with grassy weed species (Fujii *et al.*, 1992; Anaya, 1999).

Sorgoleone

Sorghum spp. are known to exude sorgoleone (2-hydroxy-5-methoxy-3-[(8′2z,11′2z)-8′2,11′2,14′2-pentadecatriene]-*p*-benzoquinone) (Fig. 5.2) in the form of hydrophobic drops from the roots of all the sorghum species evaluated to date (Fig. 5.3). Mixtures of bioactive long-chain hydroquinones are contributed to the rhizosphere by living root hairs of *Sorghum bicolor* (2 mg exudates/g fresh root weight), with sorgoleone being the major component (85–90%) (Czarnota *et al.*, 2001, 2003). This natural product is produced by living root hairs only and is released in particularly high

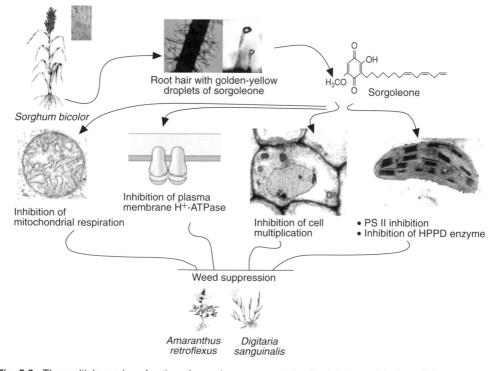

Fig. 5.3. The multiple modes of action of sorgoleone suggest that it might be an ideal candidate for continued use as a natural herbicide.

concentrations by seedling sorghum produced under various laboratory conditions (Yang *et al.*, 2004a). The pathway for biosynthesis of this compound is relatively complex and suggests that four or more genes are responsible for its production in living root hairs (Yang *et al.*, 2004b; Dayan, 2006). Environmental conditions also directly influence the production of this allelochemical, with drought and moderate levels of oxygen being important for enhanced production (Yang *et al.*, 2004a). In Africa, sorghum is used successfully as a living cover crop to suppress annual weeds when interplanted with other crops in drought conditions, particularly when densely planted (Weston, 2005).

5.5 Mechanism of Action of Allelochemicals

Sorgoleone is one example of a natural plant product or allelochemical which exhibits great specific activity as a plant growth inhibitor in broadleaved and grassy weed species grown in hydroponic bioassays at very low concentrations, e.g. 10 μM (Nimbal *et al.*, 1996). Further studies have shown that sorgoleone probably possesses multiple modes of action, and affects the chloroplastic, mitochondrial and cell replication functions in higher plants, exhibiting variable binding capacities within various organelles. For instance, (i) it binds the D_1 protein coupled with electron transfer between Q_A and Q_B within photosystem II; (ii) it mimics cyanide, potentially disrupting respiration; (iii) it inhibits hydroxyphenyl pyruvate dioxygenase (HPPD) activity, disrupting carotenoid biosynthesis, which causes bleaching; and (iv) it halts the cell replication cycle by arresting cells in prophase, metaphase and anaphase stages (Gonzalez *et al.*, 1997; Gattás Hallak *et al.*, 1999; Czarnota *et al.*, 2001; Meazza *et al.*, 2002). The multiple molecular targets of sorgoleone suggest that it could have a greater potential for use as a natural herbicide, with a limited possibility of sensitive plant species developing resistance (Fig. 5.3).

Although sorgoleone is one example of a plant-derived natural product with multiple target sites which have recently been well characterized, we know very little about specific molecular targets of most allelochemicals. Duke *et al.* (2002)

have suggested that natural products may offer opportunities for discovery and development of novel herbicides and also novel molecular target sites for herbicide activity in higher plant systems. Although research on this topic has been relatively limited, some laboratories, including those of Cutler, Dayan, Duke, Fujii, Macias and Weston, have attempted to elucidate the mode of action and structural activity relationships of numerous bioactive natural products and their related chemistries. Funding for this work is often unavailable, but linkages with the agrochemical and pharmaceutical industries have also provided support for continued research into this area. The few examples of successful herbicides and insecticides developed from bioactive natural product templates offer considerable hope that other novel families of natural products with unique pesticidal properties still remain to be discovered.

5.6 Fate of Allelochemicals in Soil

We have presented numerous examples suggesting the potential for utilization of allelopathic crop residues or the selection of allelopathic crop cultivars with enhanced suppressive ability, but one problem that surrounds the further development of this concept for long-term weed management is that of limited soil activity or persistence. Most allelochemicals, including the most frequently described systems involving simple phenolics in the rhizosphere, exhibit very limited persistence in microbially active soils (Blum *et al.*, 2000). Although allelopathy has been suggested for non-chemical weed management (Wu *et al.*, 1999), the temporal suppression by allelopathic plants and selective control of only the most sensitive species may be problematic. For example, Macias *et al.* (2005a) thoroughly studied the transformation of the hydroxamic acid DIBOA into BOA in soils containing rye residues. Although the presence of BOA is only fleeting, it rapidly undergoes biotransformation into a related structure, APO (2-aminophenoxazin-3-one). APO was noted to be the only allelochemical in this system which exhibited a relatively long half-life in the soil (3 months), and is now thought to be important in cereal allelopathy. Furubayashi *et al.* (2005) have recently reported on the adsorption, chemical transformation and microbial degradation of

L-DOPA in different soil environments. Additional systematic studies will need to be funded and conducted over time in different soil settings in order to evaluate the potential of other allelochemicals to persist and contribute to long-term weed or pest suppression in the soil rhizosphere. These studies are difficult and expensive to conduct, and require the collaboration of weed scientists, plant physiologists, chemists and microbiologists, among others, but are sorely needed to fully determine whether long-term weed suppression may be enhanced by utilization of plant residues or perennial plants in managed settings.

The production of perennial crops suggests another possibility for the utilization of allelopathic crops themselves to limit weed spread and provide some long-term control benefits. Asparagus, fescue, walnut, herbaceous ground covers such as catmint and others are known to release allelochemicals into the environment over time, in perennial production systems. Although we know that environmental fluctuations which impact crop growth rate will also impact allelochemical production, we do not understand much about the seasonality of production or how daylength, drought stress or light quality impacts the direct release of these compounds over time into the soil rhizosphere. Recent research suggests that volatile compounds which are foliarly produced may also have the potential to contribute to soil rhizosphere interactions among plants and other species, by binding to soil particles and through uptake by developing seedlings (Barney *et al.*, 2005; Weston, 2005; Eom *et al.*, 2006) Their fluxes in production and release also need to be studied in greater depth.

5.7 Laboratory versus Field Studies

Inderjit and Weston (2000) suggested that evidence for allelopathic interference could be generated under controlled laboratory conditions. In many cases, it is not possible to address physiological processes and the isolation and chemical transformation of allelochemicals in soil settings, so controlled laboratory settings are necessary in order to ensure successful completion and isolation of these metabolites. Mattice *et al.* (1997) identified the following phenolic compounds: 4-hydroxyben-

zoic, 4-hydroxyhydrocinnamic and 3,4-dihydroxyhydrocinnamic acids in aqueous production systems containing allelopathic rice cultivars. These compounds were not detected in similar systems containing non-allelopathic rice cultivars. Olofsdotter *et al.* (2002) reported that the maximum release rate of *p*-hydroxybenzoic acid from rice during the first month of growth was 10 μg/ plant per day. Using controlled laboratory conditions, it is possible to demonstrate the inhibition of weed species using higher concentrations of *p*-hydroxybenzoic acid. However, under field situations, the maximum release rate could be estimated to be 1 mg/m^2 per day at conventional densities. One cannot be certain of these figures and this predicted concentration is probably not sufficient to cause inhibition of weed species in the field. In many situations, the allelopathic potential of a crop species has only been demonstrated using artificial bioassay systems, including filter paper, agar or modified soil bioassays, and these findings are then linked to the weed-suppressive abilities of a crop in field settings (Kato-Noguchi and Ino, 2005b). Unfortunately, the current evidence generated in support of many suspected examples of allelopathic interference is often not particularly convincing.

5.8 Future of Allelopathy in Weed Management

A strong potential exists for the additional development and use of plant products or allelochemicals in medicine and agriculture for a variety of purposes. Firstly, the additional search for new plant species and cultivars and novel chemistries must be funded and this involves exploration and purification strategies, requiring sensitive and repeatable bioassays for the detection and accurate assessment of activity. In addition, close collaboration with trained synthesis chemists is needed to further alter and study structural activity relationships in an attempt to design pesticides with greater specific activity, soil persistence and selectivity, as well as environmental safety. The design of new products possessing novel modes of action for weed management will no doubt facilitate the continued use of herbicides for future weed management systems. In addition, the development of crop germplasm expressing enhanced competitive and allelopathic traits for weed management will also be of future interest to breeders and crop and landscape managers for the development of management systems with reduced chemical and labour inputs. Many of these possibilities have only been touched upon in current research efforts, and certainly the possibility for continued improvement of pesticide offerings and successful crop germplasm exists. However, funding for this research and development effort has never been easy to procure and continues to be severely limited. One can only hope that additional funds for development of biorational pest management will be considered by public and private agencies and will rise to the forefront of plant-based research.

5.9 References

Anaya, A.L. (1999) Allelopathy as a tool in the management of biotic resources in agroecosystems. *Critical Reviews in Plant Sciences* 18, 697–739.

Anderson, L.A., Harris, A. and Phillipson, J.D. (1983) Production of cytotoxic canthin-6-one alkaloids by *Ailanthus altissima* plant cell cultures. *Journal of Natural Products* 46, 374–378.

Barney, J.N., DiTommaso, A. and Weston, L.A. (2005) Differences in invasibility of two contrasting habitats and invasiveness of two mugwort (*Artemisia vulgaris*) populations. *Journal of Applied Ecology* 42, 567–576.

Bertholdsson, N.-O. (2004) Variation in allelopathic activity over one hundred years of barley selection and breeding. *Weed Research* 44, 78–86.

Bertholdsson, N.-O. (2005) Early vigour and allelopathy: two useful traits for enhanced barley and wheat competitiveness against weeds. *Weed Research* 45, 94–102.

Bertin, C., Yang, X. and Weston, L.A. (2003) The role of root exudates and allelochemicals in the rhizosphere. *Plant and Soil* 256, 67–83.

Bertin, C., Weston, L.A., Huang, T., Jander, G., Owens, T., Meinwald, J. and Schroeder, F.C. (2007). Grass root chemistry: *meta*-tyrosine, an herbicidal non-protein aminoacid. *PNAS*, Early Edition, 1–6.

Bhowmik, P.C. and Inderjit (2003) Challenges and opportunities in implementing allelopathy for natural weed management. *Crop Protection* 22, 661–671.

Blum, U., Staman, K.L., Flint, L.J. and Shafer, S.R. (2000) Induction and/or selection of phenolic acid-utilizing bulk soil and rhizosphere bacteria and their influence on phenolic acid phytotoxicity. *Journal of Chemical Ecology* 26, 2059–2078

Casinovi, C.G., Ceccherelli, P., Fardella, G. and Grandolini, G. (1983) Isolation and structure of a quassinoid from *Ailanthus glandulosa*. *Phytochemistry* 22, 2871–2873.

Czarnota, M.A., Paul, R.N., Dayan, F.E., Nimbal, C.I. and Weston, L.A. (2001) Mode of action, localization of production, chemical nature, and activity of sorgoleone: a potent PSII inhibitor in *Sorghum* spp. root exudates. *Weed Technology* 15, 813–825.

Czarnota, M., Rimando, A.M. and Weston, L.A. (2003) Evaluation of root exudates of seven sorghum accessions. *Journal of Chemical Ecology* 29, 2073–2083.

Dayan, F.E. (2006) Factors modulating the levels of the allelochemical sorgoleone in *Sorghum bicolor*. *Planta* 224, 339–346.

Dilday, R.H., Nastasi, P., Lin, J. and Smith, R.J., Jr (1991) Allelopathic activity in rice (*Oryza sativa* L.) against ducksalad (*Heteranthera limosa* (Sw.) Willd.). In: *Sustainable Agriculture for the Great Plains: Symposium Proceedings*. USDA, pp. 193–201.

Duke, S.O. (1988) Glyphosate. In: Kearney, P.C. and Kaufman, D.D. (eds) *Herbicide, Chemistry, Degradation and Mode of Action*. Marcel Dekker, New York, pp. 1–69.

Duke, S.O., Vaughn, K.C., Croom, E.M. and Elsohly, H.N. (1987) Artemisinin, a constituent of annual wormwood (*Artemisia annua*) is a selective phytotoxin. *Weed Science* 35, 499–505.

Duke, S.O., Dayan, F.E., Rimando, A.M., Schrader, K.K., Aliotta, G., Oliva, A. and Romagni, J.G. (2002) Chemicals from nature for weed management. *Weed Science* 50, 138–151.

Einhellig, F.A. and Leather, G.R. (1988) Potential for exploiting allelopathy to enhance crop production. *Journal of Chemical Ecology* 14, 1829–1844.

Eom, S.H., Yang, H.S. and Weston, L.A. (2006) An evaluation of the allelopathic potential of selected perennial groundcovers: foliar volatiles of catmint (*Nepeta* × *faasenii*) inhibit seedling growth. *Journal of Chemical Ecology* 32, 1835–1848.

Ferreira, J.F.S., Simon, J.E. and Janick, J. (1995) Developmental studies of *Artemisia annua*: flowering and artemisinin production under greenhouse and field conditions. *Planta Medica* 61, 167–170.

Fujii, Y. and Hiradate, S. (2005) A critical survey of allelochemicals in action: the importance of total activity and the weed suppression equation. In: Harper, J.D., An, M., Wu, H. and Kent, J.H. (eds) *Establishing the Scientific Base: Proceedings and Selected Papers of the Fourth World Congress on Allelopathy*. Charles Sturt University, NSW, Australia, pp. 73–76.

Fujii, Y., Shibuya, T. and Yasuda, T. (1992) Allelopathy of velvetbean: its discrimination and identification of L-DOPA as a candidate of allelopathic substances. *Japan Agricultural Research Quarterly* 25, 238–247.

Furubayashi, A., Hiradate, S. and Fujii, Y. (2005) Adsorption and transformation reactions of L-DOPA in soils. *Soil Science and Plant Nutrition* 51, 819–826.

Gattás Hallak, A.M., Davide, L.C. and Souza, I.F. (1999) Effects of sorghum (*Sorghum bicolor* L.) root exudates on the cell cycle of the bean plant (*Phaseolus vulgaris* L.) root. *Genetics and Molecular Biology* 22, 95–99.

Gonzalez, V.M., Kazimir, J., Nimbal, C., Weston, L.A. and Cheniae, G.M. (1997) Inhibition of a photosystem II electron transfer reaction by the natural product sorgoleone. *Journal of Agricultural and Food Chemistry* 45, 1415–1421.

Hassan, S.M., Aidy, I.R., Bastawisi, A.O. and Draz, A.E. (1998) Weed management in rice using allelopathic rice varieties in Egypt. In: Olofsdotter, M. (ed.) *Allelopathy in Rice*. International Rice Research Institute, Manila, Philippines, pp. 27–43.

Heisey, R.M. (1990) Allelopathic and herbicidal effects of extracts from tree-of-heaven (*Ailanthus altissima*). *American Journal of Botany* 77, 662–670.

Heisey, R.M. and Heisey, T.K. (2003) Herbicidal effects under field conditions of *Ailanthus altissima* bark extract, which contains ailantheone. *Plant and Soil* 256, 85–99.

Hoagland, R.E. (2001) Microbial allelochemicals and pathogens as bioherbicidal agents. *Weed Technology* 15, 835–857.

Inderjit and Bhowmik, P.C. (2002) Importance of allelochemical in weed invasiveness and their natural control. In: Inderjit and Mallik, A.U. (eds) *Chemical Ecology of Plants: Allelopathy in Aquatic and Terrestrial Ecosystems*. Birkhauser-Verlag, Basel, Switzerland, pp. 188–197.

Inderjit and Keating, K.I. (1999) Allelopathy: principles, procedures, processes, and promises for biological control. *Advances in Agronomy* 67, 141–231.

Inderjit and Weiner, J. (2001) Plant allelochemical interference or soil chemical ecology? *Perspectives in Plant Ecology, Evolution and Systematics* 4, 4–12.

Inderjit and Weston, L.A. (2000) Are laboratory bioassays for allelopathy suitable for prediction of field responses? *Journal of Chemical Ecology* 26, 2111–2118.

Ishibashi, M., Murae, T., Hirota, H., Naora, H., Tsuyuki, T., Takahashi, T., Itai, A. and Iitaka, Y. (1981) Shinjudilactone, a new bitter principle from *Ailanthis altissima* Swingle. *Chemistry Letters*, 1597–1598.

Jose, S. (2002) Black walnut allelopathy: current state of the science. In: Inderjit and Mallik, A.U. (eds) *Chemical Ecology of Plants: Allelopathy in Aquatic and Terrestrial Ecosystems*. Birkhauser-Verlag, Basel, Switzerland, pp. 149–172.

Jose, S. and Gillespie, A.R. (1998) Allelopathy in black walnut (*Juglans nigra* L.) alley cropping. I. Spatio-temporal variation in soil juglone in a black walnut–corn (*Zea mays* L.) alley cropping system in the midwestern USA. *Plant and Soil* 203, 191–197.

Kato-Noguchi, H. and Ino, T. (2005a) Possible involvement of momilactone B in rice allelopathy. *Journal of Plant Physiology* 162, 718–721.

Kato-Noguchi, H. and Ino, T. (2005b) Concentration and release level of momilactone B in the seedlings of eight rice cultivars. *Journal of Plant Physiology* 162, 965–969.

Lee, C.W., Yoneyama, K., Takeuchi, Y., Konnai, M., Tamogami, S. and Kodama, O. (1999) Momilactones A and B in rice straw harvested at different growth stages. *Bioscience, Biotechnology and Biochemistry* 63, 1318–1320.

Lewis, W.J., van Lenteren, J.C., Phatak, S.C. and Tumlinson, J.H. (1997) A total system approach to sustainable pest management. *Proceedings of the National Academy of Sciences of the USA* 94, 12243–12248.

Macias, F.A. (1995) Allelopathy in search for natural herbicide models. In: Inderjit, Dakshini, K.M.M. and Einhellig, F.A. (eds) *Allelopathy: Organisms, Processes, and Applications*. American Chemical Society, Washington, DC, USA, pp. 310–329.

Macias, F.A., Oliveros-Bastidas, A., Marin, D., Castellano, D., Simonet, A.M. and Molinillo, J.M.G. (2005a) Degradation studies on benzoxazinoids: soil degradation dynamics of (2R)-2-O-β-D-glucopyranosyl-4-hydroxy-(2H)-1,4-benzoxazin-3(4H)-one (DIBOA-Glc) and its degradation products. *Journal of Agricultural and Food Chemistry* 53, 554–561.

Macias, F.A., Marin, D., Oliveros-Bastidas, A., Castellano, D., Simonet, A.M. and Molinillo, J.M.G. (2005b) Structure–activity relationships (SAR) studies of benzoxazinones, their degradation products and analogues: phytotoxicity on standard target species (STS). *Journal of Agricultural and Food Chemistry* 53, 538–548.

Matsumoto, H., Sunohara, Y. and Hachinohe, M. (2004) Absorption, translocation and metabolism of L-DOPA in barnyardgrass and lettuce: their involvement in species-selective phytotoxic action. *Plant Growth Regulation* 43, 237–243.

Mattice, J., Lavy, B., Skulman, B. and Dilday, R. (1997) Searching for allelochemicals in rice that control ducksalad. In: Olofsdotter, M. (ed.) *Allelopathy in Rice.* International Rice Research Institute, Manila, Philippines, pp. 81–98.

Meazza, G., Scheffler, B.E., Tellez, M.R., Rimando, A.M., Romagni, J.G., Duke, S.O., Nanayakkara, D., Khan, I.A., Abourashed, E.A and Dayan, F.E. (2002) The inhibitory activity of natural products on plant *p*-hydroxyphenylpyruvate dioxygenase. *Phytochemistry* 60, 281–288.

Miller, D.A. (1996) Allelopathy in forage crop systems. *Agronomy Journal* 88, 854–859.

Mitchell, G., Bartlett, D.W., Fraser, T.E.M., Hawkes, T.R., Holt, D.C., Townson, J.K. and Wichert, R.A. (2001) Mesotrione: a new selective herbicide for use in maize. *Pesticide Management Science* 57, 120–128.

Nair, M.G., Whiteneck, C.J. and Putnam, A.R. (1990) 2, 2'-oxo-1, 1'-azobenzene, a microbially transformed allelochemical from 2, 3-benzoxazolinone: I. *Journal of Chemical Ecology* 16, 353–364.

Niemeyer, H.M. and Perez, F.J. (1995) Potential of hydroxamic acids in the control of cereal pests, diseases, and weeds. In: Inderjit, Dakshini, K.M.M. and Einhellig, F.A. (eds) *Allelopathy: Organisms, Processes and Applications.* American Chemical Society, Washington, DC, USA, pp. 260–270.

Nimbal, C.I., Yerkes, C.N., Weston, L.A. and Weller, S.C. (1996) Herbicidal activity and site of action of the natural product sorgoleone. *Pesticide Chemistry and Physiology* 54, 73–83.

Nishihara, E., Parvez, M.M, Araya, H. and Fujii, Y. (2004) Germination growth response of different plant species to the allelochemical L-3,4-dihydroxyphenylalanine (L-DOPA). *Plant Growth Regulation* 42, 181–189.

Nishihara, E., Parvez, M.M., Araya, H., Kawashima, S. and Fujii, Y. (2005) L-3-(3,4-Dihydroxyphenyl)alanine (L-DOPA), an allelochemical exuded from velvetbean (*Mucuna pruriens*) roots. *Plant Growth Regulation* 45, 113–120.

Olofsdotter, M. (1998) *Allelopathy in Rice.* International Rice Research Institute, Manila, Philippines.

Olofsdotter, M. and Navarez, D. (1996) Allelopathic rice in *Echinochloa crus-galli* control. *Proceedings of the Second International Weed Control Conference* 4, 1175–1182.

Olofsdotter, M., Rebulanan, M., Madrid, A., Dali, W., Navarez, D. and Olk, D.C. (2002) Why phenolic acids are unlikely primary allelochemicals in rice. *Journal of Chemical Ecology* 28, 229–242.

Polonsky, J. (1985) Quassinoid bitter principles. *Fortschritte der Chemie Organischer Naturstoffe* 47, 221–264.

Rice, E.L. (1995) *Biological Control of Weeds and Plant Diseases: Advances in Applied Allelopathy.* University of Oklahoma Press, Norman, OK, USA.

Tsao, R., Romanchuk, F.E., Peterson, C.J. and Coats, J.R. (2002) Plant growth regulatory effect and insecticidal activity of the extracts of the tree of heaven (*Ailanthus altissima* L.). *BMC Ecology* 2, 1.

Weston, L.A. (1990) Cover crop and herbicide influence on row crop seedling establishment in no-tillage culture. *Weed Science* 38, 166–171.

Weston, L.A. (1996) Utilization of allelopathy for weed management in agroecosystems. *Agronomy Journal* 88, 860–866.

Weston, L.A. (2005) History and current trends in the use of allelopathy for weed management. In: Harper, J.D., An, M., Wu, H. and Kent, J.H. (eds) *Allelopathy: Establishing the Scientific Base. Proceedings of the 4th World Congress on Allelopathy.* Wagga Wagga, Australia, pp. 15–21.

Weston, L.A. and Duke, S.O. (2003) Weed and crop allelopathy. *Critical Reviews in Plant Sciences* 22, 367–389.

Weston, L.A., Nimbal, C.I. and Jeandet, P. (1999) Allelopathic potential of grain sorghum (*Sorghum bicolor* [L.] Moench.) and related species. In: Inderjit, Dakshini, K.M.M. and Foy, C.L. (eds) *Principles and Practices in Plant Ecology: Allelochemical Interactions.* CRC Press, Boca Raton, FL, USA, pp. 467–477.

Weston, L.A., Senesac, A.F., Allaire-Shagensky, J., Eom, S.H., Weston, P.A. and Cardina, J. (2006) A safe bet for the landscape: herbaceous perennial ground covers are fairly easy to establish and maintain, making them a perfect alternative for turfgrass in a low-maintenance landscape. *American Nurseryman* February, 35–38.

Wichert, R.A., Townson, J.K., Bartlett, D.W. and Foxon, G.A. (1999) Technical review of mesotrione, a new maize herbicide. In: *Proceedings of the Brighton Crop Protection Conference,* pp. 105–110.

Willis, R.J. (2000) *Juglans* spp., juglone and allelopathy. *Allelopathy Journal* 7, 1–55.

Wu, H., Pratley, J., Lemerle, D. and Haig, T. (1999) Crop cultivars with allelopathic capability. *Weed Research* 39, 171–180.

Yang, X., Owens, T.G., Scheffler, B.E. and Weston, L.A. (2004a) Manipulation of root hair development and sorgoleone production in sorghum seedlings. *Journal of Chemical Ecology* 30, 199–213.

Yang, X., Scheffler, B.E. and Weston, L.A. (2004b) *SOR1*, a gene associated with bioherbicide production in sorghum root hairs. *Journal of Experimental Botany* 55, 2251–2259.

6 Biological Control of Weeds Using Arthropods

B. Blossey

Cornell University, Department of Natural Resources, Ithaca, NY 14853, USA

6.1 Introduction

The annual fluctuations in abundance of plant species are influenced by abiotic and biotic factors determining adult survival, dispersal, germination and recruitment. In agricultural environments, abiotic and biotic factors are manipulated to achieve high yields of desired crops, and production techniques (mechanical, physical, chemical) are designed to minimize competition by weeds that may reduce yield below economic thresholds. Each crop appears to 'select' for its own suite of associated weeds, which are considered undesirable if their abundance exceeds economic or aesthetic thresholds. In natural environments, managers try to manipulate plant community composition through fire, grazing, flooding, mowing or herbicide application. As in agricultural systems, certain species emerge as persistent problems in natural areas. While yield is of primary concern in agricultural systems, natural area managers use various metrics, including diversity (plant and animals) and rarity, in guiding their management. Use of non-selective mechanical, physical or chemical means is designed to reduce the abundance of undesirable species, while protecting or increasing the abundance of desirable species. Experience shows that non-selective management techniques often fail to reduce the abundance of non-desirable species or may need to be maintained in perpetuity, often at high cost. Biological weed control represents an economically attractive and ecologically sound alternative management technique.

Biological weed control, in the broadest sense, is the use of living organisms to manage problem plants. In theory, all herbivores could potentially be biological control agents. A near-complete list would include viruses, bacteria, insects, snails, slugs, crustaceans, nematodes, birds, fish, marsupials and mammals. In practice, the database of biocontrol programmes implemented worldwide shows that the vast majority of the >350 herbivores released against 133 plant species are insects (plus mites and several pathogens) (Julien and Griffiths, 1998). The list includes a few fish species such as grass carp (using sterile triploid individuals) to control aquatic vegetation (Bain, 1993), but few introductions into new countries have occurred in the past 40 years. No other organisms are reported to have been released to control weeds, a potentially surprising fact considering the large number of herbivore taxa and the reported impact that their feeding may have on individual plants and plant community composition (Parker *et al.*, 2006).

The effectiveness of large mammals as weed control agents is part of the folklore of invasive plant management and hundreds of websites promote the use of mammals for weed control. In contrast, the few reports in the peer-reviewed literature are not necessarily encouraging (Stanley *et al.*, 2000; Holst *et al.*, 2004). Lack of quantitative evidence for success and lack of specificity of generalist herbivores usually prevents their use as biological control agents and their long-term effects on plant invasions may be more complicated than anticipated (Parker et al., 2006). The remainder of this chapter focuses on developing the framework of biological weed control using specialist invertebrate herbivores, exclusively insects and mites (see Chapter 7 for the use of pathogens).

6.2 Definitions

Biological control is the use of specialized herbivorous natural enemies of plants considered problematic in agricultural or natural environments. The case developed further in this chapter, classical biological control, concerns the use of specialized natural enemies of non-indigenous plants from the native range to manage a target plant. Control of native plant species with indigenous or introduced herbivores is rarely attempted due to the large potential for conflicts of interests (Julien and Griffiths, 1998).

Many people use terms such as non-native, naturalized, exotic, colonizer, weed, invader, invasion, naturalization or alien interchangeably, and there is an ongoing debate about terminology. The terms used in this chapter are therefore defined below:

• A weed is any plant species 'out of place'; obviously this definition is dependent upon human attitudes and not on a set of quantifiable characters. Weeds can be native or introduced. Which species is a weed is in the eye of the beholder: one person's weed could be another person's ornamental plant.
• Non-indigenous refers to a species out of the natural range (which includes the natural zone of dispersal) where it has evolved. For example, a species introduced from southern France to Sweden (where it did not previously occur) is non-indigenous to Sweden but indigenous to Europe. Large-scale climate change, such as glaciation events, shift species range boundaries, sometimes over thousands of kilometres. A species should be considered indigenous if it is simply expanding its range, and if the

range expansion is contiguous without interference by humans.
• A naturalized species is a non-indigenous species able to establish and persist in the absence of human care in natural or agricultural environments.
• An invader is any species (indigenous or not) able to establish, increase and maintain populations to a point where the species is a major contributor (>10%) to plant biomass produced by a particular plant community. Functional characteristics of the dominant biomass producers are assumed to determine ecosystem properties (Grime, 1998). The term invader or invasive will be used indiscriminately in this chapter to represent all plant species that can become dominant, regardless of their status as native or introduced to a particular ecosystem.
• Introduced species are often referred to as aliens (or exotics). The Scientific Committee on Problems of the Environment (SCOPE) has adopted the term 'invasive alien species' to recognize introduced species (Mooney et al., 2005). This designation is unfortunate due to the typically negative connotations associated with the term 'alien' in the English language. The landmark work on introduced species in the USA (US Congress, 1993) adopted the term 'non-indigenous', recognizing that a neutral term is preferable.

6.3 Assumptions and Ecological Theory in Weed Biocontrol

Biological control of introduced plants rests on two main assumptions. The first is the enemy release hypothesis (ERH), which proposes that introduced plants become invasive because they lack natural enemies in the introduced range (Keane and Crawley, 2002); the second is that introducing specialized natural enemies of non-indigenous invasive plants from the native range results in suppression of the target weed.

While we have evidence for significant reductions in the diversity of natural enemies on introduced plants (Mitchell and Powers, 2003), the role of natural enemies providing biotic resistance continues to be debated (Colautti et al., 2004; Levine et al., 2004; Parker et al.,

2006). However, successful control of intro-duced plants with specific herbivores has been achieved in many cases (Julien and Griffiths, 1998), supporting the notion that herbivores are regulators of plant abundance.

A major problem in generalizing roles of natural enemies in plant invasions is the fact that the vast majority (90%) of introduced and naturalized plants never become invasive (Williamson, 1996). In general, we do not under-stand why certain plant species become weeds. Attempts to identify traits associated with weedi-ness (Baker, 1965, 1974) have not produced satisfactory results and rapid growth, high seed output, or short generation time cannot identify a species as a potential weed (Williamson, 1996). Unfortunately, Baker's characters con-tinue to be used routinely to declare why certain plants are weeds, yet a plant can become a weed simply through introduction into a new habitat (Williamson, 1996). An additional complication is the possibility that plants respond to changes in selection pressures by shifting their resource allo-cation to increase their competitive ability, the *evolution of increased competitive ability* (EICA) hypothesis (Blossey and Nötzold, 1995; Bossdorf *et al.*, 2005). Both the enemy release–biotic resistance framework and the EICA hypothesis are unable to explain why only a subset of naturalized plants become invasive. The role of the recipient communities in allowing invaders to gain a foothold or become dominant has received increased attention in recent decades but all plant communities appear vulnerable to plant invasions (Crawley, 1987; Lonsdale, 1999).

Historically, about one-third of all weed control programmes achieve success (Crawley, 1989), although much higher success rates have also been reported (Hoffmann, 1995; Fowler *et al.*, 2000). This indicates that herbi-vores may control their host plant abundance but the outcome is context-specific and differs within and between plant species. Our ability to predict outcomes remains poor, and reductions in host plant populations to very low levels due to attack by specialized natural enemies is not a general phenomenon of all plant–herbivore systems. However, much progress could be made by following the outline provided below. In the past, few programmes have followed these guidelines (often due to a lack of funding)

but these are considered essential components in order to achieve higher success and predictability of weed biocontrol.

6.4 Developing a Biological Weed Control Programme

The following section outlines five basic stages of an idealized classical weed biocontrol pro-gramme. Contributions may come from many different audiences and scientific disciplines and not solely from biocontrol scientists. Each stage is outlined and details of programme com-ponents are described. Different stages are not necessarily separated temporally and may over-lap significantly.

Stage 1: Identification of a problem

Duration: 1–2 years to decades.
Essential components:

- Assess distribution and rate of spread of a non-indigenous species.
- Assess impact on native plant and animal communities.
- Assess cost : benefit ratio of biocontrol.
- Survey for natural enemies in introduction area.

Plants are not introduced simultaneously across a large area but usually spread from distinct entry locations through suitable habitats (Lonsdale, 1993; Blossey *et al.*, 2001b). As species spread, the areas with the longest colonization history usually have the highest frequency of occurrence as well as the highest abundance. Many plant species appear to show lag phases (Williamson, 1996), i.e. a period of relatively slow dispersal or 'non-invasive behaviour'; thus models are of little help and herbarium or other observational records are needed in order to produce past and current distribution and abundance maps.

While distribution of the target is mapped, information on impacts in agricultural or natural environments need to be assembled. In most instances a plant becomes the target of a weed biocontrol programme because impacts are suspected due to high abundance. Quantification of impacts, although highly desirable, is rarely available (Blossey, 1999).

Often impact studies are conducted in parallel or even after weed biocontrol has been initiated. Ideally, control of a species should be attempted before impacts materialize (a precautionary principle). However, at present, the *status quo* in invasive plant control is to try traditional control measures for extended periods before initiating weed biocontrol.

Assessment of cost : benefit ratios (ecological and economical) should accompany initiation of every weed biocontrol programme. The most meaningful approach would be to begin implementation of (bio)control before widespread impacts are realized. That makes a cost : benefit analysis significantly more difficult, since impacts are assumed or realized within a limited area. Costs for a weed biocontrol programme from initiation to initial introduction of control agents are at least US$1 million, but large differences between programmes, countries and agencies may exist. Large differences in estimated costs are associated with the way costs are calculated (McConnachie *et al.*, 2003; Sheppard *et al.*, 2003; Culliney, 2005) and no standard method of valuation of ecosystem services or of individual species (such as endangered species) currently exists (Culliney, 2005). Published estimated benefit : cost ratios for weed control programmes before they were initiated ranged from 2.4:1 to 140:1 (mean 26.9:1; n = 12 programmes) compared with the realized estimated benefits after control programmes were implemented of 2.3:1 to 4000:1 (mean 293:1; n = 25 programmes) (Culliney, 2005). The estimated benefits for weed biocontrol are considerably higher than for traditional control largely due to the fact that costs are mostly independent of the area treated (control agents disperse on their own) and benefits accumulate into the future (control agents are introduced at the beginning and then maintain their populations and services) (Culliney, 2005). Despite large uncertainties in estimating benefits, all published analyses show that a strong commitment to biocontrol research and maintenance of trained personnel and well-equipped facilities can deliver high rates of return on investments (Culliney, 2005).

Classical biological control uses highly specialized natural enemies from the native range of an introduced weed. So why should herbivore surveys in the new range be conducted? In addition to plants and plant propagules, increased commerce and rapid transport mechanisms are responsible for the introduction of many herbivore species. Occasionally large numbers of accidentally introduced herbivores can be found on invasive plants (Tewksbury *et al.*, 2002) but these herbivores rarely show promise as control agents.

Stage 2: Initiation of a biocontrol programme (feasibility assessment)

Duration: 2–5 years.
Essential components:

- Identify natural enemies in the native range (literature and field surveys).
- Assess life history, distribution, impact and potential specificity.
- Develop a host-specificity screening plant list
- Proposal to target weed species for biocontrol to regulatory agency (differs by country).

Many of the objectives during stage 2 of a biocontrol programme require work in the native range of an introduced weed. Different countries such as the USA (USDA) or Australia (CSIRO) have overseas laboratories where government scientists work on development of weed biocontrol programmes. Other countries contract with researchers associated with universities or organizations such as CABI Biosciences to provide feasibility assessments. An important aspect of stage 2 is to confirm the taxonomic identity of the target species and its origin (Müller and Schroeder, 1989). Biocontrol programmes targeting leafy spurge (*Euphorbia esula*) and spotted knapweed (*Centaurea maculosa*) faced difficulties in matching plant genotype in the native and introduced ranges, with important consequences for herbivore preferences and performance (Gassmann and Schroeder, 1995). The availability of advanced genetic tools has highlighted the importance of genetic changes, including hybridization, in creating invasive species (Ellstrand and Schierenbeck, 2000). Once taxonomic problems are resolved, the native range of a species can be identified and the search for potential biological control agents can begin in earnest.

Previously, scientists had to rely on entomological collections or published accounts to create an initial list of herbivores found on a target species. With the increasingly available information on the Internet, time to discover published records has decreased. In places with a long entomological tradition and abundant knowledge of the insect fauna, a search of references (CAB Abstracts) may come up with a fairly accurate list of herbivore species; in other regions (i.e. the tropics) many unknown species may exist, often requiring extensive surveys and taxonomic work. If the fauna is fairly well known, published accounts on host-plant records, distribution, or even impact may be available for individual species and can be summarized. It is not unusual to find large numbers of herbivores (occasionally >100) associated with a plant species (Tewksbury *et al.*, 2002). Typically, the vast majority of these herbivore species are not specialized (they attack multiple plant species) and many can be excluded from a list of potential control agents. In most instances, a preliminary list of potential control agents contains 5–10 herbivores.

Host specificity of control agents (see section on Host specificity) is the most important safety feature associated with weed biocontrol programmes. Standardized procedures for selection of plants to be tested have been developed. These lists are unique to each target pest plant species and are tailored to the potential control agents and approved by regulatory agencies.

Stage 3: Detailed investigations (pre-release)

Duration: 3–10 years (varies widely depending on number of agents studied and funding levels).
Essential components:

- Assess life history, distribution and impact of potential biocontrol agents.
- Model impact of single versus multiple agents on plant demography.
- Assess host specificity.
- Propose release to regulatory agency if results of tests are satisfactory.
- Develop standardized monitoring protocol and begin monitoring plants in future release areas.
- Develop mass rearing techniques.
- Ship and release control agents.

Based on literature and field surveys conducted during stage 2, herbivores are selected for further study in the native range (or in quarantine). The first task in the native range is to locate target plant and herbivore populations. Not all herbivore species may occur in the same area, and occasionally additional species are encountered. Invasive plants are often much less abundant in the native than in the introduced range; consequently herbivores are often difficult to locate and not necessarily abundant. Different herbivores may differ in their phenology, requiring scientists to revisit field sites frequently during a growing season to assemble a full species list. Once sufficient information is available, detailed evaluations of life history and impact on target-plant performance by the different herbivore species should follow. In addition to information on herbivore biology and impact, detailed studies of target-plant biology, ecology and demography in the native and introduced range will help to identify the most promising agents through demographic modelling. This work programme should not be assumed to be a core responsibility of biocontrol scientists, but instead an opportunity for collaboration with scientists in other disciplines.

The largest expense in current weed biocontrol is evaluation of host specificity. In many instances, well over 70% of total expenditure is dedicated to raising plants and testing feeding preferences of herbivores (Blossey *et al.*, 1994b; Blossey and Schroeder, 1995). The focus on host specificity versus a focus on success has resulted in the unfortunate situation that we are able to introduce safe control agents but not necessarily effective ones (McEvoy and Coombs, 1999; Denoth *et al.*, 2002). A potential reason for lack of predictability of success is lack of pre-release impact studies which measure effects of herbivore feeding on growth, survival, biomass production and reproductive output of the target plant (McClay and Balciunas, 2005). Impact studies, whether conducted in the field through exclusion, or in common gardens through addition of herbivores, particularly in combination with demographic modelling (Davis *et al.*, 2006), may reduce scientific and monetary resources spent on ultimately unsuccessful agents (McClay and Balciunas, 2005). It is particularly important to design experiments that assess outcomes using single and multiple herbivores and then use the available data to try to predict impacts on plant demography. While such modelling is not routinely done, and we have insufficient data on the reliability of such approaches (Davis *et al.*, 2006), at least models can provide useful predictions that can then be tested after control agents are released.

In addition to host specificity, a major issue in weed biocontrol is the question of whether release of control agents results in the desired outcome (see below). Measuring outcomes is far from trivial, may require sophisticated experimental designs, and always requires long-term commitments. The development of standardized monitoring protocols to assess the development of control agent populations, their impact on the target plant, as well as the response of associated plant and animal populations, is an extremely useful tool in documenting outcomes of biocontrol (Blossey, 1999, 2004). Such monitoring protocols (see Monitoring section) should be developed and implemented before control agents are released, so as to provide baseline data of plant community composition and structure. Standardization allows comparison of data collected from across the range of an invasive plant targeted with biocontrol.

Stage 4: Detailed investigations (post-release)

Duration: 10–20+ years.
Essential components:

- Mass rearing, transport and release of control agents.
- Monitor and evaluate direct impacts on target plant, and indirect effects (food webs, trophic links) on associated communities (plants and animals).
- Determine whether additional control or restoration efforts are needed.

Assuming that the results of host-specificity investigations and assessments of impacts prove satisfactory to regulatory agencies (Sheppard *et al.*, 2003), control agents are transported from overseas into quarantine. Great care is taken not to introduce potential entomopathogenic diseases or parasitoids. In most instances, organisms go through a generation in quarantine. At the beginning of an introduction programme, control agents are usually

in extreme demand but in very short supply. Many limitations for field collections in the native range exist (low plant abundance, parasitoids, predators, etc) and, if reared in quarantine, numbers are equally low. Traditionally, insects were often field-released at selected sites and then redistribution programmes were established once the field populations had grown to collectable sizes. It appears more effective to develop mass production techniques (Blossey and Hunt, 1999; Blossey *et al.*, 2000) that allow rapid dissemination of control agents; an approach taken by the control programme targeting *L. salicaria* (Blossey *et al.*, 2001a). Not only are insects distributed over a larger range, but stakeholders urgently awaiting arrival of control agents have the opportunity to participate, including in rearing programmes – a distinct advantage in maintaining the 'momentum' of a programme.

After the initial releases are made, mass production will probably continue for 5–10 years until populations of control agents are widely available for field collections. Long-term monitoring (see below for details) is essential to assess whether control can be achieved and under what circumstances. Often it takes 3–5 years before impacts of control agents become apparent; occasionally it may take longer, or control programmes may fail completely. Are insects unable to control their host plant in certain habitats? Are single or multiple agent combinations more successful in suppressing the target weed? Is control, or lack thereof, a function of latitude, longitude or other climatic influences? How are replacement communities assembled? Is there a need to maintain other control techniques or should restoration of native plant communities be attempted? In many instances, local seed banks may not provide sufficient propagules for colonization once the target plant is declining. To prevent other invasive plants from gaining a foothold, active reseeding or replanting of desirable species may be necessary. Research will potentially provide different answers for different programmes and in different areas of the introduced range. Depending on the outcome of these investigations, additional restoration measures (reseeding, replanting, and control of other invasive species) may be needed in order to create the desired outcome.

Stage 5: Ecological and economic assessment

Essential components:

- Review literature.
- Assess opinions of stakeholders.

Control programmes can last for several decades before a final assessment of the overall outcome may be feasible. A final analysis should assess costs and benefits through a literature review assessing published economic and ecological implications. In addition, stakeholder surveys may provide useful feedback on whether the programme was a success in the eyes of those charged with managing a particular area or resource. Such surveys should be conducted in a timely manner; turnover in personnel and new managers may not remember how problematic certain species were if no quantitative records have been kept.

6.5 A Historical Overview

The following sections describe the history of weed biological control, beginning with an overview of successes, an examination of what constitutes success, and an exploration of host specificity, monitoring and conflicts of interest (which include non-target effects).

Success and failure in weed biocontrol programmes

A number of previous reviews have summarized much of the history of weed biocontrol programmes (Crawley, 1989; Lawton, 1990; McFadyen, 1998) and, due to the long duration of these programmes, not much additional summary information can be added to these previous analyses, but they basically agree that at least one-third of all control programmes achieve substantial or complete suppression of the target weed. Other scientists have used more restricted analyses and, depending on how success is defined, success increases to well over 50% (Hoffmann, 1995; Fowler *et al.*, 2000). While major findings are outlined below, readers should refer to these and additional references for details.

The most active programmes in the 200-year history of weed biocontrol have been located in the USA, Australia, South Africa, Canada and New Zealand (Julien and Griffiths, 1998), although many more countries have tried biocontrol. Weed biocontrol claimed its first success with spectacular control of prickly pear cacti (*Opuntia* spp.) in Australia (Dodd, 1940). This initial success was repeated in other countries, and among the best-known historic examples of successful control are species such as *Lantana camara*, *Hypericum perforatum*, *Eichornia crassipes*, *Alternanthera philoxeroides* and *Salvinia molesta* (Crawley, 1989). Success in one country often results in attempts to repeat results in other areas of the introduced range – sometimes these attempts succeed, occasionally they fail: an indication that success can be a function of interaction of invasive plant with local biotic and abiotic conditions. The context-specific outcomes of weed biocontrol are illustrated by the interaction of nodding thistle (*Carduus nutans*) and the seed head feeding weevil *Rhinocyllus conicus*. This weevil is one of the few successful seed-feeding biocontrol agents, and rapid declines followed its release in the USA (Kok and Surles, 1975). In New Zealand, under nearly identical herbivore attack rates, no control was reported, and therefore it appears that characteristics of the invaded community determine both the invasive success of *C. nutans* and the ability of *R. conicus* to control its host plant (Shea *et al.*, 2005). Among more recent examples of successful control are *Euphorbia esula*, *Lythrum salicaria*, *Senecio jacobaea*, *Sesbania punicea*, *Acacia saligna* and *Azolla filiculoides* (McFadyen, 2000; McConnachie *et al.*, 2004). However, biological weed control has also encountered a number of plant species (*Cirsium arvense*, *Cyperus rotundus* and *Chromolaena odorata*) that appear close to impossible to get under control using herbivores (Crawley, 1989).

Success in classical biological weed control relies on the availability of host-specific control agents able to suppress target weed populations. While predictions about the realized host specificity of herbivorous biocontrol agents based on pre-release evaluations are sophisticated and reliable (McFadyen, 1998; Pemberton, 2000; Blossey *et al.*, 2001c; Louda et al., 2003), forecasting control success has been much less successful and is often compared to a lottery (Crawley, 1989; Lawton, 1990; McEvoy and Coombs, 1999; Denoth et al., 2002). Scoring systems (Goeden, 1983), traits of successful control agents (Crawley, 1986, 1989), and climate matching (Wapshere, 1985) have been tried in an attempt to improve agent selection; however, different approaches provide contradictory results when applied to the same system (Blossey, 1995). More recently, demographic models (Shea and Kelly, 1998; McEvoy and Coombs, 1999; Davis et al., 2006) have been used to evaluate the promise of single or multiple control agents. While it is too early to assess the long-term success of demographic models, traditional approaches have failed to provide the desired success rates (Blossey and Hunt-Joshi, 2003) or allow unequivocal prioritization of control agents. This leaves weed biocontrol scientists and practitioners at the present time with little else but informed trial and error.

A major shortcoming is a lack of pre-release impact studies and follow-up monitoring (Blossey, 2004). Historically, too many agents with little impact on plant demography were released, wasting valuable resources and increasing the risk for food-web effects (see sections on What constitutes success? and Conflicts of interest). As detailed for stage 3, impact studies measuring effects of different herbivore combinations and densities on plant performance and plant demography can yield valuable information. This information can be used to develop predictive models (Davis et al., 2006) and allows prioritization of control agents (McClay and Balciunas, 2005). Reducing the number of introductions of species that ultimately contribute little to control will have the additional benefit of reducing potential risks to non-target plants (McEvoy and Coombs, 2000). Such studies can be done at various venues (field, common, garden) and under different designs (herbivore exclusion or addition) and may provide a new and improved framework for predicting success. It is also crucial to quantitatively assess the role of different herbivores after they have been released. Visual or anecdotal assessments, which form the basis of many programme evaluations (Julien and Griffiths, 1998), are notoriously unreliable and tend to

overestimate the contributions of highly visible defoliators while less visible species such as root feeders go unnoticed. A recent evaluation of programme outcomes demonstrated that despite their 'invisibility', root-feeders are more successful in controlling their host plant than their above-ground counterparts (Blossey and Hunt-Joshi, 2003).

What constitutes success?

Success in weed biological control is measured in many different ways (Crawley, 1989; Lawton, 1990; Hoffmann, 1995; McFadyen, 1998; Fowler et al., 2000; McFadyen, 2000). In the most spectacular cases – such as control of Opuntia cacti in Australia; the floating tropical water lettuce, Salvinia molesta; and alligator weed, Alternanthera philoxeroides, in the south-eastern USA – there is little dispute about outcomes. Similarly, if control agents do not establish or do not harm individual plants, there is little controversy is assigning the programme a failure. However, traditionally, most programmes use simple observational reports (complete, marked, or no control) without quantifications (Crawley, 1989; Julien and Griffiths, 1998). If herbivores successfully control a plant species in some countries, but not in others (Crawley, 1989), it is difficult to assign success or failure to entire programmes. Regardless of these context-specific outcomes (all control is local), there are some very clear desirable or undesirable outcomes. The important question is whether we have reliable and defensible data to assess whether they have been achieved. Desirable outcomes in the order in which they may occur are listed below.

1. *Establishment*: Control agent(s) establish self-sustaining populations.
2. *Population growth*: Control agent(s) populations increase.
3. *Impact on plant individuals*: Control agent(s) impact performance (growth, seed output, survival) of host plant individuals.
4. *Impact on plant populations*: Control agent(s) affect population dynamics, resulting in declines of the host plant. These abundance declines can be measured in various ways (e.g. biomass, cover, number of stems), and may

vary depending on growth form and architecture of the plant. Declines should be quantified in proportional reductions from the abundance levels the weed obtained before control agents were released, and thus may range from small (5–10%) to substantial reductions (>80%).

5. *Reduced costs for other control measures*: Success of a biocontrol programme can be measured in terms of cost savings if other control measures are less often used or are entirely superfluous (Hoffmann, 1995; Fowler *et al.*, 2000).

6. *Reduction or elimination of negative impacts by the target weed*: Ecosystem impacts of a weed vary widely depending on the ecosystem and traits of invaders. For example, the impact of rangeland weeds can be measured using stocking or livestock growth rates. Some species may impact wildlife abundance or threaten rare and endangered species. Some species have more subtle influences on ecosystem processes such as changes in fire frequency or soil biota changes associated with cheatgrass (*Bromus tectorum*) invasion (Sperry *et al.*, 2006); changes in amphibian community composition associated with *Lythrum salicaria*, purple loosestrife (Brown *et al.*, 2006); or disruption of mycorrhizal mutualisms in native plant species by garlic mustard, *Alliaria petiolata* (Stinson *et al.*, 2006). Ideally, all impacts are known and quantified before biocontrol agents are released, and potential reductions in these negative ecosystems effects associated with the release of biocontrol agents can be measured. Knowledge of impacts allows assessments of whether release of biocontrol agents reduces or eliminates negative ecosystem consequences.

Host specificity

Releases of specialized herbivores from the home range of an invasive plant are often met with concerns that: (i) biocontrol agents may attack non-target plants; and (ii) biocontrol agents may, over evolutionary time, become less host-specific and attack non-target species (Simberloff and Stiling, 1996; Louda *et al.*, 1997). Over the past 60 years, weed biocontrol scientists have developed sophisticated procedures to assess the dietary restrictions of

control agents (Manly, 1993; Clement and Cristofaro, 1995; USDA, 1999; Briese, 2004, 2005) and several reviews of the literature provided little evidence for host-shifts among released insect herbivores after their introduction (Marohasy, 1998; Blossey *et al.*, 2001c; van Klinken and Edwards, 2002).

When weed biocontrol scientists refer to host specificity or host range of a potential biocontrol agent, they refer typically to the 'realized' host range (van Klinken, 2000), while tests actually measure the fundamental (physiological) host range. The former describes the plant species actually used in the field by a herbivore, the latter includes all plant species that a herbivore is potentially able to utilize. The realized host range is a subset of the fundamental host range and is influenced largely by behavioural responses to constraints of different developmental stages of a herbivore. For example, an internally feeding larva may be able to feed and complete development on a certain plant; however, female oviposition choice and lack of larval mobility may prevent attack in the field (Blossey *et al.*, 1994b).

Host specificity testing starts with a selection of plants (typically 50–80 species) to be tested, and this selection is unique to each plant species and each country. To aid selection, biocontrol scientists have devised a system referred to as the centrifugal phylogenetic method (Wapshere, 1974), emphasizing the need to focus on plant species closely related to the target (i.e. the same genus, followed by family, order etc). This is a fundamental shift from earlier procedures testing agricultural or ornamental species. Current selection procedures include species that occur in the same habitat as the target, species with similar chemistry, and species attacked by close relatives of potential control agents, plus rare and endangered species (USDA, 1999; Briese, 2004).

Once a test plant list is assembled and approved by regulatory authorities, tests are either conducted in the native range of the target plant or in quarantine. Typically scientists use no-choice or multiple-choice tests using cut plant parts or potted plants. Tests are conducted in very confined conditions (Petri dishes), small cages, or field enclosures, or even open field tests. Testing conditions have to be adapted to feeding mode and behaviour of potential agents. Both female oviposition choice and larval acceptance of test plants and their ability to survive and complete development is measured. Difficulties in interpreting contradictory results obtained under different circumstances have long been recognized (Cullen, 1990). While the realism of testing conditions increases from no-choice to multiple-choice, and from Petri dishes, to cages, to field cage or open field tests, every test can have problems in interpretation and may result in false positives or negatives (Marohasy, 1998). No-choice tests are very conservative and are prone to false positives, but are able to reduce the number of plant species needed in more sophisticated (and expensive) tests. Species not attacked in no-choice conditions are usually considered safe.

The fact that an introduced biological control agent may attack native species is, in itself, not necessarily a reason for concern (Blossey et al., 2001c). In many instances, survival and recruitment is low (Turner, 1985; Willis and Ash, 1996) and is reduced as distance from the original host increases (Schooler and McEvoy, 2006). Only four insect species used as biocontrol agents are known to have established self-sustaining populations on non-target species in the absence of the original host (Blossey et al., 2001c) and this potential was known at the time of introduction.

The release of weed biocontrol agents is often permitted (after environmental assessment), even if the potential for non-target attack exists, when potential harm caused by the herbivore is significantly less than the harm caused by the uncontrolled spread of the weed or by other control methods (Blossey et al., 1994a, 2001a). When introducing biocontrol agents, we must be concerned about their impact if this affects a non-target species' population, distribution or abundance, not just the fact that individual plants are attacked (Louda et al., 1997).

Published and anecdotal evidence suggests that, with the exception of the two high-profile cases of Rhinocyllus conicus Fröhlich attacking native North American Cirsium species (Louda et al., 1997) and Cactoblastis cactorum Berg. attacking native Opuntia species in North America, weed biocontrol is safe and does not cause extended non-target effects. But critics of

weed biocontrol can point to the lack of follow-up monitoring and the fact that lack of evidence does not indicate an absence of non-target effects. Overall, host specificity screening has consistently provided the best assurance for the safety of non-target species (McFadyen, 1998; Pemberton, 2000; Gassmann and Louda, 2001). Non-target impacts of R. conicus and C. cactorum can be traced to poor decision-making processes before 1970 that permitted the release of non-specific herbivores (Pemberton, 2000; Gassmann and Louda, 2001). Contemporary regulations incorporate measures to avoid similar mistakes (USDA, 1999), but the need for improved monitoring to develop a reliable database remains.

Conflicts of interest

Objections to the control of non-indigenous species can be grouped into five categories: (i) economic; (ii) ecological; (iii) aesthetic; (iv) ethical; and (v) risks associated with the development of biological weed control (Blossey, 1999). Although potential negative impacts of herbicides on applicators, ecosystem function, and on non-target species are widely recognized, traditional techniques (mechanical, physical, chemical) used to control invasive plants usually meet with little resistance. Many people assume that local control efforts have localized impacts and can be discontinued if unwanted side-effects occur; and that control at one site does not affect populations of the target at other sites. However, biological control is irreversible after control agents are established, and release at one location has local, regional and potential continental implications when control agents spread beyond their initial release sites. Concerns over safety of weed biocontrol focus on two topics:

1. the attack of non-target species (discussed above under Host specificity);
2. undesirable non-target effects through 'ripple effects' in the food web after introduction of biocontrol agents.

Changes in food webs have received recent attention (Pearson and Callaway, 2005, 2006). For example an increase in mouse populations carrying hanta virus as a result of a biocontrol agent attacking *Centaurea maculosa* without causing declines in their host plant. A major problem with this debate is that the invasion of *C. maculosa* has probably altered species composition and ecosystem processes (Callaway and Aschehoug, 2000; Callaway *et al.*, 2004), yet we have no idea about the extent of changes due to *C. maculosa* invasion. Again, the problem is the lack of sophisticated data on ecological communities before invasion or before biocontrol agent introductions (Blossey, 1999). Nevertheless, the introduction of additional trophic links in ecological communities will result in changes in local food webs. In many instances, additional links in food webs have been interpreted as positive (Blossey, 2003), but concerns over indirect effects of such redirected resource flows are raised (Pearson *et al.*, 2000). Of particular concern is the effect of new food subsidies provided by biocontrol agents to generalist predator populations. If predator populations increase, alternative native prey populations may decline as a consequence (Pearson and Callaway, 2003).

We have only just started to address potential food web implications of biocontrol agent releases and much more work needs to be done. Most importantly, such assessments need to use appropriate reference scenarios to assess the potential positive or negative impacts of biological control (Blossey, 2003). Any comparison of populations, energy flows, and food web linkages after biocontrol implementation should use pre-plant invasion data to assess changes and the effectiveness of restoration attempts. Such data are usually not available and most researchers rely on comparisons between invaded and uninvaded habitats (see Blossey, 2003, for a detailed discussion of the pitfalls of such approaches). Ultimately, quantitative long-term evaluations of changes associated with invasions or release of biological control programmes at various trophic levels can provide much-needed information about associated ecological benefits or costs.

Monitoring

Scepticism concerning the need, safety and effectiveness of insect introductions for weed control is, at least in part, the result of a lack of quantitative data on the impacts of the release of weed biocontrol agents on native ecosystems. This lack of scientific certainty about impacts has led to harsh criticism of attitudes towards non-indigenous species and control efforts (Hager and McCoy, 1998; Sagoff, 1999) and exposes a fundamental weakness in our ability to assess the impacts of introduced species on individual species or ecosystem processes. Lack of evidence does not necessarily imply a lack of impacts (Blossey *et al.*, 2001a) and impacts of invasive plant species may be greatly underestimated; yet we lack clear and convincing evidence due to the lack of long-term data. Biocontrol practitioners identified the need for long-term follow-up work decades ago (Huffaker and Kennett, 1959; Schroeder, 1983), but little progress has been made in collecting quantitative data on the effects of biocontrol agents on target plant performance or in documenting responses of associated plant and animal communities (Blossey, 1999). A better understanding of successes (and failures!) should over time lead to better selection and release strategies (Malecki *et al.*, 1993; McEvoy and Coombs, 1999).

The responsibility for developing monitoring protocols rests with biocontrol practitioners familiar with target plants and their response to control agents. Monitoring protocols should be completed and implemented before control agents are actually field-released. Well-executed, long-term monitoring offers exciting opportunities to merge basic and applied ecological research using teams of investigators. However, to allow widespread adoption of protocols and participation by non-academic personnel, monitoring protocols should balance scientific sophistication with ease of application (Blossey and Skinner, 2000). Incorporating long-term monitoring will certainly increase costs associated with biocontrol, but these costs may be offset by saving resources currently wasted on unsuccessful agents.

At a minimum, assessments should include measures of control agent abundance and impact on host plants and host-plant populations, and performance measures for associated plant communities at release and control sites (where no releases are made). These measures would capture many of the direct effects associated with the release of biological control agents. However, indirect effects are prevalent in ecological communities and many species and trophic levels are linked through such indirect interactions. The type of biological inventories will be case-specific. For example monitoring for biocontrol effects on rangelands will be different from monitoring in forests or wetlands, yet overall goals are similar. It is at the discretion of those implementing biological control programmes to select, justify and defend the most appropriate metrics and scope of monitoring.

6.6 Summary

Weed biocontrol could have a bright future as an environmentally sound pest management option if weed biocontrol scientists are able to quantitatively document the benefits (economically and ecologically) of their activities. The use of specialized insect herbivores as biocontrol agents will most probably focus on natural areas and more permanent plant communities (forests, rangelands, orchards) and have limited applicability in croplands with short rotations. However, even after a 200-year history, the science of weed biocontrol has not matured sufficiently, and large gaps in sophisticated programme implementation and follow-up remain. Open questions on how to select successful herbivores and the true ecosystem impacts of herbivore releases on ecological communities require urgent attention. Pre-release impact studies and follow-up monitoring after control agent releases could greatly improve the science of weed biocontrol.

6.7 References

Bain, M.B. (1993) Assessing impacts of introduced aquatic species: grass carp in large systems. *Environmental Management* 17, 211–224.

Baker, H.G. (1965) Characters and modes of origin of weeds. In: Baker, H.G. and Stebbins, G.L. (eds) *The Genetics of Colonizing Species*. Academic Press, New York, USA, pp. 147–172.

Baker, H.G. (1974) The evolution of weeds. *Annual Review of Ecology and Systematics* 5, 1–24.

Blossey, B. (1995) A comparison of various approaches for evaluating potential biological control agents using insects on *Lythrum salicaria*. *Biological Control* 5, 113–122.

Blossey, B. (1999) Before, during, and after: the need for long-term monitoring in invasive plant species management. *Biological Invasions* 1, 301–311.

Blossey, B. (2003) A framework for evaluating potential ecological effects of implementing biological control of *Phragmites australis*. *Estuaries* 26, 607–617.

Blossey, B. (2004) Monitoring in weed biological control programs. In: Coombs, E., Clark, J., Piper, G.L. and Cofrancescor, A.F., Jr (eds) *Biological Control of Invasive Plants in the United States*. Oregon State University Press, Corvallis, OR, USA, pp. 95–105.

Blossey, B. and Hunt, T.R. (1999) Mass rearing methods for *Galerucella calmariensis* and *G. pusilla* (Coleoptera: Chrysomelidae), biological control agents of *Lythrum salicaria* (Lythraceae). *Journal of Economic Entomology* 92, 325–334.

Blossey, B. and Hunt-Joshi, T.R. (2003) Belowground herbivory by insects: influence on plants and aboveground herbivores. *Annual Review of Entomology* 48, 521–547.

Blossey, B. and Nötzold, R. (1995) Evolution of increased competitive ability in invasive nonindigenous plants: a hypothesis. *Journal of Ecology* 83, 887–889.

Blossey, B. and Schroeder, D. (1995) Host specificity of three potential biological weed control agents attacking flowers and seeds of *Lythrum salicaria* (purple loosestrife). *Biological Control* 5, 47–53.

Blossey, B. and Skinner, L.C. (2000) Design and importance of post-release monitoring. In: Spencer, N.A. (ed.) *Proceedings of the X International Symposium on Biological Control of Weeds*. Montana State University, 4–10 July 1999. Bozeman, MT, USA, pp. 693–706

Blossey, B., Schroeder, D., Hight, S.D. and Malecki, R.A. (1994a) Host specificity and environmental impact of two leaf beetles (*Galerucella calmariensis* and *G. pusilla*) for biological control of purple loosestrife (*Lythrum salicaria*). *Weed Science* 42, 134–140.

Blossey, B., Schroeder, D., Hight, S.D. and Malecki, R.A. (1994b) Host specificity and environmental impact of the weevil *Hylobius transversovittatus*, a biological control agent of purple loosestrife (*Lythrum salicaria*). *Weed Science* 42, 128–133.

Blossey, B., Eberts, D., Morrison, E. and Hunt, T.R. (2000) Mass rearing the weevil *Hylobius transversovittatus* (Coleoptera: Curculionidae), biological control agent of *Lythrum salicaria*, on semiartificial diet. *Journal of Economic Entomology* 93, 1644–1656.

Blossey, B., Skinner, L.C. and Taylor, J. (2001a) Impact and management of purple loosestrife (*Lythrum salicaria*) in North America. *Biodiversity and Conservation* 10, 1787–1807.

Blossey, B., Nuzzo, V., Hinz, H. and Gerber, E. (2001b) Developing biological control of *Alliaria petiolata* (M. Bieb.) Cavara and Grande (garlic mustard). *Natural Areas Journal* 21, 357–367.

Blossey, B., Casagrande, R., Tewksbury, L., Landis, D.A., Wiedenmann, R.N. and Ellis, D.R. (2001c) Nontarget feeding of leaf-beetles introduced to control purple loosestrife (*Lythrum salicaria* L.). *Natural Areas Journal* 21, 368–377.

Bossdorf, O., Auge, H., Lafuma, L., Rogers, W.E., Siemann, E. and Prati, D. (2005) Phenotypic and genetic differentiation between native and introduced plant populations. *Oecologia* 144, 1–11.

Briese, D.T. (2004) Weed biological control: applying science to solve seemingly intractable problems. *Australian Journal of Entomology* 42, 304–317.

Briese, D.T. (2005) Translating host-specificity test results into the real world: the need to harmonize the yin and yang of current testing procedures. *Biological Control* 35, 208–214.

Brown, C.J., Blossey, B., Maerz, J.C. and Joule, S.J. (2006) Invasive plant and experimental venue affect tadpole performance. *Biological Invasions* 8, 327–338.

Callaway, R.M. and Aschehoug, E.T. (2000) Invasive plants versus their new and old neighbors: a mechanism for exotic invasion. *Science* 290, 521–523.

Callaway, R.M., Thelen, G.C., Barth, S., Ramsey, P.W. and Gannon, J.E. (2004) Soil fungi alter interactions between the invader *Centaurea maculosa* and North American natives. *Ecology* 85, 1062–1071.

Clement, S.L. and Cristofaro, M. (1995) Open field tests in host-specificity determination of insects for biological control of weeds. *Biocontrol Science and Technology* 5, 395–406.

Colautti, R.I., Ricciardi, A., Grigorovich, I.A. and MacIsaac, H.J. (2004) Is invasion success explained by the enemy release hypothesis? *Ecology Letters* 7, 721–733.

Crawley, M.J. (1986) The population biology of invaders. *Philosophical Transactions of the Royal Society of London B* 314, 711–731.

Crawley, M.J. (1987) What makes a community invasible? In: Gray, A.J., Crawley, M.J. and Edwards, P.J. (eds) *Colonization, Succession and Stability*. Blackwell Scientific, Oxford, UK, pp. 429–453.

Crawley, M.J. (1989) The successes and failures of weed biocontrol using insects. *Biocontrol News and Information* 10, 213–223.

Cullen, J.M. (1990) Current problems in host-specificity screening. In: Delfosse, E.S. (ed.) *Proceedings of the VII International Symposium on Biological Control of Weeds*. Istituto Sperimentale per la Patologia Vegetale, MAF Rome, Rome, Italy, pp. 27–36.

Culliney, T.W. (2005) Benefits of classical biological control for managing invasive plants. *Critical Reviews in Plant Sciences* 24, 131–150.

Davis, A.S., Landis, D.A., Nuzzo, V.A., Blossey, B., Gerber, E. and Hinz, H.L. (2006) Demographic models inform selection of biocontrol agents for garlic mustard. *Ecological Applications* 16, 2399–2410.

Denoth, M., Frid, L. and Myers, J.H. (2002) Multiple agents in biological control: improving the odds? *Biological Control* 24, 20–30.

Dodd, A.P. (1940) *The Biological Campaign against Prickly-Pear*. Commonwealth Prickly Pear Board, Brisbane, Australia.

Ellstrand, N.C. and Schierenbeck, K. (2000) Hybridization as a stimulus for the evolution of invasiveness in plants? *Proceedings of the National Academy of Sciences* 97, 7043–7050.

Fowler, S.V., Syrett, P. and Hill, R.L. (2000) Success and safety in the biological control of environmental weeds in New Zealand. *Austral Ecology* 25, 553–561.

Gassmann, A. and Louda, S.M. (2001) *Rhinocyllus conicus*: initial evaluation and subsequent ecological impacts in North America. In: Wajnberg, E., Scott, J.K. and Quimby, P.C. (eds) *Evaluating Indirect Ecological Effects of Biological Control*. CABI Publishing, Wallingford, UK, pp. 147–183.

Gassmann, A. and Schroeder, D. (1995) The search for effective biological control agents in Europe: history and lessons from leafy spurge (*Euphorbia esula* L.) and cypress spurge (*Euphorbia cyparissia* L.). *Biological Control* 5, 466–477.

Goeden, R.D. (1983) Critique and revision of Harris' scoring system for selection of insect agents in biological control of weeds. *Protection Ecology* 5, 287–301.

Grime, J.P. (1998) Benefits of plant diversity to ecosystems: immediate, filter and founder effects. *Journal of Ecology* 86, 902–910.

Hager, H.A. and McCoy, K.D. (1998) The implications of accepting untested hypotheses: a review of the effects of purple loosestrife (*Lythrum salicaria*) in North America. *Biodiversity and Conservation* 7, 1069–1079.

Hoffmann, J.H. (1995) Biological control of weeds: the way forward, a South African perspective. In: Waage, J.K. (ed.) *British Crop Protection Council Symposium no. 64*. The British Crop Protection Council, Farnham, UK, pp. 77–98

Holst, P.J., Allan, C.J., Campbell, M.H. and Gilmour, A.R. (2004) Grazing pasture weeds by goats and sheep. 2. Scotch broom (*Cytisus scoparius* L.). *Australian Journal of Experimental Agriculture* 44, 553–557.

Huffaker, C.B. and Kennett, C.E. (1959) A ten-year study of vegetational changes associated with biological control of Klamath weed. *Journal of Range Management* 12, 69–82.

Julien, M.H. and Griffiths, M.W. (1998) *Biological Control of Weeds: A World Catalogue of Agents and their Target Weeds*. CABI Publishing, Wallingford, UK.

Keane, R.M. and Crawley, M.J. (2002) Exotic plant invasions and the enemy release hypothesis. *Trends in Ecology and Evolution* 17, 164–170.

Kok, L.T. and Surles, W.W. (1975) Successful biocontrol of musk thistle by an introduced weevil, *Rhinocyllus conicus*. *Environmental Entomology* 4, 125–127.

Lawton, J.H. (1990) Biological control of plants: a review of generalisations, rules and principles using insects as agents. In: Basset, C., Whitehouse, L.J. and Zabkiewicz, J.A. (eds) *Alternatives to the Chemical Control of Weeds*. FRI Bulletin 155, Rotorua, New Zealand, pp. 3–17

Levine, J.M., Adler, P.B. and Yelenik, S.G. (2004) A meta-analysis of biotic resistance to exotic plant invasions. *Ecology Letters* 7, 975–989.

Lonsdale, W.M. (1993) Rates of spread of an invasive species: *Mimosa pigra* in northern Australia. *Journal of Ecology* 81, 513–521.

Lonsdale, W.M. (1999) Global patterns of plant invasions and the concept of invasibility. *Ecology* 80, 1522–1536.

Louda, S.M., Kendall, D., Connor, J. and Simberloff, D. (1997) Ecological effects of an insect introduced for the biological control of weeds. *Science* 277, 1088–1090.

Louda, S.M., Pemberton, R.W., Johnson, M.T. and Follett, P.A. (2003) Nontarget effects: the Achilles' heel of biological control: retrospective analyses to reduce risk associated with biocontrol introductions. *Annual Review of Entomology* 48, 365–396.

Malecki, R.A., Blossey, B., Hight, S.D., Schroeder, D., Kok, L.T. and Coulson, J.R. (1993) Biological control of purple loosestrife. *Bioscience* 43, 680–686.

Manly, B.F.J. (1993) Comments on design and analysis of multiple-choice feeding preference experiments. *Oecologia* 93, 149–152.

Marohasy, J. (1998) The design and interpretation of host-specificity tests for weed biological control with particular reference to insect behavior. *BioControl* 19, 13–20.

McClay, A.S. and Balciunas, J.K. (2005) The role of pre-release efficacy assessment in selecting classical biological control agents for weeds: applying the Anna Karenina principle. *Biological Control* 35, 197–207.

McConnachie, A.J., de Wit, M.P., Hill, M.P. and Byrne, M.J. (2003) Economic evaluation of the successful biological control of *Azolla filiculoides* in South Africa. *Biological Control* 28, 25–32.

McConnachie, A.J., Hill, M.P. and Byrne, M.J. (2004) Field assessment of a frond-feeding weevil, a successful biological control agent of red waterfern, *Azolla filiculoides*, in southern Africa. *Biological Control* 29, 326–331.

McEvoy, P.B. and Coombs, E.M. (1999) Biological control of plant invaders: regional patterns, field experiments, and structured population models. *Ecological Applications* 9, 387–401.

McEvoy, P.B. and Coombs, E.M. (2000) Why things bite back: unintended consequences of biological weed control. In: Follett, P.A. and Duan, J.J. (eds) *Non-target Effects of Biological Control*. Kluwer Academic, Boston, MA, USA, pp. 167–194.

McFadyen, R.E.C. (1998) Biological control of weeds. *Annual Review of Entomology* 43, 369–393.

McFadyen, R.E.C. (2000) Successes in biological control of weeds. In: Spencer, N.A. (ed.) *Proceedings of the X International Symposium on Biological Control of Weeds*. Montana State University, 4–10 July 1999. Bozeman, MT, USA, pp. 3–14

Mitchell, C.E. and Powers, A.G. (2003) Release of invasive plants from fungal and viral pathogens. *Nature* 421, 625–627.

Mooney, H.A., Mack, R.N., McNeely, J.A., Neville, L.E., Schei, P.J. and Waage, J.K. (2005) *Invasive Alien Species*. Island Press, Washington, DC, USA.

Müller, H. and Schroeder, D. (1989) The biological control of diffuse and spotted knapweed in North America: what did we learn? In: Fay, P.K. and Lacey, J.R. (eds) *Proceedings of the 1989 Knapweed Symposium*. Montana State University, Bozeman, MT, USA, pp. 151–169

Parker, J.D., Burkepile, D.E. and Hay, M.E. (2006) Opposing effects of native and exotic herbivores on plant invasions. *Science* 311, 1459–1461.

Pearson, D.E. and Callaway, R.M. (2003) Indirect effects of host-specific biological control agents. *Trends in Ecology and Evolution* 18, 456–461.

Pearson, D.E. and Callaway, R.M. (2005) Indirect nontarget effects of host-specific biological control agents: implications for biological control. *Biological Control* 35, 288–298.

Pearson, D.E. and Callaway, R.M. (2006) Biological control agents elevate hantavirus by subsidizing deer mouse populations. *Ecology Letters* 9, 443–450.

Pearson, D.E., McKelvey, K.S. and Ruggiero, L.F. (2000) Non-target effects of an introduced biological control agent on deer mouse ecology. *Oecologia* 122, 121–128.

Pemberton, R.W. (2000) Predictable risk to native plants in weed biocontrol. *Oecologia* 125, 489–494.

Sagoff, M. (1999) What's wrong with alien species? *Report for the Institute for Philosophy and Public Policy* 19, 16–23.

Schooler, S.S. and McEvoy, P.B. (2006) Relationship between insect density and plant damage for the golden loosestrife beetle, *Galerucella pusilla*, on purple loosestrife (*Lythrum salicaria*). *Biological Control* 36, 100–105.

Schroeder, D. (1983) Biological control of weeds. In: Fletcher, W.E. (ed.) *Recent Advances in Weed Research*. Commonwealth Agricultural Bureau, Farnham, UK, pp. 41–78.

Shea, K. and Kelly, D. (1998) Estimating biocontrol agent impact with matrix models: *Carduus nutans* in New Zealand. *Ecological Applications* 8, 824–832.

Shea, K., Kelly, D., Sheppard, A.W. and Woodburn, T.L. (2005) Context-dependent biological control of an invasive thistle. *Ecology* 86, 3174–3181.

Sheppard, A.W., Hill, R.L., DeClerck-Floate, R.A., McClay, A., Olckers, T., Quimby, P.C. and Zimmermann, H.G. (2003) A global review of risk-benefit-cost analysis for the introduction of classical biological control agents against weeds: a crisis in the making? *Biocontrol News and Information* 24, 77N–94N.

Simberloff, D. and Stiling, P. (1996) How risky is biological control? *Ecology* 77, 1965–1974.

Sperry, L.J., Belnap, J. and Evans, R.D. (2006) *Bromus tectorum* invasion alters nitrogen dynamics in an undisturbed arid grassland and ecosystem. *Ecology* 87, 603–615.

Stanley, D.F., Holst, P.J. and Allan, C.J. (2000) The effect of sheep and goat grazing on variegated thistle (*Silybum marianum*) populations in annual pasture. *Plant Protection Quarterly* 15, 116–118.

Stinson, K.A., Campbell, S.A., Powell, J.R., Wolfe, B.E., Callaway, R.M., Thelen, G.C., Hallett, S.G., Prati, D. and Klironomos, J.N. (2006) Invasive plant suppresses the growth of native tree seedlings by disrupting belowground mutualisms. *Public Library of Science Biology* 4(5), 140 [doi:10.1371/journal.pbio.0040140].

Tewksbury, L., Casagrande, R., Blossey, B., Häfliger, P. and Schwarzländer, M. (2002) Potential for biological control of *Phragmites australis* in North America. *Biological Control* 23, 191–212.

Turner, C.E. (1985) Conflicting interests and biological control of weeds. In: Delfosse, E.S. (ed.) *Proceedings of the VI International Symposium on Biological Control of Weeds*. Agriculture Canada, Ottawa, 19–24 August. Vancouver, Canada, pp. 203–224.

US Congress (1993) *Harmful Non-Indigenous Species in the United States*. Office of Technology Assessment, OTA-F-565, Washington, DC, USA.

USDA (1999) *Reviewers Manual for the Technical Advisory Group for Biological Control Agents of Weeds*. Manual Unit of Plant Protection and Quarantine, Animal Plant Health Inspection Service (APHIS), United States Department of Agriculture, Annapolis, MD, USA.

van Klinken, R.D. (2000) Host specificity testing: why do we do it and how can we do it better? In: van Driesche, R., Heard, T., McClay, A. and Reardon, R. (eds) *Host Specificity of Exotic Arthropod Biological Control Agents: The Biological Basis for Improvements in Safety*. US Forest Service, Forest Health Technology Enterprise Team, Morgantown, WV, USA, pp. 54–68

van Klinken, R.D. and Edwards, O.R. (2002) Is host specificity of weed biocontrol agents likely to evolve rapidly following establishment? *Ecology Letters* 5, 590–595.

Wapshere, A.J. (1974) A strategy for evaluating the safety of organisms for biological weed control. *Annals of Applied Biology* 77, 200–211.

Wapshere, A.J. (1985) Effectiveness of biological control agents for weeds: present quandaries. *Agriculture, Ecosystems and Environment* 13, 261–280.

Williamson, M. (1996) *Biological Invasions*. Chapman and Hall, London.

Willis, A.J. and Ash, J.E. (1996) Combinations of stress and herbivory by a biological control mite on the growth of target and non-target native species of *Hypericum* in Australia. In: Moran, V.C. and Hoffmann, J.H. (eds) *Proceedings of the IX International Symposium on Biological Control of Weeds*, 19–26 January 1996, University of Cape Town, Stellenbosch, South Africa, pp. 93–100.

7 Bioherbicides for Weed Control

M.A. Weaver, M.E. Lyn, C.D. Boyette and R.E. Hoagland

*United States Department of Agriculture, Agricultural Research Service,
PO Box 350, Stoneville, MS 38776, USA*

7.1 Introduction

Management of weeds is a necessary but expensive challenge. Chemical weed control accounts for over $14 bn spent annually (Kiely et al., 2004), excluding immense indirect costs to producers, consumers and the environment, and resulting also in the development of resistant weed biotypes. While chemical herbicides effectively control unwanted vegetation, many herbicides are no longer available due to lack of re-registration, competition from other products, and development of numerous genetically modified crops with resistance to broad-spectrum herbicides, namely glyphosate and glufosinate. The implementation of conservation tillage practices to promote soil quality, to minimize erosion, or to simplify crop management has increased reliance on 'burn-down' herbicides and placed additional selection pressure on weeds to develop resistance. After years of applying herbicides, often in the presence of high weed pressure, 180 species of herbicide-resistant weeds have been identified (WeedScience, 2006). The majority of herbicide usage is for agronomic areas or turf, but few herbicides are registered for, or are being developed for, smaller markets or niche weed problems, such as invasive weeds in noncropland areas. Furthermore, chemical weed control is not an option in organic cropping systems and near to sensitive natural habitats. The high costs involved in developing and registering chemical herbicides, and recent trends in environmental awareness concerning pesticides in general, have prompted researchers to develop additional weed control tools, such as biological weed control using plant pathogens.

A review of pathogen-based weed control prospects by Charles Wilson (1969) noted that 'the idea of using plant pathogens to control weeds is almost as old as the science of plant pathology itself', but that the 'seeds of the idea … have lain dormant since their sowing'. Since that review, almost 40 years ago, numerous pathogens for weed control have been identified and a few have enjoyed limited commercial success (Hoagland, 1990, 2001).

Classical pathogen-mediated biocontrol of weeds generally employs an exotic pest to manage a weed population. This is an effective weed management strategy in many systems (Bedi et al., 2002; also see Blossey, Chapter 6, this volume). An alternative method is to overwhelm the target weed with direct pathogen application, or multiple applications of a pathogen. Because this tactic uses biological agents in an application similar to chemical herbicidal applications, it is often called the 'bioherbicidal' approach. When the plant pathogens are fungi, these bioherbicides are often called 'mycoherbicides'.

7.2 Discovery of Bioherbicides

A scientific strategy should be utilized in the discovery of classical biocontrol agents (Berner and Bruckart, 2005). Often the weed to be controlled is exotic, so a search is made near its region of origin to find potential biocontrol organisms that have co-evolved with the given weed host. These potential biocontrol agents are then screened for possible undesirable, non-target effects, and studied to evaluate environmental parameters for efficacy. High levels of host specificity are desirable and often unavoidable, as with the rust pathogens. An example of this ongoing work is the effort to control yellow starthistle (*Centaurea solstitialis*) by *Puccinia jaceae* var. *solstitialis* (Bruckart, 2006).

The bioherbicidal approach, in contrast with classical biocontrol, more commonly relies on indigenous pathogens. A common assumption is that highly virulent pathogens always make the most effective bioherbicides, but this concept has been effectively challenged (Hallett, 2005). Some host–pathogen interactions produce dramatic symptoms, but may not meaningfully reduce the weed population. In contrast, one of the most commercially successful bioherbicides to date, *Colletotrichum gloeosporioides* f. sp. *aeschynomene*, does not produce impressive symptoms or particularly rapid mortality of northern jointvetch (*Aeschynomene indica*), the target weed. Instead, its success has been due to: (i) its low cost of production; (ii) its comparatively simple formulation requirements; and (iii) its rapid and efficient secondary spread in the field (Bowers, 1986; Smith, 1991; D.O. TeBeest, personal communication). This might not be the only model for success with a bioherbicide, but it provides a useful benchmark for comparing candidate bioherbicides.

Other practical considerations for the commercial success of a bioherbicide include the ease of obtaining both patent and product registration. Registration of a chemical herbicide is a lengthy and expensive process, but the US Environmental Protection Agency has recognized potential environmental benefits for biopesticides, simplifying the process, so that registration can be achieved in as little as 1 year (EPA, 2006; Slininger *et al.*, 2003). Furthermore, some of the expense of bioherbicide registration can be reduced for 'low-risk' applications.

Finally, the discovery of candidate bioherbicides must consider economics, and several questions need to be answered.

- Can propagules be generated on low-cost substrates?
- Will they remain viable in storage after production?
- Will production require lengthy incubation in expensive bioreactors?
- Will the inoculum require extensive processing before application?
- What conventional herbicides will it compete with?
- Is the market large and stable enough to recoup development and registration costs?
- Will the product generate sufficient profit?

It would be beneficial if markets could be identified where low-cost chemical inputs are not available, such as in organic crop production or small-acreage, but potentially high-value, crops that have been neglected by the chemical industry. Invasive weeds in natural or low-input managed ecosystems, such as public lands, forests and conservation easements may also be attractive targets for bioherbicides, as they may be perceived as more compatible than chemical herbicides in these sensitive areas. Bedi *et al.* (2002) described a semi-quantitative means of measuring the commercial potential of a bioherbicide candidate, incorporating many of these traits.

7.3 Mycoherbicide Production Technology

For practical and economic reasons, the propagules of a bioherbicide must be rapidly and inexpensively produced. Asexually produced fungal spores, or conidia, are generally the most cost-effective and easiest to produce under laboratory conditions (Templeton *et al.*, 1979). Since spores provide the most common method for natural dispersal and typically have longevity, they should logically serve as the best candidates as infective units of mycoherbicides. For fungi that that do not produce spores, or do not produce them readily or efficiently, the production and use of mycelial fragments may be possible (Boyette *et al.*, 1991b). Mycelial formulations present challenges because they are generally more difficult to quantify, less readily separated from the culture medium, and are less infective than conidia. Spores often have longer shelf-lives and are more tolerant of suboptimal storage conditions (Churchill, 1982). The recent development of *Mycoleptodiscus terrestris* for the control of hydrilla (*Hydrilla verticillata*) has identified microsclerotia as a readily produced, desiccation-tolerant inoculum (Shearer and Jackson, 2006).

Selection of culture medium

Defined media, vegetable juice or agar culture can be used in the production of inoculum for experimental bioherbicidal systems (Boyette, 2000; Gressel, 2002). Researchers must recognize the unique nutritional requirements of each candidate bioherbicide with regard to carbon and nitrogen sources, pH, inorganic and trace elements and, in some cases, vitamins, amino acids or essential oils. However, to mass-produce mycoherbicides on a large scale for pilot tests or industry evaluation, these requirements must be met within the context of economic constraints. These requirements are most often realized using crude agricultural or industrial products that are readily available at low cost. Nitrogen sources, such as soybean flour, corn steep liquor, distillers solubles, brewers yeast, autolysed yeast, milk solids, cottonseed flour and linseed meal, are some of the materials that have been used to produce mycoherbicides. Carbon sources commonly tested include cornstarch, cornflour, glucose, hydrolysed-corn-derived materials, glycerol and sucrose (Churchill, 1982). Additionally, some fungi require light for sporulation, which may add complexity and increase production costs.

Carbon sources that do not maximize vegetative growth may enhance sporulation. The carbon, nitrogen and mineral levels that lead to optimal growth and sporulation may require precise and empirical balancing (Jackson, 1997). In addition to the effect on growth and sporulation, the carbon : nitrogen (C:N) ratio may affect viability, longevity and virulence of the fungus. For example, the vegetative growth of *Fusarium solani* f. sp. *phaseoli* was increased by a high C:N ratio, while virulence of the fungus on its host plant, *Phaseolus vulgaris*, was decreased.

Conversely, a low C:N growth medium resulted in decreased vegetative growth and increased virulence (Toussoun *et al.*, 1960). In contrast, Phillips and Margosan (1987) found the spore volume, nuclear number and virulence of *Botrytis cinerea* against the hybrid rose (*Rosa* spp.) increased linearly in response to increasing glucose concentration.

Slininger *et al.* (2003) reviewed a systematic approach taken to evaluate both the commercial potential of biocontrol strains and media selection. Media selection involves producing large quantities of inoculum quickly and inexpensively while simultaneously maintaining pathogen virulence. In the case of *Colletotrichum truncatum* for control of hemp sesbania (*Sesbania exaltata*), it was discovered that the optimum C:N ratio for production of conidia did not yield the most virulent conidia. Therefore, a balance point was found to achieve these two goals (Jackson, 1997; Jackson *et al.*, 1996; Wraight *et al.*, 2001).

Inoculum density may also affect fungal sporulation. Slade *et al.* (1987) found that a high inoculum density (2.5×10^6 spores/ml) of *Colletotrichum gloeosporioides* resulted in slimy masses of conidia, called 'slime spots', when grown on several commonly used growth media. Slime spots are associated with microcyclic conidiation, where sporulation occurs directly after spore germination in the absence of mycelial growth. Conversely, reduced inoculum concentrations or concentrated growth media resulted in dense, vegetative mycelial growth; and microcyclical conidiation did not occur (Hildebrand and McCain, 1978; Slade *et al.*, 1987).

Solid substrate fermentation

Solid substrate fermentation may be the only practical method for spore production if spores cannot be produced using submerged fermentation. Various cereal grains and vegetative residues have been used to produce simple, inexpensive inocula for a number of plant pathogenic fungi (Tuite, 1969). Hildebrand and McCain (1978) used wheat straw infested with *Fusarium oxysporum* f. sp. *cannabis*, to control marijuana (*Cannabis sativa*). Boyette *et al.* (1984) used oat seed infested with *F. solani* f. sp. *cucurbitae* to control Texas gourd (*Cucurbita texana*). These types of bulky substrates are difficult to sterilize, inoculate and store until they are ready for use in the field. Some of these problems can be overcome by separation of spores from the substrate for subsequent drying, formulation and storage. These processes add cost and complexity to product development.

Combined solid substrate and submerged fermentations

Several mycoherbicides have been produced using combined solid and submerged fermentation techniques. *Alternaria macrospora*, a pathogen for control of spurred anoda (*Anoda cristata*), was first mass-produced by culturing fungal mycelium for 48 h in a vegetable-juice-based liquid medium. Fungal biomass was then collected, blended, mixed with vermiculite, spread into foil-lined pans, and exposed to either fluorescent light or direct sunlight to induce sporulation. After air-drying, the mixture was sieved, packaged, and stored at 4°C (Walker, 1981). This procedure has also been used to produce inoculum of *Colletotrichum malvarum* for control of prickly sida (*Sida spinosa*) and *F. lateritium* for control of spurred anoda, velvetleaf (*Abutilon theophrasti*) and prickly sida.

A modification of this technique was used to produce spores of *A. cassiae* for use as a mycoherbicide against sicklepod (*Senna obtusifolia*) (Walker and Riley, 1982). Fungal mycelium was grown in submerged culture for 24 h, collected, homogenized, poured into foil-lined trays, and then exposed to 10 min of ultraviolet light every 12 h for 5 days to induce sporulation. The mycelia sporulated prolifically as the medium dried. After 72 h, the spores were collected by vacuum, dried over calcium sulphate, and stored at 4°C. Approximately 8 g of spores were produced per litre of growth medium with this simple technique, to yield a product density of 10^8 spores/g (Walker and Riley, 1982). Sufficient quantities of *A. cassiae* spores were produced using this technique for a 2-year pilot study. This technique was also used to produce spores of *A. crassa* for jimsonweed (*Datura stramonium*) control (Quimby, 1989); *A. helianthi* for cocklebur (*Xanthium strumarium*) and wild sunflower (*Helianthus annuus*) control (Van Dyke and

Winder, 1985); and *Bipolaris sorghicola* for johnsongrass (*Sorghum halepense*) control (Bowers, 1982).

Submerged culture fermentations

From both practical and economic perspectives, biocontrol fungi that sporulate in liquid culture are favoured over those that require additional steps to induce sporulation. This factor alone has proved to be advantageous for the commercial development of a fungus as a mycoherbicide (Bowers, 1982).

For early developmental studies using small-scale experiments, inoculum can be produced in shake-flasks. However, with shake-flasks it is difficult to maintain and adjust parameters that affect growth, such as the correct pH, temperature and aeration essential for optimal growth and sporulation. For larger quantities of inoculum or systems that require more precise control, laboratory-scale fermenters are essential. Some fermenters monitor and provide programmed control of environmental factors including temperature, agitation, dissolved oxygen and pH.

Slade *et al.* (1987) developed a simple method to assess inoculum production of *Colletotrichum gloeosporioides* in liquid culture using microplate assays of the fungus on various solid media. This system could possibly be used to provide an accurate, rapid and inexpensive means of screening various growth media for spore yield and virulence.

Systematic approaches to growth medium development can yield significant economic returns. Mitchell (2003) examined 47 carbon sources in an effort to maximize spore production of *Septoria polygonorum*, a pathogen of smartweed (*Polygonum* spp.). After identifying pea brine as the best carbon source, 38 factors (numerous inorganic amendments, fatty acids, complex nitrogen sources, surfactants, etc) were screened to find the best combination. The final composition yielded production of more than 10^8 spores/ml. A similar stepwise, surface-response modelling approach to medium selection was used by Mitchell *et al.* (2003) to maximize production of *Gloeocercospora sorghi*, a bioherbicide of johnsongrass (*Sorghum halepense*). By methodically testing numerous components, at several concentrations, in many combinations, much higher production levels can be achieved, with concomitant economic returns.

Two registered mycoherbicides – Collego (Encore Technologies, Minnentoka, MN) and DeVine (Abbott Laboratories, North Chicago, IL) were both produced using submerged liquid culture techniques. The formulation of these bioherbicides is discussed later.

7.4 Bioherbicide Formulation

Bioherbicide formulations are engineered or strategically developed materials consisting of spores or other propagules of one or more microbial agent(s) previously identified as a weed pathogen. Various bioherbicide formulation types are possible and many have been explored on an experimental basis and are discussed throughout this chapter. Suspension concentrates, wettable powders, dry flowables, water-dispersible or wettable granules, and non-disintegrable granules are some of the possible formulations.

Formulations are developed for different reasons associated with manipulation (including handling and application), stabilization or shelf-life, and efficacy. The importance of formulation as it relates to each of these factors is discussed below.

Manipulation

Loose particles ranging in size from sub-micron up to several tens of microns (micrometres) are prone to aerosolization, i.e. becoming suspended in air for long periods of time (Griffin, 2004). Atmospheric aerosols that are microorganisms, plant material, and associated cell-wall materials and metabolites are specifically referred to as bioaerosols (Kuske, 2006; El-Morsy, 2006; Reoun An et al., 2006). Exposure to such aerosols can pose health risks (WHO, 2000) to developers and users of bioherbicides. Furthermore, other issues arising from aerosolized agents include loss of applied active ingredient and deposition to non-target or off-sites (Brown and Hovmøller, 2002;

Griffin et al., 2003). This process of aerosolization and the associated hazards are similar to the risks with liquid spray application of synthetic pesticides (Tsai et al., 2005). Incorporation of bioherbicide propagules into macroscopic solids or other steps to minimize aerosolization may be an important technique to promote safe handling and application of agents.

Stabilization and storage

Stabilization and long-term storage of a mycoherbicide is dependent on formulation composition and the water content of the dried product. For long-term storage of agents, cellular metabolism can be controlled by lowering either the water activity (Aw) or the storage temperature of the product. Commercially, one would prefer to store the agent at ambient temperatures; therefore, the shelf-life of mycoherbicides should preferentially be extended by lowering the Aw. Reduced Aw implies the agent could be in either a solid-state formulation such as a granule or dispersed in oil. Oils can be phytotoxic to non-target plants (Tworkoski, 2002) and may undergo lipid oxidation upon extended storage. Thus, careful selection of oil type and inclusion of antioxidants as stabilizing agents may be necessary to prevent unwanted chemical changes to the formulation and to extend the shelf-life of oil-based mycoherbicides.

The science of choosing formulation ingredients to improve the stability of solid-state bioherbicides is not fully understood. However, the significance of formulation composition as it relates to storage stability of bioherbicide propagules has been demonstrated (Silman et al., 1993; Connick et al., 1996, 1997; Amsellem et al., 1999; Shabana et al., 2003; Müller-Stöver et al., 2004; Friesen et al., 2006). For example, in research on the shelf-life of either conidia or conidia plus mycelium of *Fusarium arthrosporioides* and *F. oxysporum*, 'Stabileze' (a mixture of starch, sucrose, corn oil and silica) was found to be superior to alginate bead formulations in preserving the viability of each weed control agent (Amsellem et al., 1999). Shelf-life of *C. truncatum* spores, formulated in a solid/perlite–cornmeal–agar mixture, at 15°C was longer than in a liquid formulation or a solid/vermiculite mixture (Silman *et al.*, 1993). In wheat flour–kaolin 'pesta' granules, *C. truncatum* spores germinated in 87% of granules that were stored for 1 year at 25°C (Connick *et al.*, 1996). Viable pesta granules containing *C. truncatum* microsclerotia were observed after 10 years of storage at 4°C (Boyette *et al.*, 2007b). In other studies, significantly different trends were observed in the viability of *F. oxysporum* f. sp. *orthoceras* microconidia in ten differently amended pesta formulations, each containing various adjuvants (Shabana *et al.*, 2003). In all these pesta formulations, glycerol imparted a negative effect on shelf-life, but stillage (an alcohol manufacturing by-product) and Water Lock (a 'super-absorbent' polymer)-amended pesta formulations exhibited the worst shelf-life among those evaluated (Shabana *et al.*, 2003).

In addition to formulation, storage stability can be improved further by maintaining the product at an optimal water content or water activity (Connick *et al.*, 1996; Shabana *et al.*, 2003). An example is the storage studies of pesta containing *C. truncatum* microsclerotia by Connick *et al.* (1997). These investigators were able to identify, for a single formulation, water activities ranging from 0.12 to 0.33 at 25°C that were conducive to long-term storage. At water activities above 0.33, the shelf-life of pesta with *C. truncatum* was inferior to granules stored in drier air environments.

Collectively, the above studies indicate that improvements in bioherbicide storage stability are possible through the choice of specific formulation ingredients and appropriate drying. Since shelf-life is dependent on optimal water activities, packaging is also an important parameter in improving the shelf-life of mycoherbicides.

Efficacy

Finally, the efficacy of weed pathogens may be enhanced through formulation. Formulations amended with particular adjuvants and nutrients can: (i) stimulate biological activity while reducing biological competition from pre-existing microorganisms; (ii) protect the agent from environmental factors such as UV light, wind, and rainwater removal from the target plant surface; (iii) facilitate and sustain propagule germination, growth and infection; and (iv) improve coverage and agent–target interaction of the formulated spray droplets on plant tissue.

The first mycoherbicide registered and marketed in the USA was based on the phytopathogenic fungus *Phytophthora palmivora*. The product, DeVine, was used to control strangler vine or milkweed vine (*Morrenia odorata*), a pest of Florida citrus orchards. The product had a shelf-life limited to a few weeks and required refrigeration during storage and shipment to preserve the live chlamydospores. Nevertheless the need to effectively control strangler vine assured commercial success for several years. The product is no longer commercially available because the market niche is too small to sustain commercial interest (Ridings, 1986). This, however, is an example of the potential for agents with short shelf-lives that might be found for target weeds that are of regional importance. For example, of the estimated 3 million hectares covered by kudzu (Forseth and Innis, 2004), several hundred thousand hectares of kudzu could be within a 500 km distribution radius of a mid-south USA bioherbicide production site.

The second US bioherbicide product was registered in 1982. It was based on *Colletotrichum gloeosporioides* f. sp. *aeschynomene* (CGA). The Upjohn Company, in collaboration with researchers at the University of Arkansas and the US Department of Agriculture, was able to mass-produce CGA and market a formulated biological control agent under the trade name Collego. It was developed for northern jointvetch control in soybean and rice fields, and approved for use in Arkansas and Louisiana. The Collego product was delivered in two packages – one of dried CGA spores in an inert carrier material, the other an inert rehydrating osmoticum for reviving spores before spraying. The components of both packages were added to the desired volume of water immediately before application (reviewed in Smith, 1986). While this product has not been available commercially for several years, changing agronomic practices have led to renewed interest in biological control of northern jointvetch. An effort is under way to bring this product back to the marketplace under the name LockDown (K. Cartwright, Agricultural Research Initiatives, personal communication).

BioMal (Philom Bios, Saskatoon, Canada) is another registered mycoherbicide based on the hydrophilic fungus *Colletotrichum gloeosporioides* f. sp. *malvae*. It was delivered as a wettable silica gel powder for control of round-leaved mallow (*Malva pusilla*). The spores, as formulated, dispersed readily in water for application, and routinely provided more than 90% control in the field (Ridings, 1986).

Granular formulations

With some exceptions, liquid mycoherbicide formulations are generally best suited for use as post-emergence sprays and are used primarily to incite leaf and stem diseases. Conversely, pathogens that infect below the soil surface are best delivered in a solid or granular formulation. Granular formulations are better suited for use as pre-planting or pre-emergence mycoherbicides than are liquid spray formulations because: (i) granules provide a buffer from environmental extremes; (ii) granules can provide a food-base for the fungus, prolonging persistence; and (iii) granules are less likely to be washed away from the treated areas (Mitchell, 1986; Wymore *et al.*, 1988).

A cornmeal–sand formulation of *Fusarium solani* f. sp. *cucurbitae* was used to produce a mixture of mycelium, microconidia, macroconidia and chlamydospores. The ratio of these spore types can be altered by the addition of various nutrients to the basal medium. Excellent control (96%) of Texas gourd was achieved using pre-planting and pre-emergence applications with granular formulations of this fungus (reviewed in Boyette, 2000).

Vermiculite has also been used effectively to prepare solid substrate mycoherbicide formulations. Walker (1981) produced mycelia of *Alternaria macrospora* in liquid shake culture and mixed the mycelium with vermiculite. The fungus sporulated profusely in the mixture and, after air-drying, applications were made both pre-emergence and post-emergence to spurred anoda. Pre-emergent application of fungus-infested vermiculite resulted in control rates equivalent to those achieved with post-emergence foliar sprays.

Granular formulations of several biocontrol fungi have also been made using sodium alginate (Walker and Connick, 1983; Weidemann and Templeton, 1988); a procedure adapted from research with time-released herbicide formulations (Connick, 1982). Fungal mycelium is mixed into a sodium alginate solution with various fillers, such as kaolin clay, and the mixture is dripped into 0.25 M calcium chloride. The Ca^{2+} ions react with the sodium alginate to form gel beads. The beads are allowed to harden briefly in the calcium chloride solution and then they are collected, rinsed and air-dried. Granules are relatively uniform in size and shape and can be used in a manner similar to pre-planting or pre-emergence herbicides, or rehydrated and exposed to UV light to induce spore production for other applications.

A pasta-like process is another approach to producing granules of several different fungi, such as *C. truncatum* for hemp sesbania control, *F. lateritium* for velvetleaf control, and *F. oxysporum* for sicklepod control. Granules are produced by mixing semolina wheat flour and kaolin clay with fungal propagules contained in a liquid component; either water or residual liquid growth medium. The mixture is kneaded into dough, rolled into thin sheets with a pasta press, and air-dried for 48 h. Sheets are then milled and sieved to obtain uniform-sized granules which are stored at 4°C. These granules, called 'Pesta', provided 90–100% weed control in glasshouse tests. In field tests, 'Pesta' granules containing *C. truncatum* provided 80–85% control of hemp sesbania in 3 years of tests (Connick *et al.*, 1993; Boyette *et al.*, 2007b). Various mycoherbicide formulations are listed in Table 7.1.

Adjuvants for liquid formulations

The simplest mycoherbicide delivery system is suspension of the agent in water for spray application. However, many weeds possess a waxy cuticle that inhibits the even spreading of droplets across the leaf surface, thus preventing uniform distribution of the agent. Surfactants facilitate distribution of an agent across the phylloplane by reducing surface tension caused by the waxy cuticle. Various non-ionic surfactants have been used in mycoherbicide research. Some surfactants may affect the growth or germination of fungal propagules.

Table 7.1. Experimental and commercial bioherbicide formulations.

Weed host	Pathogen	Formulation[a]
Liquid suspension formulations		
Spurred anoda (*Anoda cristata*)	*Fusarium lateritium*	Water + Tween-20 surfactant (0.02%)
Spurred anoda (*Anoda cristata*)	*Colletotrichum coccodes*	Water + sorbitol (0.75%)
Round-leaved mallow (*Malva pusilla*)	*Colletotrichum gloeosporioides* f. sp. *malvae*	**BioMal**
Annual bluegrass (*Poa annua*)	*Xanthamonas campestris*	**Camperico**
Spurred anoda (*Anoda cristata*)	*Alternaria macrospora*	Water + nonoxynol surfactant (0.02%) + sucrose (5% w/v)
Giant ragweed (*Ambrosia trifida*)	*Protomyces gravidus*	Water
Field bindweed (*Convolvulus arvensis*)	*Phomopsis convolvulus*	Water + gelatin (0.1%)
Jimsonweed (*Datura stramonium*)	*Alternaria crassa*	Water + nonoxynol surfactant (0.4%)
Florida beggarweed (*Desmonium tortuosum*)	*Colletotrichum truncatum*	Water + nonoxynol surfactant (0.4%)
Sicklepod (*Cassia obtusifolia*)	*Alternaria cassiae*	Water + nonoxynol surfactant (0.4%)
Common purslane (*Portulaca oleracea*)	*Dichotomophthora portulacaceae*	Water + Tween-20 (0.02%)
Horse purslane (*Trianthema portulacastrum*)	*Gibbago trianthemae*	Water + Tween-20 (0.02%)
Hemp sesbania (*Sesbania exaltata*)	*Colletotrichum truncatum*	Water + nonoxynol surfactant (0.2%); paraffin wax, mineral oil, soybean oil lecithin; unrefined corn oil
Eastern black nightshade (*Solanum ptycanthum*)	*Colletotrichum coccodes*	Water + Tween-80 surfactant (0.02%)
Stranglervine (*Morrenia odorata*)	*Phytophthora palmivora*	**DeVine** Chlamydospores in water
Solid formulations		
Velvetleaf (*Abutilon theophrasti*)	*Fusarium lateritium*	Sodium alginate, kaolin granules
Northern jointvetch (*Aeschynomene virginica*)	*Colletotrichum gloeosporioides* f. sp. *aeschynomene*	**Collego** Component A: dried spores Component B: rehydrating agent + surfactant
Spurred anoda (*Anoda cristata*)	*Alternaria macrospora*	Vermiculite
Texas gourd (*Cucurbita texana*)	*Fusarium solani* f. sp. *cucurbitae*	Cornmeal/sand; sodium alginate–kaolin granules
Marijuana (*Cannabis sativa*)	*Fusarium oxysporum* var. *cannabis*	Fungus-infested wheat straw
Hemp sesbania (*Sesbania exaltata*)	*Colletotrichum truncatum*	Fungus-infested wheat gluten/kaolin clay (Pesta)
Sicklepod (*Cassia obtusifolia*)	*Fusarium oxysporum*	Fungus-infested wheat gluten/kaolin clay (Pesta)
Sicklepod (*Cassia obtusifolia*) and others	*Alternaria cassiae*	**CASST**

[a] Names in bold are commercial formulations.
Adapted from Boyette (2000).

For example, *Alternaria cassiae* spores do not germinate consistently in Tween-20 or Tween-80 non-ionic surfactants but readily germinate in 0.02–0.04% non-ionic, nonoxynol surfactants (Walker and Riley, 1982). Tests should be conducted to measure any effect of a given surfactant on the candidate mycoherbicide. Other liquid formulations are listed in Table 7.1.

Various adjuvants and amendments have been used either to improve or to modify spore germination, increase pathogen virulence, minimize environmental constraints, or alter host preference, each of which may greatly influence the mycoherbicidal performance of a candidate bioherbicide. The addition of sucrose to aqueous suspensions of *A. macrospora* resulted in greater control of spurred anoda (Walker, 1981). Also, increased spore germination and disease severity occurred on Florida beggarweed (*Desmodium tortuosum*) when small quantities of sucrose and xanthan gum were added to aqueous spore suspensions of *C. truncatum* (Cardina *et al.*, 1988).

Disease severity on johnsongrass infected by *Bipolaris sorghicola* was significantly increased by adding 1% Soy-Dox to the fungal spray mixture. Similarly, the addition of sorbitol yielded a 20-fold increase in the number of viable spores of *C. coccodes* re-isolated from inoculated velvetleaf. When this amendment was added to *C. coccodes* for velvetleaf control, three 9-h dew periods on consecutive nights were as effective as a single 18-h dew treatment (Wymore and Watson, 1986).

Most pathogens being evaluated as mycoherbicides require high water activities (i.e. humidity greater than 80%, or dew) over a period of time in order to germinate, penetrate, infect and kill the target weed. This period of time ranges from 6 h to more than 24 h, depending upon the pathogen and the weed host (reviewed in Boyette, 2000). Invert (water-in-oil) emulsions can retard evaporation, thereby decreasing the length of time that additional free moisture is required for spore germination and for infection (Quimby *et al.*, 1988; Daigle *et al.*, 1990). In these studies, lecithin was used as an emulsifying agent, and paraffin oil and wax were used to further retard evaporation and help retain droplet size. Specialized spraying equipment was developed to deliver this viscous material (McWhorter *et al.*, 1988;

Quimby *et al.*, 1988). Glasshouse and field results indicated that excellent control (>95%) of sicklepod with *A. cassiae* could be achieved with little or no dew (Quimby *et al.*, 1988). This system was used to enhance hemp sesbania control in the field with *C. truncatum*. The control (95%) achieved was comparable to that achieved with the synthetic herbicide, acifluorfen. Less than 10% control of hemp sesbania occurred in plots treated with the fungus applied with a water-only carrier (Boyette *et al.*, 1993).

Protection against ultraviolet radiation

Solar radiation is one reason for mycoherbicides that perform well in glasshouse trials to fail in the field or exhibit sporadic field efficacy (Yang and TeBeest, 1993; Walker and Tilley, 1997; Charudattan, 2000). The transmitted solar spectrum is attenuated by the windows of glasshouses and spectrally altered by cover materials that are typically treated with ultraviolet (UV) inhibitors to prolong their lifespan. Like some synthetic herbicide formulations, UV protection may be crucial in preserving the applied active ingredient, particularly for formulations that deposit agents onto leaf surfaces, where they remain exposed to solar radiation.

Recently, the effects of sunlight on mortality, germination rate, and germ tube length for different phytopathogenic species of *Colletotrichum* were explored (Ghajar *et al.*, 2006a). Exposure to sunlight decreased germination rate and germ tube length of *C. gloeosporioides* conidia isolated from *Polystigma rubrum* subsp. *rubrum* (Stojanović *et al.*, 1999). More recently, UV-A (320–400 nm) photons were found to stimulate appressorium formation, while UV-B (280–320 nm) photons delayed conidium germination in *C. orbiculare* and *Plectosporium alismatis*. As the dose of UV-B increased, these photons deactivated conidia and also delayed germination of the survivors (Ghajar *et al.*, 2006a).

Studies on UV protection of fungal entomopathogens have indicated that protectants can prolong the viability of conidia (Burges, 1998; Burges and Jones, 1998; Leland and Behle, 2005). A calcium cross-linked lignin UV barrier produced a tenfold increase in *Beauveria bassiana* germination response time (RT_{50}) from 2.8 to 28.3 h after incubation for 48 h (Leland and Behle, 2005). In a follow-up study by Ghajar *et al.* (2006b), formulations containing water- and oil-soluble compounds that protect against UV damage were explored as formulation additives and post-UV-B-exposed germination rates increased to levels similar to the unexposed or so-called 'dark' control. In addition, these authors reported a significant increase in disease development over control levels when *C. orbiculare* was applied with particular UV protectants in leaf disc bioassays (Ghajar *et al.*, 2006b).

Formulation can alter or expand host selectivity of a bioherbicide. For example, the host selectivity of *A. crassa*, a mycoherbicide for jimsonweed, can be altered by the addition of water-soluble filtrates of jimsonweed or dilute pectin suspensions (Boyette and Abbas, 1994). Several plant species that were either resistant or which exhibited a hypersensitive reaction to the fungus alone, exhibited various degrees of susceptibility following these amendments. Among the important weed species that were highly susceptible to infection following addition of these amendments were hemp sesbania, eastern black nightshade (*Solanum ptycanthum*), cocklebur and showy crotalaria (*Crotalaria spectabilis*). Several solanaceous crop species, including tomato (*Lycopersicon esculentum*), aubergine (*Solanum melonegra*), potato (*S. tuberosum*) and tobacco (*Nicotiana tabacum*), also became susceptible to infection when these amendments were used. With proper timing of application, it is possible that these amendments could enhance the weed control spectrum of *A. crassa* (Boyette and Abbas, 1994).

Amsellem *et al.* (1991) found that the host specificities of *A. cassiae* and *A. crassa* were greatly expanded, and that a saprophytic *Cephalosporium* species became pathogenic when these fungi were formulated in an invert emulsion. Similarly, the host ranges of *C. truncatum* and *C. gloeosporioides* f. sp. *aeschynomene* (Collego) were also expanded when spores of either pathogen were formulated in an invert emulsion. In rice field plots, over 90% of hemp sesbania plants were controlled by Collego/invert emulsion treat-ments, while aqueous suspensions of Collego had no effect upon hemp sesbania (Boyette *et al.*, 1991a, 1992). A similar response occurred with *C. truncatum*. Aqueous inundative or wound inoculations with aqueous spore suspensions of *C. truncatum* had no effect on northern jointvetch, but susceptibility to infection was induced when the fungus was formulated in the invert emulsion.

Most mycoherbicides have a limited host range. For the purposes of safety and registration, this is an advantage. However, from an economic standpoint, this could preclude the practical use of a candidate mycoherbicide, since a single weed species rarely predominates in row crop situations (McWhorter and Chandler, 1982). One solution to this limitation is to apply mixtures of pathogens to mixed weed populations. For example, northern jointvetch and winged waterprimrose (*Jussiae decurrens*), two troublesome weeds in rice, were simultaneously controlled with a single application of CGA and *C. gloeosporioides* f. sp. *jussiae* (Boyette *et al.*, 1979). A mixture of these two pathogens with the addition of *C. malvarum* also effectively controlled northern jointvetch, winged waterprimrose and prickly sida (TeBeest and Templeton, 1985). Various weed pathogens may not be compatible with each other. Thus, mixtures of pathogens need to be screened prior to formulation.

7.5 Application Technology

In a recent review on the state of the art in bioherbicides, Hallett noted that application technology, especially liquid spray application, had lagged (Hallett, 2005). Citing published reports by Egley and Boyette (1993), Chapple and Bateman (1997), Bateman (1998), and his own unpublished results, he highlighted the role of droplet size and deposition patterns in weed control.

Bioherbicidal weed control research is often conducted with very high inoculum rates and unrealistically high application volumes (e.g. 'spray to runoff'). Ground-based herbicide application rates are generally less than 200 l/ha and aerial application rates are much lower. When application rates are expressed as CFU/ml, the actual number of infective units to treat an area is obfuscated. Without a clear understanding of the true application rate, any assessment of a pathogen's potential as a biocontrol agent should be considered to be preliminary.

In using native pathogens to manage indigenous weed species, the weed scientist is attempting to alter the natural balance. This is hardly an anathema – altering this balance is intrinsic to most agronomic practices – but the ecological forces at work warrant consideration. The dose–response relationship should not be assumed to be linear over a broad range of inoculum concentrations and, as reviewed (Hallett, 2005), application rates are often well beyond the linear range. Density-dependent pathogen mortality, hyperparasitism and competition for infection sites conspire to reduce efficiency at these levels (Newton *et al.*, 1997, 1998; Horn, 2006). Consequently, if inadequate control is provided at a given inoculum rate, it may be more cost-effective to reconsider adjuvants, formulation and delivery systems than to simply increase the dose.

The method of production of mycoherbicides may directly determine the method of application. Mycoherbicides are applied in much the same manner as chemical herbicides and often with the same equipment. Tanks, lines and nozzles on the spraying system must be void of chemical residues that may be detrimental to mycoherbicides. A slurry of activated charcoal and liquid detergent can be used for cleansing spray equipment (Quimby and Boyette, 1987). Similarly, pesticides, especially fungicides, applied to mycoherbicide-treated areas may reduce mycoherbicide effectiveness. For example, the fungicide benomyl and the herbicide propiconazol, applied sequentially 7 and 14 days after Collego application, suppressed disease development in northern jointvetch (Khodayari and Smith, 1988). Similarly, the efficacy of DeVine was reduced if the fungicides Aliette and Ridomil were used within 45 days following mycoherbicide application (Kenney, 1986). This result might be expected, as these fungicides are active against *Phytophthora* spp., but it highlights the need to evaluate bioherbicides for compatibility with agronomic practices and agrochemical programmes.

7.6 Compatibility of Bioherbicides with other Management Practices

While classical biocontrol is most often practised in natural or low-input ecosystems, bioherbicides generally aim to control weeds in more highly managed systems. It is unrealistic to expect major changes in cropping systems to accommodate bioherbicides, so formulation and application technology must work with, or least not interfere with, accepted agronomic practices.

Wyss *et al.* (2004) recognized that agrochemicals can interfere with biocontrol agents in distinct phases. They measured the compatibility of synthetic herbicides, fungicides, insecticides and adjuvants with the *Amaranthus* spp. biocontrol agent, *Phomopsis amaranthicola*. Spore germination was tolerant of very high rates (over $2\times$ maximum labelled rates) of some agrochemicals, such as benomyl, atrazine, imazethapyr and pendimethalin, but was completely inhibited by low rates ($0.25\times$ maximum labelled rates) of chlorothalonil, iprodione and diuron. Similar differences in compatibility were observed regarding vegetative growth and sporulation. The authors noted that this type of in vitro screening is useful, but that ultimately the agrochemical effects need to be assessed in the field so that effects on pathogenicity and weed control can be quantified. The purple nutsedge (*Cyperus rotundus*) pathogen, *Dactylaria higginsii*, was also evaluated for tolerance to several pesticides (Yandoc *et al.*, 2006). The herbicide imazapyr was well tolerated by the fungus, but all other evaluated pesticides inhibited conidial germination and/or mycelial growth.

One study examined in-field interactions between a synthetic herbicide and a bioherbicide, *C. truncatum*, for hemp sesbania control (Boyette *et al.*, 2007a). This weed is problematic in soybean and cotton fields in the southern USA, where glyphosate-based weed control management is common. In the context of that system, the authors evaluated *C. truncatum* for bioherbicidal efficacy in the field when applied before, simultaneously with, and after glyphosate applications. In these studies, the bioherbicide provided control of the weed target when applied after glyphosate, but not when applied simultaneously or before the herbicide (Fig. 7.1).

Yandoc *et al.* (2006), citing unpublished observations of Smith and Hallett, stated that glyphosate itself was not toxic to *Aposphaeria amaranthi* but the commercially formulated products were. We have observed that various commercial formulations of glyphosate have very different effects on *Myrothecium verrucaria* spore viability (unpublished data).

The addition of sublethal rates of the herbicides linuron, imazaquin and lactofen to *A. cassiae* spores significantly increased control of sicklepod when applied in an invert formulation (Quimby and Boyette, 1986). Control of velvetleaf was significantly improved by sequential applications of the herbicide 2,4 DB and spores of *F. lateritium*. However, spore germination and disease severity were greatly reduced when the fungus and herbicide were tank-mixed (reviewed in Boyette, 2000). Biocontrol of velvetleaf was also improved significantly by the addition of thiadiazuron, a cotton defoliant, to an aqueous spray mixture of *C. coccodes* (Wymore *et al.*, 1988).

The rust fungus, *Puccinia canaliculata*, does not provide consistently high control of its host weed yellow nutsedge (*Cyperus esculentus*)

when uredospores are applied alone, even under optimal environmental conditions (Bruckart *et al.*, 1988). However, sequential applications of the herbicide paraquat followed by *P. canaliculata* spores resulted in a synergistic disease interaction, with almost complete yellow nutsedge control, compared to only 10% and 60% control, respectively, for paraquat or the fungus alone (Callaway *et al.*, 1987).

Khodayari *et al.* (1987) demonstrated that the weed control spectrum of *Colletotrichum gloeosporioides* f. sp. *aeschynomene* can be extended by mixing with acifluorfen, a herbicide that is effective in controlling hemp sesbania, but is ineffective in controlling northern jointvetch. In these tests, both weeds were effectively controlled in soybeans by a single application of the mixture, providing that the microenvironment was favourable for infection.

Erwinia carotovora, a candidate bacterial bioherbicide for tropical soda apple (*Solanum viarum*) was evaluated for compatibility with commercial formulations of dicamba and triclopyr (Roberts *et al.*, 2002). In this unique pathosystem, the biocontrol agent alone did not cause any leaf injury, but it prevented the

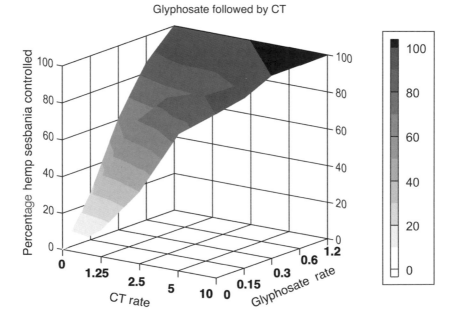

Fig. 7.1. Hemp sesbania control by *Colletotrichum truncatum* (CT) preceded by glyphosate.

regeneration of plants injured in the herbicide treatments.

In the previous examples, bioherbicides were used in concert with chemical herbicides to improve weed control. Shearer and Nelson (2002) evaluated co-application of *M. terrestris* with the herbicide endothall to improve control of hydrilla (*Hydrilla verticillata*) and also to minimize injury to non-target vegetation. At high treatment levels, both the pathogen and the herbicide have been reported to cause disease or injury to other plants (Shearer and Nelson, 2002; Shearer and Jackson, 2006). When used together, however, greater than 90% control was achieved in mesocosm studies, even with reduced rates of endothall and *M. terrestris*. Non-target plant injury still occurred, but the researchers suggested that this integrated control offered a means of effective hydrilla control while altering the dosage of bioherbicide and chemical herbicide to protect desirable vegetation (Shearer and Nelson, 2002).

7.7 Risk Analysis and Mitigation for Bioherbicides

Beyond the aspects of efficacy and economic viability, there may be other risks associated with bioherbicides. Some of the identified hazards include toxicity, infection, or allergenic responses to producers or applicators of the biocontrol agent. After application there are risks of damage to non-target plants, direct impacts on wildlife or other microorganisms, and indirect species- or community-level effects by perturbations in trophic networks (Vurro *et al.*, 2001; Delfosse 2005). While not trivializing the magnitude of these hazards, the likelihood of some of these events is generally low. Risk mitigation can be realized through lowering the intrinsic hazard or by reducing the probability of exposure. For example, *M. verrucaria* produces secondary metabolites; the trichothecene mycotoxins (Abbas *et al.*, 2001, 2002). It might be possible to make this a more acceptable bioherbicide by altering of the conditions of inoculum growth and formulation to reduce or exclude these toxins from the final product (Hoagland *et al.*, 2007a,b). Mutation, strain selection, strain improvement or genetic engineering could yield

an atoxigenic isolate. It has been reported that infected plants do not accumulate detectable levels of trichothecenes (Abbas *et al.*, 2001). Another risk factor associated with *M. verrucaria* is its extensive host range, which includes several important crop species (Walker and Tilley, 1997). Clarke *et al.* (2007) demonstrated the efficacy of *M. verrucaria* against old world climbing fern (*Lygodium microphyllum*) without significant disease symptoms on co-occurring native plant species. The potential threat posed by bioherbicides to non-target plants, which is actually less than that of many widely used herbicides, might be mitigated by restrictions on application sites and times. Furthermore, while *M. verrucaria* is an effective pathogen when formulated, it does not form infections when applied without surfactant, which effectively halts any secondary disease cycles or off-target movement.

7.8 Conclusions

As noted earlier, Charles Wilson (1969) addressed the potential of bioherbicides almost 40 years ago. Since his writing, a few of those seeds have flourished but, in the words of a more recent review (Hallett, 2005), the field 'has languished in recent years'.

In the early years of bioherbicide discovery and development, the conventional thinking was that a good agent was one that could be grown cheaply and quickly, was an aggressive pathogen, was patentable, could be easily applied, had single-weed specificity, and that if satisfactory control was not realized, then inundate the weed with the pathogen (Templeton *et al.*, 1979; TeBeest, 1991; Zidak and Quimby, 1998). Some of these principles are still valid; the marketplace has little use for slow-growing, fastidious bioherbicidal microorganisms. Other parts of the dogma have been effectively challenged. While a highly specific pathogen is desirable and intrinsically safe, economic realities may prevent the commercialization of such a pathogen. This premium on safety from specific agents may be unwarranted. Even before the advent of transgenic cropping systems, numerous successful synthetic herbicides with very broad activities were deployed. It may be possible to use broad-spectrum pathogens without intolerable non-target effects (De Jong et al., 1999; Pilgeram and Sands, 1999). Instead of replacing herbicides, synergy between pathogens and chemical herbicides may be a more successful approach to weed management (Boyette et al., 2007a).

A portion of the bioherbicide canon rejected by many is the reliance on very high inoculum levels. While mycoherbicides offer natural weed control, there is nothing 'natural' about applying mycoherbicides in extraordinarily high titres, artificially deposited onto weed foliage, in anticipation of an epidemic. Gressel's review found published application rates from 200 to 500,000 times higher than necessary based on his assumptions (Gressel, 2002). This should be considered encouraging, since this suggests there is an opportunity to profoundly lower the doses

and maintain weed control through improved formulation and application techniques.

The future of bioherbicides may lie with agents that were once considered too risky. Broad-spectrum bioherbicides such as M. verrucaria, Sclerotinia sclerotiorum and others might be made 'safe' by means of appropriate deployment strategies or development of ecologically impaired strains, respectively. Other pathogens with low virulence, but with desirable specificity or epidemiology, might be made more virulent through genetic modification (Gressel, 2002).

Through a pragmatic understanding of economic constraints and safety, the intelligent use of formulation and deployment methods, and with genetic engineering, the pathogen discoveries of the past might be harnessed as the bioherbicides of the future.

7.9 References

Abbas, H.K., Tak, H., Boyette, C.D., Shier, W.T. and Jarvis, B.B. (2001) Macrocyclic trichothecenes are undetectable in kudzu (Pueraria montana) plants treated with a high-producing isolate of Myrothecium verrucaria. Phytochemistry 58, 269–276.

Abbas, H.K., Johnson, B.B., Shier, W.T., Tak, H., Jarvis, B.B. and Boyette, C.D. (2002) Phytotoxicity and mammalian cytotoxicity of macrocyclic trichothecenes from Myrothecium verrucaria. Phytochemistry 59, 309–313.

Amsellem, Z., Sharon, A. and Gressel, J. (1991) Abolition of selectivity of two mycoherbicidal organisms and enhanced virulence of a virulent by an invert emulsion. Phytopathology 81, 985–988.

Amsellem, Z., Zidack, N.K., Quimby, P., Jr and Gressel, J. (1999) Long-term dry preservation of viable mycelia of two mycoherbicidal organisms. Crop Protection 18, 643–649.

Bateman, R.P. (1998) Delivery systems and protocols for biopesticides. In: Hall, F.R. and Menn, J.J. (eds) Biopesticides: Use and Delivery. Humana Press, Totowa, NJ, USA, pp. 509–528.

Bedi, J.S., Muller-Stover, D. and Sauerborn, J. (2002) Status of plant pathogens for the biological control of weeds. In: Koul, O., Dhaliwal, G.S., Marwaha, S.S. and Arora, J.K. (eds) Biopesticides and Pest Management. Campus Books International, New Delhi, India, pp. 316–331.

Berner, D.K. and Bruckart, W.L. (2005) A decision tree for evaluation of exotic plant pathogens for biological control of introduced invasive weeds. Biological Control 34, 222–232.

Bowers, R.C. (1982) Commercialization of microbial biological control agents. In: Charudattan, R. and Walker, H.L. (eds) Biological Control of Weeds with Plant Pathogens. John Wiley and Sons, New York, pp. 157–173.

Bowers, R.C. (1986) Commercialization of Collego: an industrialist's view. Weed Science 34 (Suppl. 1), 24–25.

Boyette, C.D. (2000) The bioherbicide approach: using phytopathogens to control weeds. In: Cobb, A.H. and Kirkwood, R.C. (eds) Herbicides and Their Mechanisms of Action. Sheffield Academic Press, Sheffield, UK, pp. 134–152.

Boyette, C.D. and Abbas, H.K. (1994) Host range alteration of the bioherbicidal fungus Alternaria crassa with fruit pectin and plant filtrates. Weed Science 42, 487–491.

Boyette, C.D., Templeton, G.D. and Smith, R.J., Jr (1979) Control of winged waterprimrose (Jussiaea decurrens) and northern jointvetch (Aeschynomene virginica) with fungal pathogens. Weed Science 27, 497–501.

Boyette, C.D., Templeton, G.E. and Oliver, L.R. (1984) Texas gourd (Cucurbita texana) control with Fusarium solani f. sp. cucurbitae. Weed Science 32, 497–501.

Boyette, C.D., Abbas, H.K. and Smith, R.J., Jr (1991a) Invert emulsions alter host specificity of biocontrol fungi. *Phytopathology* 71, 126.

Boyette, C.D., Weidemann, G.J., TeBeest, D.O. and Quimby, P.C., Jr (1991b) Biological control of jimsonweed (*Datura stramonium*) with *Alternaria crassa. Weed Science* 39, 673–667.

Boyette, C.D., Smith, R.J., Jr, Abbas, H.K. and McAlpine, J.R. (1992) Bioherbicidal control of hemp sesbania (*Sesbania exaltata*) and northern jointvetch (*Aeschynomene virginica*) in rice with anthracnose pathogen-invert formulations. *Proceedings of the Southern Weed Science Society* 45, 293.

Boyette, C.D., Quimby, P.C., Jr, Bryson, C.T., Egley, G.H. and Fulgham, R.E. (1993) Biological control of hemp sesbania (*Sesbania exaltata*) under field conditions with *Colletotrichum truncatum* formulated in an invert emulsion. *Weed Science* 41, 497–500.

Boyette, C.D., Hoagland, R.E. and Weaver, M.A. (2007a) Interaction of a bioherbicide and glyphosate for controlling hemp sesbania in glyphosate-resistant soybean. *Weed Biology and Management* (in press).

Boyette, C.D., Jackson, M.A., Bryson, C.T., Hoagland, R.E., Connick, W.J. and Daigle, D.J. (2007b) *Sesbania exaltata* biocontrol with *Colletotrichum truncatum* microsclerotia formulated in 'Pesta' granules. *BioControl* 52, 413–426.

Brown, J.K.M. and Hovmøller, M.S. (2002) Aerial dispersal of pathogens on the global and continental scales and its impact on plant disease. *Science* 297, 537–541.

Burges, H.D. (1998) Formulation of mycoinsecticides. In: Burges, H.D. (ed.) *Formulations of Microbial Biopesticides, Beneficial Microorganisms, Nematodes and Seed Treatments*, Kluwer Academic, London, pp. 132–185.

Burges, H.D. and Jones, K.A. (1998) Formulation of bacterial, viruses, and protozoa to control insects. In: Burges, H.D. (ed.) *Formulations of Microbial Biopesticides, Beneficial Microorganisms, Nematodes and Seed Treatments*, Kluwer Academic, London, pp. 33–127.

Bruckart, W.L. (2006) Supplemental risk evaluations of *Puccinia jaceae* var. *solstitialis* for biological control of yellow starthistle. *Biological Control* 37, 359–366.

Bruckart, W.L., Johnson, D.R. and Frank, J.R. (1988) Bentazon reduces rust-induced disease in yellow nutsedge, *Cyperus esculentus. Weed Science* 2, 299–303.

Callaway, M.B., Phatak, S.C. and Wells, H.D. (1987) Interactions of *Puccinia canaliculata* (Schw.) Lagerh. with herbicides on tuber production and growth of *Cyperus esculentus* L. *Tropical Pest Management* 33, 22–26.

Cardina, J., Littrell, R.H. and Hanlin, R.T. (1988) Anthracnose of Florida beggarweed (*Desmodium tortuosum*) caused by *Colletotrichum truncatum. Weed Science* 36, 329–334.

Chapple, A.C. and Bateman, R.P. (1997) Application systems for microbial pesticides: necessity not novelty. *British Crop Protection Council Mongraphs* 89, 181–190.

Charudattan, R. (2000) Current status of biological control of weeds. In: Kennedy, G.G. and Sutton, T.B. (eds) *Emerging Technologies for Integrated Pest Management: Concepts, Research and Implementation*. APS Press, St Paul, MN, USA, pp. 269–288.

Churchill, B.W. (1982) Mass production of microorganisms for biological control. In: Charudattan, R. and Walker, H.L. (eds) *Biological Control of Weeds with Plant Pathogens*. John Wiley and Sons, New York, pp. 139–156.

Clarke, T.C., Shetty, K.G., Jayachandran, K. and Norland, M.R. (2007) *Myrothecium verrucaria*: a potential biological control agent for the invasive 'Old World climbing fern' (*Lygodium microphyllum*). *BioControl* [doi:10.1007/s10526-006-9035-3].

Connick, W.J., Jr (1982) Controlled release of the herbicides 2,4-D and dichlobenil from alginate gels. *Journal of Applied Polymer Science* 27, 3341–3348.

Connick, W.J., Jr, Nickle, W.R. and Boyette, C.D. (1993) Wheat flour granules containing mycoherbicides and entomogenous nematodes In: Lumsden, R.D. and Vaughn, J.L. (eds) *Pest Management: Biologically Based Technologies: Proceedings of the 28th BARC Symposium Series*. Beltsville, MD, USA, pp. 238–240.

Connick, W.J., Daigle, D.J., Boyette, C.D., Williams, K.S., Vinyard, B.T. and Quimby, P.C., Jr (1996) Water activity and other factors that affect the viability of *Colletotrichum truncatum* conidia in wheat flour–kaolin granules ('Pesta'). *Biocontrol Science and Technology* 6, 277–284.

Connick, W.J., Jr, Jackson, M.A., Williams, K.S. and Boyette, C.D. (1997) Stability of microsclerotial inoculum of *Colletotrichum truncatum* encapsulated in wheat flour–kaolin granules. *World Journal of Microbiology and Biotechnology* 13, 549–554.

Daigle, D.J., Connick, W.J., Jr, Quimby, P.C., Jr, Evans, J.P., Trask-Merrell, B. and Fulgham, F.E. (1990) Invert emulsions: carrier and water source for the mycoherbicide, *Alternaria cassiae. Weed Technology* 4, 327–331.

De Jong, M.D., Aylor, D.E. and Bourdot, G.W. (1999) A methodology for risk analysis of pluvivorous fungi in biological control: *Sclerotina sclerotiorum* as a model. *BioControl* 43, 397–419

Delfosse, E.S. (2005) Risk and ethics in biological control. *Biological Control* 35, 319–329.

Egley, G.H. and Boyette, C.D. (1993) Invert emulsion droplet size and mycoherbicidal activity of *Colletotrichum truncatum*. *Weed Technology* 7, 417–424.

El-Morsy, E.-S.M. (2006) Preliminary survey of indoor and outdoor airborne microfungi at coastal buildings in Egypt. *Aerobiologia* 22, 197–210.

EPA [Environmental Protection Agency] (2006) *Pesticides: Regulating Pesticides* [see http://www.epa.gov/pesticides/regulating/index.htm Accessed 26 May 2006].

Forseth, I.N., Jr and Innis, A.F. (2004) Kudzu (*Pueraria montana*): history, physiology, and ecology combine to make a major ecosystem threat. *Critical Reviews in Plant Sciences* 23, 401–413.

Friesen, T.J., Holloway, G., Hill, G.A. and Pugsley, T.S. (2006) Effect of conditions and protectants on the survival of *Penicillium bilaiae* during storgage. *Biocontrol Science and Technology* 16, 89–98.

Ghajar, F., Holford, P., Cother, E. and Beattie, A. (2006a) Effects of ultraviolet radiation, simulated or as natural sunlight, on conidium germination and appressorium formation by fungi with potential as mycoherbistats. *Biocontrol Science and Technology* 16, 451–469.

Ghajar, F., Holford, P., Cother, E. and Beattie, A. (2006b) Enhancing survival and subsequent infectivity of conidia of potential mycoherbistats using UV protectants. *Biocontrol Science and Technology* 16, 451–469.

Gressel, J. (2002) *Molecular Biology of Weed Control*. Taylor and Francis, New York.

Griffin, D.W. (2004) Terrestrial microorganisms at an altitude of 20,000 m in Earth's atmosphere. *Aerobiologia* 20, 135–140.

Griffin, D.W., Kellogg, C.A., Garrison, V.H., Lisle, J.T., Borden, T.C. and Shinn, E.A. (2003) African dust in the Caribbean atmosphere. *Aerobiologia* 19, 143–157.

Hallett, S.G. (2005) Where are the bioherbicides? *Weed Science* 53, 404–415.

Hildebrand, D.C. and McCain, A.H. (1978) The use of various substrates for large scale production of *Fusarium oxysporum* f. sp. *cannabis* inoculum. *Phytopathology* 68, 1099–1101.

Hoagland, R.E. (1990) *Microbes and Microbial Products as Herbicides: ACS Symposium Series 439*. ACS Books, Washington, DC, USA.

Hoagland, R.E. (2001) Microbial allelochemicals and pathogens as bioherbicidal agents. *Weed Technology* 15, 835–857.

Hoagland, R.E., Boyette, C.D., Weaver, M.A. and Abbas, H.K. (2007a) Bioherbicides: research and risks. *Toxin Reviews* (in press).

Hoagland, R.E., Weaver, M.A. and Boyette, C.D. (2007b) *Myrothecium verrucaria* as a bioherbicide, and strategies to reduce its non-target risks. *Allelopathy Journal* 19, 179–192.

Horn, B.W. (2006) Relationship between soil densities of *Aspergillus* species and colonization of wounded peanut seeds. *Canadian Journal of Microbiology* 52, 951–960.

Jackson, M.A. (1997) Optimizing nutritional conditions for the liquid culture production of effective fungal biological control agents. *Journal of Industrial Microbiology and Biotechnology* 19, 180–187.

Jackson, M.A., Shasha, B.S. and Schisler, D.A. (1996) Formulation of *Colletotrichum truncatum* microsclerotia for improved biocontrol of the weed hemp sesbania (*Sesbania exaltata*). *Biological Control* 7, 107–113.

Kenney, D.S. (1986) DeVine – the way it was developed: an industrialist's view. *Weed Science* 34 (Suppl. 1), 15–16.

Khodayari, K. and Smith, R.J., Jr (1988) A mycoherbicide integrated with fungicides in rice, *Oryza sativa*. *Weed Technology* 2, 282–285.

Khodayari, K., Smith, R.J., Jr, Walker, J.T. and TeBeest, D.O. (1987) Applicators for a weed pathogen plus acifluorfen in soybean. *Weed Technology* 1, 37–40.

Kiely, T., Donaldson, D. and Grube, A. (2004) *Pesticide Industry Sales and Usage: 2000 and 2001 Market Estimates*. Environmental Protection Agency (EPA), Washington, DC, USA.

Kuske, C.R. (2006) Current and emerging technologies for the study of bacteria in the outdoor air. *Current Opinion in Biotechnology* 17, 291–296.

Leland, J.E. and Behle, R.W. (2005) Coating *Beauveria bassiana* with lignin for protection from solar radiation and effects on pathogenicity to *Lygus lineolaris* (*Heteroptera*: Miridae). *Biocontrol Science and Technology* 15, 309–320.

McWhorter, C.G. and Chandler, J.M. (1982) Conventional weed control technology. In: Charudattan, R. and Walker, H.L. (eds) *Biological Control of Weeds with Plant Pathogens*. John Wiley and Sons, New York, pp. 5–27.

McWhorter, C.G., Fulgham, F.E. and Barrentine, W.L. (1988) An air-assist spray nozzle for applying herbicides in ultra-low volume. *Weed Science* 36, 118–121.

Mitchell, J.K. (1986) *Dichotomopthora portulacae* causing black stem rot on common purslane in Texas. *Plant Disease* 70, 603.

Mitchell, J.K. (2003) Development of a submerged-liquid sporulation medium for the potential smartweed bioherbicide *Septoria polygonorum*. *Biological Control* 27, 293–299.

Mitchell, J.K., Njalamimba-Bertsch, M., Bradford, N.R. and Birdsong, J.A. (2003) Development of a submerged-liquid sporulation medium for the johnsongrass bioherbicide *Gloeocercospora sorghi*. *Journal of Industrial Microbiology and Biotechnology* 30, 599–605.

Müller-Stöver, D., Thomas, H., Sauerborn, J. and Kroschel, J. (2004) Two granular formulations of *Fusarium oxysporum* f.sp. *orthoceras* to mitigate sunflower broomrape *Orobanche cumana*. *BioControl* 49, 595–602.

Newton, M.R., Kinkel, L.L. and Leonard, K.J. (1997) Competition and density-dependent fitness in a plant parasitic fungus. *Ecology* 78, 1774–1784.

Newton, M.R., Kinkel, L.L. and Leonard, K.J. (1998) Determinants of density- and frequency-dependent fitness in competing plant pathogens. *Phytopathology* 88, 45–51.

Phillips, D.J. and Margosan, D.A. (1987) Size, nuclear number, and aggressiveness of *Botrytis cinerea* spores produced on media of varied glucose concentrations. *Phytopathology* 77, 1606–1612.

Pilgeram, A.L. and Sands, D.C. (1999) Mycoherbicides. In: Hall, F.R. and Menn, J.J. (eds) *Biopesticides: Use and Delivery*. Humana Press, Totowa, NJ, USA.

Quimby, P.C., Jr (1989) Response of common cocklebur (*Xanthium strumarium*) to *Alternaria helianthi*. *Weed Technology* 3, 177–181.

Quimby, P.C., Jr and Boyette, C.D. (1986) Pathogenic control of prickly sida and velvetleaf: an alternate technique for producing and testing *Fusarium lateritium*. *Proceedings of the Annual Meeting of the Southern Weed Science Society* 38, 365–371.

Quimby, P.C., Jr and Boyette, C.D. (1987) Production and application of biocontrol agents. In: McWhorter, C.G. and Gebhardt, M.R. (eds) *Methods of Applying Herbicides*. Monograph Series, No. 4, Weed Science Society, Champaign, IL, USA, pp. 265–280.

Quimby, P.C., Jr, Fulgham, F.E., Boyette, C.D. and Connick, W.J., Jr (1988) An invert emulsion replaces dew in biocontrol of sicklepod: a preliminary study. In: Hovde, D.A. and Beestman, G.B. (eds) *Pesticide Formulations and Application Systems*. ASTM-STP 980, American Society for Testing and Materials, Philadelphia, PA, USA, pp. 264–270.

Reoun An, H., Mainelis, G. and White, L. (2006) Development and calibration of real-time PCR for quantification of airborne microorganisms in air samples. *Atmospheric Environment* 40, 7924–7939.

Ridings, W.H. (1986) Biological control of stranglervine in citrus: a researcher's view. *Weed Science* 34 (Suppl. 1), 31–32.

Roberts, P.D., Urs, R.R., Wiersma, H.I. and Mullahey, J.J. (2002) Effect of bacterium–herbicide combinations on tropical soda apple. *Biological Control* 24, 238–244.

Shabana, M.S., Müller-Stöver, D. and Sauerborn, J. (2003) Granular Pesta formulation of *Fusarium oxysporum* f. sp. *orthoceras* for biological control of sunflower broomrape: efficacy and shelf-life. *Biological Control* 26, 189–201.

Shearer, J.F. and Jackson, M.A. (2006) Liquid culturing of microsclerotia of *Mycoleptodiscus terrestris*, a potential biological control agent for the management of hydrilla. *Biological Control* 38, 298–306.

Shearer, J.F. and Nelson, L.S. (2002) Integrated use of endothall and a fungal pathogen for management of the submersed aquatic macrophyte *Hydrilla verticillata*. *Weed Technology* 16, 224–230.

Silman, R.W., Bothast, R.J. and Schisler, D.A. (1993) Production of *Colletotrichum truncatum* for use as a mycoherbicide: effects of culture, drying and storage on recovery and efficacy. *Biotechnology Advances* 11, 561–575.

Slade, S.J., Harris, R.F., Smith, C.S. and Andrews, J.H. (1987) Microcycle conidiation and spore-carrying capacity of *Colletotrichum gloeosporioides* on solid media. *Applied and Environmental Microbiology* 53, 2106–2110.

Slininger, P.J., Behle, R.W., Jackson, M.A. and Schisler, D.A. (2003) Discovery and development of biological agents to control crop pests. *Neotropical Entomology* 32, 183–195.

Smith, J.R., Jr (1991) Integration of biological control agents with chemical pesticides. In: TeBeest, D.O. (ed.) *Microbial Control of Weeds*. Chapman and Hall, New York, pp. 189–208.

Smith, R.J. (1986) Biological control of northern jointvetch (*Aeschynomene virginica*) in rice (*Oryza sativa*) and soybeans (*Glycine max*): a researcher's view. *Weed Science* 34 (Suppl. 1), 17–23.

Stojanović, S., Starović, M. and Matijević, D. (1999) Factors affecting conidial germination of *Colletotrichum gloeosporioides* isolated from *Polystigma rubrum* subsp. *rubrum* stromata. *Acta Phytopathologica et Entomologica Hungarica* 34, 63–73.

TeBeest, D.O. (1991) *Microbial Control of Weeds*. Chapman and Hall, New York.

TeBeest, D.O. and Templeton, G.E. (1985) Mycoherbicides: progress in the biological control of weeds. *Plant Disease* 69, 6–10.

Templeton, G.E., TeBeest, D.O. and Smith, R.J. (1979) Biological weed control with mycoherbicides. *Annual Review of Phytopathology* 17, 301–310.

Toussoun, T.A., Nash, S.N. and Snyder, W.C. (1960) The effect of nitrogen sources and glucose on the pathogenesis of *Fusarium solani* f. *phaseoli*. *Phytopathology* 50, 137–140.

Tsai, M.-Y., Elgethun, K., Ramaprasad, J., Yost, M.G., Felsot, A.S., Herbert, V.R., and Fenske, R.A. (2005) The Washington aerial spray drift study: modeling pesticide spray drift deposition from an aerial application. *Atmospheric Environment* 39, 6194–6203.

Tuite, J. (1969) *Plant Pathological Methods: Fungi and Bacteria*. Burgess Publications, Minneapolis, MN, USA.

Tworkoski, T. (2002) Herbicide effects of essential oils. *Weed Science* 50, 425–431.

Van Dyke, C.G. and Winder, R.S. (1985) *Cercospora dubia* (Reiss) Wint. on *Chenopodium album* L.: greenhouse efficacy and biocontrol potential. *Proceedings of the Annual Meeting of the Southern Weed Society* 38, 373.

Vurro, M., Gressel, J., Butt, T., Harman, G.E., Pilgeram, A., St. Leger, R.J. and Nuss, D.L. (2001) *Enhancing Biocontrol Agents and Handling Risks*. IOS Press, Amsterdam, The Netherlands.

Walker, H.L. (1981) *Fusarium lateritium*: a pathogen of spurred anoda (*Anoda cristata*), prickly sida (*Sida spinosa*) and velvetleaf (*Abutilon theophrasti*). *Weed Science* 29, 629–631.

Walker, H.L. and Connick, W.J., Jr (1983) Sodium alginate for production and formulation of mycoherbicides. *Weed Science* 33, 333–338.

Walker, H.L. and Riley, J.A. (1982) Evaluation of *Alternaria cassiae* for the biocontrol of sicklepod (*Cassia obtusifolia*). *Weed Science* 30, 651–654.

Walker, H.L. and Tilley, A.M. (1997) Evaluation of an isolate of *Myrothecium verrucaria* from sicklepod (*Senna obtusifolia*) as a potential mycoherbicide agent. *Biological Control* 10, 104–112.

WeedScience (2006) *International Survey of Herbicide Resistant Weeds* [see http://www.weedscience.org/in.asp Accessed 26 May 2006].

Weidemann, G.J. and Templeton, G.E. (1988) Control of Texas gourd, *Curcurbitia texana*, with *Fusarium solani* f.sp. *cucurbitae*. *Weed Technology* 2, 271–274.

WHO [World Health Organization] (2000) *Guidelines for Concentration and Exposure-Response Measurements of Fine and Ultra-fine Particulate Matter for Use in Epidemiological Studies*. World Health Organization, Geneva, Switzerland.

Wilson, C.L. (1969) Use of plant pathogens in weed control. *Annual Review of Phytopathology* 7, 411–433.

Wraight, S.P., Jackson, M.A. and DeKock, S.L. (2001) Production, stabilization and formulation of fungal biocontrol agents. In: Butt, T., Jackson, C. and Magan, N. (eds) *Fungal Biocontrol Agents: Progress, Problems, and Potential*. CAB International, Wallingford, UK.

Wymore, L.A. and Watson, A.K. (1986) An adjuvant increases survival and efficacy of *Colletotrichum coccodes*: a mycoherbicide for velvetleaf (*Abutilon theophrasti*). *Phytopathology* 76, 1115–1116.

Wymore, L.A., Poirier, C., Watson, A.K. and Gotlieb, A.R. (1988) *Colletotrichum coccodes*, a potential bioherbicide for control of velvetleaf (*Abutilon theophrasti*). *Plant Disease* 72, 534–538.

Wyss, G.S., Charudattan, R., Rosskopf, E.N. and Littell, R.C. (2004) Effects of selected pesticides and adjuvants on germination and vegetative growth of *Phompsis amaranthicola*, a biocontrol agent for *Amaranthus* spp. *Weed Research* 44, 469–492.

Yandoc, C.B., Rosskopf, E.N., Pitelli, R.L.C.M. and Charudattan, R. (2006) Effect of selected pesticides on conidial germination and mycelial growth of *Dactylaria higginsii*, a potential bioherbicide for purple nutsedge (*Cyperus rotundus*). *Weed Technology* 20, 255–260.

Yang, X.B. and TeBeest, D.O. (1993) Epidemiological mechanisms of mycoherbicides effectiveness. *Phytopathology* 83, 891–893.

Zidak, N.K. and Quimby, P.C. (1998) Formulation and application of plant pathogens for biological weed control. In: Hall, F.R. and Menn, J.J. (eds) *Biopesticides: Use and Delivery*. Humana Press, Totowa, NJ, USA, pp. 371–382.

8 Mechanical Weed Management

D.C. Cloutier,[1] R.Y. van der Weide,[2] A. Peruzzi[3] and M.L. Leblanc[4]

[1]Institut de Malherbologie, 102 Brentwood Rd, Beaconsfield, Québec,
QC, Canada, H9W 4M3; [2]Applied Plant Research, PO Box 430, 8200 AK Lelystad,
The Netherlands; [3]Sezione Meccanica Agraria e Meccanizzazione Agricola (MAMA),
DAGA – University of Pisa, via S. Michele degli Scalzi, 2 – 56124, Pisa, Italy;
[4]Institut de Recherche et de Développement en Agroenvironnement, 3300,
rue Sicotte, C.P. 480, Saint-Hyacinthe, Québec, QC, Canada, J2S 7B8

8.1 Introduction

Weed control has always been closely associated with farming. It is very likely that the first weeding action was by hand-pulling. This was followed by using a stick which became a hand-hoe. As agriculture became more mechanized, fields were successfully kept weed-free with mechanical weed management techniques and tools pulled first by animals and eventually by tractors (Wicks et al., 1995). The appearance of herbicides in the mid-20th century contributed to a decreased reliance on mechanical weeders on farms. Nevertheless, these implements have continued to evolve and are very efficient and versatile in controlling weeds in a variety of cropping systems.

Mechanical weed management consists of three main techniques: the use of tillage, cutting weeds and pulling weeds. These three techniques are presented separately in this chapter.

8.2 Tillage

According to the American Society of Agricultural Engineers (ASAE, 2005), tillage generally refers to the changing of soil conditions for the enhancement of crop production. It can be further subdivided into three categories: primary tillage, secondary tillage and cultivating tillage (Wicks et al., 1995; ASAE, 2004). This section is further subdivided into two subsections: with or without soil inversion.

Cropping systems with soil inversion

Primary tillage

Primary tillage is the first soil-working operation in soil-inversion-based cropping systems. Its objective is to prepare the soil for planting by reducing soil strength, covering plant material, and by rearranging aggregates (ASAE, 2005). In these cropping systems, primary tillage techniques are always aggressive and usually carried out at considerable depth, leaving an uneven soil surface. In other cases, primary tillage may leave a more even soil surface, e.g. when soil packers are used in association with ploughs. For weed species that are propagated by seeds, primary tillage can contribute to control by burying a portion of the seeds at depths from which they are unable to emerge (Kouwenhoven, 2000). Primary tillage can also play a role in controlling perennial weeds by burying some of their propagules deep, thereby preventing or slowing down their emergence. Some of the propagules can be brought up to the soil surface, where they will be exposed directly to cold or warm temperatures or desiccation conditions (Cloutier and Leblanc, 2001; Mohler, 2001). The tools used to perform primary tillage in soil-inversion-based cropping systems are mainly mouldboard ploughs, but disc ploughs, powered rotary ploughs, diggers and chisel ploughs can also be used for this purpose (Barthelemy et al., 1987; Peruzzi and Sartori, 1997; Cloutier and Leblanc, 2001; ASAE, 2005).

Secondary tillage

In secondary tillage, the soil is not worked as aggressively or as deeply as in primary tillage. The purpose of secondary tillage is to further pulverize the soil, mix various materials such as fertilizer, lime, manure and pesticides into the soil, level and firm the soil, close air pockets, and control weeds (ASAE, 2004). Seedbed preparation is the final secondary tillage operation except when used in the stale or false seedbed technique (Leblanc and Cloutier, 1996). The equipment used to perform secondary tillage are different types of cultivators, harrows (disc, spring tine, radial blade and rolling) and power take-off (PTO)-powered machines. Several of these implements may also be used instead of common primary tillage implements (ploughing, digging, etc) to prepare fields. In these cases, the soil is tilled (crumbled and stirred) down to a depth of 10–15 cm, which is beneficial in conserving or increasing soil organic matter content, and in saving time, fuel and money (Barthelemy et al., 1987; Peruzzi and Sartori, 1997). Initially there might be some problems with weeds when using reduced-tillage techniques, since they are not effective against the potential flora and they might even stimulate weed seed germination. Consequently, mechanical weed management has to be intensive and performed with particular care using secondary tillage, seedbed preparation and the false seedbed technique. Optimally, farmers will alternate ploughing with chiselling and use reduced tillage to optimize soil management, till it at different depths, and change the mechanical actions year after year in order to conserve organic matter and to increase fertility, to save time, fuel and money and, last but not least, to improve annual and perennial weed species control and crop development and yield (Peruzzi and Sartori, 1997; Mohler, 2001; Bàrberi, 2002).

In the following subsections, seedbed preparation is presented first, followed by some noteworthy techniques such as tilling in the dark, using the false or stale seedbed technique, and raised bed cultivation.

SEEDBED PREPARATION. The cultivators are always equipped with rigid or flexible tines working at a depth (on ploughed soil) ranging from 15–25 cm when heavy cultivators are used with the aim of reducing clod size, lifting the soil, increasing soil roughness and controlling perennial weeds, down to 5–10 cm when light cultivators are used to prepare the seedbed. The tines may be rigid or flexible. The rigid tines are often partly or completely curved, work at greater depths, reduce clod size, have a good weeding action on actual weed flora by uprooting them, and may also partly control the vegetative and reproductive structures of perennial species that are brought to the soil surface where they may be exposed to the elements. The flexible tines are usually curved, work at a shallow depth, require a lower drawbar pull, and crumble and intensively stir the tilled soil layer. The tines vibrate with the forward movement of the tractor, which helps in incorporating crop and weed residues into the soil.

The tip of the tines can be equipped with teeth which may be of different shapes; large tools (e.g. goose foot) enhance the uprooting effect on actual weed flora. Any cultivator passage has a weeding action, but it might also stimulate weed seed germination and emergence (Barthelemy et al., 1987; Peruzzi and Sartori, 1997; Mohler, 2001; Bàrberi, 2002; ASAE, 2004).

TILLAGE IN DARKNESS. The technique of doing the final seedbed preparation in darkness has proved to be a valid preventive method of weed control under some conditions. Tillage in darkness has also been referred to as photocontrol of weeds by several authors (Hartmann and Nezadal, 1990; Juroszek and Gerhards, 2004). This technique relies on the fact that many weed species require light to germinate (Hartmann and Nezadal, 1990). The technique consists of doing the last tillage operation for the seedbed preparation in darkness, either during the night or by covering the tillage implement with an opaque material that prevents light from reaching the soil being tilled. In a recent literature review of over 30 different studies, Juroszek and Gerhards (2004) reported that, according to one study, this technique caused a decrease in weed ground cover of over 97% compared with daylight tillage, while, in another study, the same technique caused an increase of 80%. Admittedly this technique gives inconsistent results, but it was found to decrease or delay weed emergence sufficiently to provide a decrease in weed ground cover of slightly less than 30% on average (Juroszek and Gerhards, 2004). Although small, this decrease could be advantageous for the crop by decreasing the intensity of weed control required.

STALE SEEDBED AND FALSE SEEDBED. Tillage can increase weed emergence from the potential weed flora (seed/bud bank) as mentioned previously in this chapter. Consequently, a set of related techniques have been developed to take advantage of this phenomenon, namely stale seedbed and false seedbed techniques. The general procedure consists of ploughing and tilling the field to prepare the crop seedbed while promoting the maximum emergence of weeds. To this end, the soil could even be firmed to promote a greater emergence of weeds by improving soil contact with weed seeds. Once the seedbed has been prepared, crop seeding or planting is delayed in order to allow sufficient time for weeds to emerge and be destroyed (Mohler, 2001).

The stale seedbed technique involves preparing the seedbed as above and, prior to planting the crop, or crop emergence (particularly when crop seeds are characterized by a slow germination: e.g. onion, carrot, spinach, etc), the emerged weeds are destroyed without disturbing the soil in order to minimize further emergence (Mohler, 2001). Traditionally, herbicides have been utilized, but propane flamers can also be used to control weeds. Rasmussen (2003) reports that a stale seedbed where weed flaming was used had a 30% decrease in weed density compared with a control without flaming. A 2- or 4-week delay in planting (stale seedbed with flaming) resulted in 55% and 79% less weeds, respectively, than in the control with no delay and no flaming. Balsari *et al.* (1994) reported a 60% decrease in weed density and percentage of ground cover 16 days after flaming compared with an untreated control.

The false seedbed technique is similar to the stale seedbed technique except that the seedbed is cultivated instead of being left undisturbed. After a period of time sufficient for the weeds to emerge but not develop too much (approximately 1 week), the soil is cultivated as shallowly as possible. The cultivation depth for subsequent operations should not exceed the depth of the first operation, otherwise new weed seeds might be brought to the surface from lower soil levels. When soil conditions and time permit, this procedure can be repeated several times prior to sowing or planting the crop. The false seedbed technique has not been well documented; however, this practice is widespread on organic farms (Mohler, 2001) and a reduction in weed density of 63–85% has been observed in some situations (Gunsolus, 1990; Leblanc and Cloutier, 1996). Riemens *et al.* (2006) report that, depending on location and year, false seedbed prior to planting of lettuce decreased the number of weeds observed during crop growth by 43–83%. In silage maize, a false seedbed created 3 weeks before sowing decreased the density of early-emerging weed species but had an inconsistent effect on late-emerging species depending on the year and/or sowing times (van der Weide and Bleeker, 1998). These techniques result in delayed planting, which can decrease yields (Rasmussen, 2004).

The false seedbed technique is often carried out by means of flex-tine harrows, but it is also possible to use a rolling harrow developed by researchers at the University of Pisa, Italy (Peruzzi *et al.*, 2005a). This implement can effectively control weeds, even under unfavourable soil conditions, by tilling superficially and causing significant crumbling of the soil. This harrow is equipped with spike discs placed at the front and cage rolls mounted at the rear. Ground-driven by the movement of the tractor, the front and rear tools are connected to one another by a chain drive with a ratio equal to 2 (Fig. 8.1).

The discs and the rolls can be arranged in two different ways on the axles. They can be tightly placed together to superficially till (3–4 cm) the whole treated area for non-selective mechanical weed control in a false seedbed operation (Fig. 8.2), or they can be widely spaced to perform precision inter-row weeding in a row crop (Fig. 8.3) (Peruzzi *et al.*, 2005a).

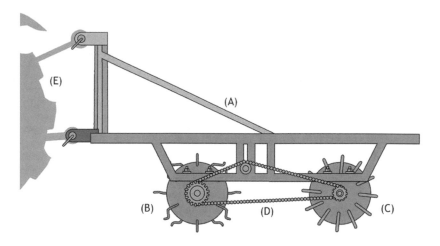

Fig. 8.1. Schematic diagram of the rolling harrow: (A) frame; (B) front axle equipped with spike discs; (C) rear axle equipped with cage rolls; (D) chain drive; (E) three-point linkage. (Drawn by Andrea Peruzzi.)

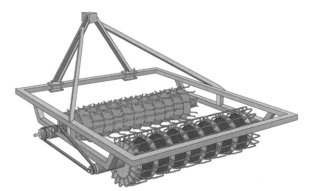

Fig. 8.2. Close arrangement of the tools of the rolling harrow for non-selective treatments. (Drawn by Andrea Peruzzi.)

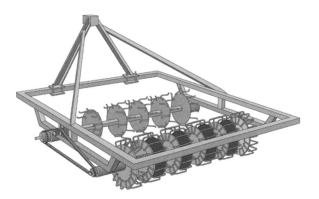

Fig. 8.3. Spaced arrangement of the tools of the rolling harrow for precision inter-row weeding. (Drawn by Andrea Peruzzi.)

The action of the rolling harrow is characterized by the passage of the spike discs that till the top 3–4 cm of the soil followed by the passage of the cage rolls that work at a higher peripheral speed, tilling and crumbling the first 1–2 cm of the soil. These two actions separate weeds from the soil, achieving an excellent level of control. The harrow also stimulates the germination and emergence of new weeds, making it very suitable for the false seedbed technique where the objective is to reduce the weed seedbank and, consequently, the potential weed flora.

The efficacy of the rolling harrow was determined in spinach fields in the Serchio Valley in Tuscany, central Italy, and in carrot, fennel and chicory fields in the Fucino Valley in Abruzzo, southern Italy. When used in a false seedbed, the rolling harrow decreased weed density by approximately 20% more than the flex-tine harrow when assessed after crop emergence. When used for inter-row precision weeding, the steerage hoe equipped with rigid tines decreased weed density by 30–50% more when evaluated 15 days after the treatment (Peruzzi *et al.*, 2005a).

RAISED BED CULTIVATION. In many geographical areas of Europe, the cultivation of vegetables is carried out on raised beds or on strips that are formed every year just before planting. The general intention is to improve growing conditions either by increasing drainage of water for crops susceptible to excess water (e.g. spinach) or by loosening the soil for root crops such as carrot. This technique facilitates the formation of permanent traffic lanes where the tractor wheels always pass, confining soil compaction to a small area (Peruzzi *et al.*, 2005b,c).

In these production systems, primary and secondary tillage are generally carried out on the whole cultivated surface. After seedbed preparation, raised beds are formed by specialized equipment or the cultivated area is divided into strips. In conventional cropping systems where herbicides are used, these operations coincide or are immediately followed by planting. In cropping systems based on non-chemical weed control, all field operations preceding planting are performed only on the raised beds, or on the cultivated strips (Peruzzi *et al.*, 2005b,c).

Cultivating tillage

Previously referred to as tertiary tillage, cultivating tillage is the term suggested by the ASAE (2004). Cultivating tillage equipment is used after crop planting to carry out shallow tillage to loosen the soil and to control weeds (Cloutier and Leblanc, 2001). These implements are commonly called cultivators.

Soil loosening by cultivators has been proven to improve crop yield even in the absence of weeds (Buckingham, 1984; Leblanc and Cloutier, 2001a,b). This positive contribution to yield could be ascribed to the fact that soil loosening breaks the soil crust when one is present, possibly improving crop development and growth; it also breaks up soil capillaries, preventing water evaporation under warm and dry growing conditions; it can enhance mineralization of organic matter; and improves water infiltration in the soil (Blake and Aldrich, 1955; Souty and Rode, 1994, Buhler *et al.*, 1995; Leblanc *et al.*, 1998; Cloutier and Leblanc, 2001; Steinmann, 2002).

Cultivating tillage can destroy weeds in several different ways. After a cultivator passes over a field, complete or partial burial of weeds can be an important cause of mortality (Cavers and Kane, 1990; Rasmussen, 1991; Kurstjens and Perdok, 2000). Another mode of action is by uprooting and breakage of the weed root contact with the soil (Cavers and Kane, 1990; Rasmussen, 1992; Weber and Meyer, 1993; Kurstjens *et al.*, 2000; Kurstjens and Kropff, 2001). Mechanical tearing, breaking or cutting the plant can also result in mortality (Toukura *et al.*, 2006). Cultivation is more effective in dry soils because weeds often die by desiccation and mortality is severely decreased under wet conditions. Cultivating when the soil is too wet will damage the soil structure and possibly spread perennial weeds (Cloutier and Leblanc, 2001).

Cultivators can be classified according to where they are used in a crop. These categories are: broadcast cultivators, which are passed both on and between the crop rows; inter-row cultivators, which are only used between crop rows; and finally, intra-row cultivators, which are used to remove weeds from the crop rows (Cloutier and Leblanc, 2001; Leblanc and Cloutier, 2001b; Melander *et al.*, 2005).

BROADCAST: PASSED BOTH ON THE CROPS AND BETWEEN THE CROP ROWS. Broadcast, sometimes referred to as full-field or blind cultivation, consists of cultivating with the same intensity both on and between the crop rows. The cultivations can be done before or after crop emergence. There are several types of harrow that can be used for this type of cultivation but the most common implements used are chain harrows, flex-tine harrows in Europe, and rotary hoes in North America.

Rotary hoes and most other cultivators have often been accused of promoting new weed germination, because it has been observed that there is sometimes a flush of weed emergence immediately after cultivation. Often, on soils that are subject to crust formation, this phenomenon could be better explained by the breaking of the soil crust with the cultivator passage rather than by new weed germination. In fact, Cloutier *et al.* (1996) observed in a field experiment that less than 5% of all germinated weeds in the soil emerged. A large proportion of these weeds were unable to emerge because of the presence of the soil crust (Leblanc *et al.*, 1998). Soil crust presence also explains the big flush of weed emergence on a day following rain. This is rarely caused by sudden and rapid weed germination. A more likely explanation is that the soil crust

becomes more plastic when moist, offering less resistance to weed emergence. Although breaking the soil crust could be seen as promoting weed emergence, it is generally more beneficial to the crops.

Chain harrows

Chain harrows have short shanks fitted on chains rather than a rigid frame, so that they hug the ground. They are especially effective on light soils and prior to crop emergence, or in short crops.

Flex-tine harrows

The flex-tine harrow is the most commonly used implement in this category in Europe (Fig. 8.4). Because these harrows are rear-mounted on the tractor and not pulled on the soil, they can be used in taller crops and on the top of ridges. Flex-tine harrows have rigid frames and a variety of different tines. They have fine, flexible tines which destroy weeds by vibrating in all directions. Rigid-tine harrows, best for heavy soils, consist of several sets of spikes or rigid blades angled at the tip; the spikes or blades are mounted on a rigid frame or a floating section. The spikes or blades vibrate perpendicularly to the direction in which the tractor is moving. Depending on the model,

Fig. 8.4. Flex-tine harrow in silage maize. (Photo by Wageningen UR Applied Plant Research.)

tension on the tines can be adjusted individually or collectively to change the intensity of the treatment. The working width ranges from 1.5 to over 24 m, but the most common width is 6 m. The driving speed with these harrows varies between 3 and 12 km/h (4 km/h for sensitive crops or stages, with 8 km/h being the more commonly used speed). Cultivation depth can be adjusted by depth wheels on the harrow (when present) or by the tractor's hydraulic system, since they are attached to the three-point hitch. Depth can also be adjusted by changing the tine angles and driving speed.

Pre-crop emergence cultivation with harrows is selective because the crop seeds are planted more deeply than the weed seeds or are larger than the weed seeds, and are therefore not affected or only slightly affected by cultivation (Dal Re, 2003; Peruzzi *et al.*, 2005b). In general, this is a benign treatment that destroys only weeds that are at the white thread stage (weeds that have germinated but not emerged), dicotyledonous weed seedlings before the two-leaf stage, and monocotyledonous weeds at the one-leaf stage. However, where crop seeds are planted deep enough, tines can be adjusted to be more aggressive (angled forward) and driving speed can be increased to destroy more developed weeds such as small-seeded dicotyledonous weeds with 2–4 true leaves. Fairly aggressive harrowing is possible with deep-sown crops such as beans, peas and maize. However, care is needed with shallow-sown crops such as spring-sown onions and sugarbeet, where cultivation depth is of great importance.

Post-crop emergence broadcast treatment is selective, given the fact that the crop is better rooted. Since the crop has larger seeds (and therefore more energy reserves) or is transplanted, it becomes established faster than the weeds. Harrowing weeds in their earlier stages of development (e.g. until the first true leaves are visible), can result in excellent levels of control. However, harrowing might have to be repeated several times to maintain acceptable weed control levels during the growing season (Rasmussen, 1993). Spring-tine harrows can be used post-emergence in cereals, maize, potatoes, peas, beans, many planted vegetables and relatively sensitive crops such as sugarbeet. In sensitive crops, harrows cannot be used in the early crop growth stages such as before four

true leaves in sugarbeet. Cultivation speed should be decreased (e.g. 3–4 km/h) when a crop is at a sensitive development stage such as the two-leaf stage in maize. In this particular case, the tines should be at the vertical setting. Information has been compiled where suggestions are made concerning which harrow or other equipment to choose for various crop growth stages (Fig. 8.5; example taken from van der Schans *et al.*, 2006).

Rotary hoes

The rotary hoe is a harrow with two gangs of hoe wheels that are rolled on the ground (Fig. 8.6). The wheel axles are horizontal and the two sets of wheels are offset for maximum soil contact. High-residue models have a greater distance between the two gangs to prevent the accumulation of plant residues. The hoe wheels have several rigid and curved teeth that are sometimes referred to as 'spoons' because they have a wider point at their tip, similar to a spoon. The teeth penetrate almost straight down but lift the soil as they emerge, pulling young weed seedlings. The selectivity of the rotary hoe is attributed to the crop seeds being deeper than the working depth or of the crop being better rooted than the weeds.

Rotary hoes are implements that are widely used in North America, even by growers who utilize herbicides. They use the rotary hoe to incorporate herbicides into the soil and to break the soil crust when one is present. Rotary hoes can be used to cultivate a field relatively quickly and cheaply (Buckingham, 1984; Bowman, 1997). Their width varies from 3 to 12 m and the optimal speed at which they should be operated varies from 8 to 24 km/h. Extra weights might have to be added because teeth penetration decreases as speed increases. The ideal working depth of the rotary hoes varies between 2 and 5 cm. They can be passed before or after crop emergence. They are most effective against weeds at the white-thread stage but, with the exception of monocotyledonous weeds, will control many weed species at the two-leaf stage. Crops such as maize, soybean and various field beans tolerate one or several cultivations with the rotary hoe (Bowman, 1997; Leblanc and Cloutier, 2001a,b). It is often recommended to increase the seeding rate of crops that receive multiple cultivations with the rotary hoe to compensate for some of the crop uprooting that can occur.

INTER-ROW CULTIVATION.   The use of inter-row cultivators is generally widespread and well mastered. These implements are used in row crops by conventional growers as well as growers who do not use herbicides. There is minimal risk to the crop and weed control is generally

excellent. The only limitations are crop height and growth stage because of tractor and cultivator ground clearance and potential damage to crop foliage. Also, because of critical periods of weed interference, it is preferable to carry out inter-row cultivations early rather than late in the season. Another problem with late cultivations is that when weeds are well developed, cultivators could easily get plugged with plant material. Cultivator shields can be used early in

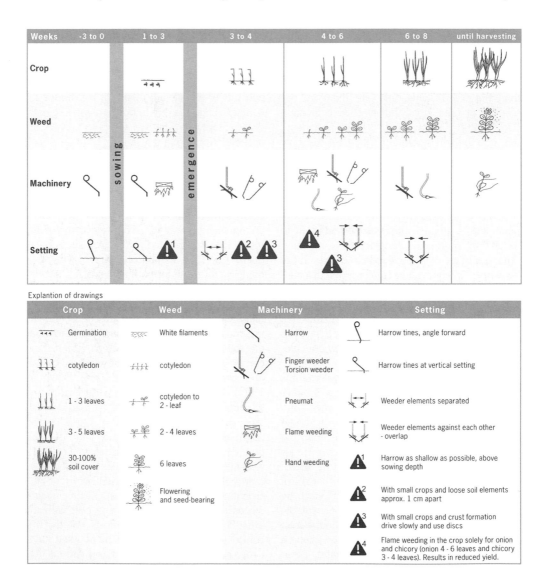

Fig. 8.5. Example taken from van der Schans *et al.* (2006). Possibilities and machine settings for weed control in crop rows for small seed crops, tightly spaced, such as carrots, chicory, onions, red beet and spinach.

Fig. 8.6. Rotary hoe in action in cereals. (Photo by Maryse Leblanc.)

the season to prevent accidental burying or breaking of crop seedlings by soil or plant residues thrown by the cultivator. Cultivator shields come in a variety of forms. There are rolling, panel, tent and wheel shields that move along each side of the crop row (Bowman, 1997; Cloutier and Leblanc, 2001).

Inter-row cultivation can be carried out by inter-row cultivators, discs, brush weeders, rotary cultivators, rolling cultivators, basket weeders and rolling harrows.

Inter-row cultivators

Inter-row cultivators were the earliest and are the most widespread type of cultivators used in row crops. In general, mechanical weed control between crop rows is carried out with a group of cultivating tools (usually on three to five shanks, called a gang) mounted on a toolbar, one gang per inter-row. Ideally, the implement should cultivate the same number of inter-rows that were represented by one planter pass, or a whole fraction, because adjacent planter passages are seldom totally parallel and equidistant. The width of the toolbar and the number of blades depends on the width of the working passage and on the row distance of the planting or sowing machine. Cultivating as much of the inter-row area as possible without damaging the crop should be the objective of inter-row cultivation. The distance between the crop rows and

the precision of the implement determine the working width of the gangs. Some accidental crop damage might occur when working very close to the crop rows, in the presence of soil crust, or when high tractor speeds are used (Bowman, 1997; Mohler, 2001).

The gangs mounted on the toolbar can either have rigid or vibrating shanks to which various types of points (shovels, sweeps and weed knives) can be attached. These points vary in width from a few centimetres to 76 cm. Each tool consists of a shank which is typically long and narrow. The shank ends in a point and connects to the toolbar or the frame (Cloutier and Leblanc, 2001). An assortment of different cultivating tools can be fitted between two crop rows when the distance between the rows is 25 cm or more. The major benefit offered by this approach is the ability to adjust gangs to fit any inter-row width. Alongside the working width of the cultivating tools, the type of soil cultivation attained is also of importance. The ideal cultivation depth is less than 4 cm because there is a risk of crop root pruning if cultivating too close and too deep. Inter-row cultivators have been classified as being adapted for low (up to 20%), moderate (up to 30%), high (up to 60%) and maximum residue levels (up to 90%) (Bowman, 1997).

Cultivators using vibrating shanks are usually considered light-duty cultivators. These shanks can be C-curved or S-shaped; commonly referred to as Danish S-tines. Various types of sweeps (duckfoot, goosefoot, triangular, backland, etc) can be attached to the shanks. The S-shaped shanks vibrate vigorously, shatter the soil and kill weeds, while the C-curved shanks vibrate considerably less (Bowman, 1997). The greater the speed, the more aggressive the tools with vibrating shanks will cultivate and the more they could stimulate the germination of new weeds.

Inter-row cultivators with rigid shanks are considered heavy cultivators, better used in fields with high residues. These cultivators cut off weeds and disturb the soil to a lesser extent than cultivators with vibrating shanks. An implement with rigid shanks will disturb soils the least when passed at an approximate speed of 6 km/h. Wide, sharp sweeps can be attached to these shanks. This type of cultivator is more effective against bigger weeds than the ones with vibrating shanks.

Mounting gangs on a parallelogram with a gauge wheel ensures that soil contours are followed closely. The best uniform hoeing depth can be achieved with a minimum distance between the gang and the gauge. The location of the toolbar relative to the tractor and steering systems is discussed in a separate section below.

Discs

Although discs alone are sometimes used to replace shanks and points on gangs, they are usually mounted on gangs with shanks and points to cultivate very close to the crop row while other weeding tools cultivate the rest of the inter-row. Some implements might require a second operator to guide the gangs in order to increase cultivation precision. Discs can be adjusted to throw soil towards the crop row or to remove soil and weeds away from the row.

Brush weeders

There are several different types of brush weeder. In general, the brushes are made of fibreglass and are flexible (Fig. 8.7). There are horizontal-axis brushes and vertical-axis brushes that are either driven by the tractor's power take-off (PTO) shaft, electric motors or by hydraulics. Working very superficially, these weeders mainly uproot, but do also bury or break weeds. A protective shield panel or tent

can be used to protect the crop. In the case of rear-mounted implements, a second operator might be needed to steer the brushes so as to cultivate as close as possible to the crop without damaging it.

When using horizontal-axis brushes, their rotation speed should be only slightly faster than the tractor speed, otherwise too much dust will be generated. A higher rotational speed will not improve the effect; however, the bristles will wear out more rapidly. The soil must not be too hard or too fine. When the soil is too hard, the brush weeder will remove only the part of the weeds above the soil, and the weeds will readily regrow. When the soil is too hard for hoeing, brush weeders can be used to remove the part of the weeds above the soil. When used on moist soil, the effect will diminish as a result of soil sticking to the bristles.

Some models of vertical-axis brushes can have the angle, rpm and rotating direction of the brushes adjusted. Vertical-axis brushes can be adjusted to throw soil towards the crop row or to remove soil and weeds away from the row (Fogelberg and Kritz, 1999).

Rotary cultivators

Rotary cultivators refer to rotary tilling cultivators which have multiple heads (one per inter-row) and rotary tillers which have a single head that covers several inter-rows. Driven by the tractor's PTO, the cultivators have a vertical, horizontal or oblique axis. Designed for shallow tillage, the inter-row gangs are made of blades, points or knives that rotate at high speed just below the soil surface (Bowman, 1997). Rotary cultivators can cultivate close to the crop row and they are very effective in controlling weeds. However, implements with horizontal axes require time-consuming adjustments, or else gangs with specific working widths must be available for each inter-row distance. The working width of the Weed-fix cultivator (Fig. 8.8), a rotary cultivator with vertical axis, can be adjusted by moving the two rotors closer to or further from each other.

Rolling cultivators

Rolling cultivators have gangs of wheels that are ground-driven. The wheels can be 'spiders'

Fig. 8.7. Vertical axis brush weeder. (Photo by Wageningen UR Applied Plant Research.)

(curved teeth), notched discs, 'stars' etc. There are three to five discs per gang and gangs can be arranged to throw soil towards or away from the crop row. Because gangs are mounted diagonally from the crop row, there are generally two gangs of wheels per inter-row (Lampkin, 1990; Bowman, 1997).

Basket weeders

Basket weeders, also referred to as rolling cages, are cylindrical, made of quarter-inch spring wire, and ground-driven (Fig. 8.9). These cultivators are pulled rather than rear-mounted on a tractor. Ground-driven by the movement of the tractor, the front and rear tools are connected to one another by a chain drive with a ratio equal to 2. The first set of baskets loosens the soil and the second pulverizes it, uprooting young weed seedlings.

Rolling harrows

The disposition of the tools of the rolling harrow enables it to perform efficient selective inter-row weed control (see Fig. 8.3). Young weed seedlings are controlled because they are uprooted from the soil, even if it is very wet and plastic. To enhance the precision and the weeding action, particularly important when vegetable crops are cultivated, the implement may be equipped with a manual guidance system. Moreover, the rolling harrow may be equipped with flexible tools that can selectively control weeds in the rows. This type of precision hoeing is very effective because it

Fig. 8.8. Photo of the Weed-Fix cultivator, a vertical-axis rotary cultivator. (Photo by Wageningen UR Applied Plant Research.)

Fig. 8.9. Photo of the Buddingh basket weeder Model K used in carrots. (Photo by Daniel Cloutier.)

makes it possible to selectively perform a post-emergence treatment on the whole cultivated surface. Thus, this implement is multipurpose and versatile, as it can be used for false seedbed and for precision hoeing (Peruzzi *et al.*, 2005a).

Guidance systems

Guidance systems (mechanical or electronic) allow cultivation to be done at greater speeds and reduce the risk of crop damage. Also, it is possible to cultivate more of the inter-row area by using these systems. Hoeing 1 cm closer to the row will, in the case of onions planted in 25 cm spaced rows, keep an additional 6.5% of the field clean of weeds. This will save between 10 and 30 hours of weeding per hectare in organically grown onions. Consequently, a guidance system that steers accurately in combination with the maximum cultivation width will result in reduced manual weeding costs.

The benefit of weed control increases with every additional centimetre of cultivated intra-row space. The uncultivated strip in which the crop grows must be as narrow as possible while minimizing the amount of crop damaged by the weeding equipment. Intra-row weeders such as finger and torsion weeders achieve better results when the weeders are kept in the same position relative to the crop row. When an operator is driving and steering unassisted, an extra 3–4 cm of clearance from the crop row is required in order to prevent damage to the crop. Consequently, a strip of at least 6–8 cm will not be cultivated in the row. More accurate steering can reduce the uncultivated strip to approximately 4 cm wide.

Steering systems have been developed for the accurate control of weeding machinery. With toolbars mounted on the front of tractors, the driver has an excellent view of the toolbar and the crop rows, and there is sufficient space for the machinery. However, the disadvantage is that a minor correction in the direction in which the tractor is moving results in a much greater correction in the position of the inter-row cultivator. A toolbar mounted between the front and rear wheels of the tractor can be steered with much greater accuracy. However, most tractors offer insufficient space; moreover the driver in the cabin has an insufficient view of the underside of the tractor. Although special implement carriers have been developed for cultivating machinery, their high costs have resulted in little interest in their use. When weeding machinery is mounted on the rear of the tractor, the driver cannot see anything that is happening behind; consequently a distance of 8 cm or more often has to be maintained between the crop rows and the blades of the intra-row cultivator. This results in combined uncultivated strips of at least 16 cm wide around the crop row.

However, for horticultural crops, the accuracy required for operating near the row is sometimes achieved by having a second person seated on a rear-mounted cultivator to steer it.

Mechanical steering systems

The implement must be level to function adequately and the cultivator must also be able to move freely. Some knowledge and experience with the adjustments of the cultivator settings is required. This system will not achieve an optimum effect in the case of differences in soil structure, uneven soil or the presence of ruts.

Weeding equipment mounted rigidly to the front of the tractor responds with an amplified movement following the driver's steering correction. Consequently this makes the steering inaccurate.

The Mutsaert QI steering system resolves this shortcoming; the centre of rotation of the cultivators is located immediately behind the toolbar (van der Schans *et al.*, 2006). If a driver makes a steering correction, the result is a smaller adjustment of the cultivator position. The driver is then able to steer as accurately as with a toolbar mounted between the front and rear wheels of a tractor or implement carrier. Cultivating close to the crop rows is also possible by allowing the cultivator to be guided by the crop. This originally required a fairly strong crop (such as maize or beans at a height of about 10 cm). However, the system has now been improved to the extent that even beet plants with four leaves and the ridges of furrow drills can be used to control the guides. Unfortunately, this steering system cannot be combined with the finger or torsion weeders. Furthermore, there are also mechanical systems where guide wheels follow ridges or grooves or deep furrows created at seeding (Bowman, 1997).

Electrical systems

During the past few years, considerable progress has been made with systems using cameras and software to process images acquired live and processed in real-time. An overview of the systems available in Europe has been presented by van der Schans *et al.* (2006). These systems consist of a camera which locates the crop row(s) a few metres in front of the cultivator. The camera is mounted on a toolbar. Software uses the image to calculate the row position. The toolbar is mounted on a side-shift fitted to the tractor. A side-shift consists of two parts: a front frame attached to the tractor's three-point hitch, which is fixed, and a rear plate to which the implement is attached (Bowman, 1997). The rear plate is hydraulically moved right or left. The side-shift's controls correct the position of the camera and, in so doing, the position of the inter-row cultivators relative to the crop row.

Systems using image recognition to determine the position of individual crop plants are under development. When there is sufficient space between crop plants in the row, a cultivator or a flame weeder can then be used to cultivate or burn weeds between crop plants.

INTRA-ROW CULTIVATION

Finger weeders

The original finger weeder is the Buddingh Model C (USA). It has two pairs of truncated steel cones that are ground-driven by metal tines that point vertically. Each cone has rubber spikes or 'fingers' that point horizontally outwards. The crop row is between each pair of cones. The rubber fingers from opposite cones connect together in the row, pulling out small weeds in the process. The space between opposite cones can be widened to prevent crop damage. This type of cultivator is effective against young weed seedlings and is gentle to the crop provided that it is well rooted. Finger weeders have been imported and modified by various European manufacturers (Fig. 8.10).

Compared to the harrow, finger weeders have the disadvantage that they need very accurate steering to work as close as possible in the crop rows, and thus their working capacity is relatively low. However they are gentler to the crop and can easily be combined with inter-row

cultivation. The finger weeders operate from the sides of the crop row and beneath most of the crop leaves. As with inter-row brush weeding, finger weeders also cause relatively more weed uprooting and move them away from the crop rows. Finger weeders are more effective against weeds with true leaves, but the weeds still need to be small and/or easy to uproot. The tools can be used in many transplanted vegetables, beans, spring-seeded rape, seeded onions (from two-leaf stage and beyond), red beet and sugarbeet (from 2–4 leaves), carrots (two-leaf stage and beyond) and strawberries (Bowman, 1997; van der Schans *et al.*, 2006).

Torsion weeders

Intra-row weeding using the torsion weeder is selective because the crop is better rooted (and better anchored) than the weeds. Torsion weeders are made up of pairs of spring tines connected to a rigid frame and bent in various ways (e.g. angled downward and back towards the row) so that two short segments (only a few centimetres long) work very close together and parallel to the soil surface, even overlapping over the crop plants (Fig. 8.11). The coiled base allows tips to flex with soil contours and around established crop plants, uprooting young weed seedlings within the row (Ascard and Bellinder, 1996; Bowman, 1997; Bleeker *et al.*, 2002; Melander, 2004).

The diameter of the spring tines may vary from 5–6 mm to 9–10 mm, increasing the

Fig. 8.10. Finger weeders, manufactured in Europe, operating in transplanted cabbages. (Photo by Wageningen UR Applied Plant Research.)

aggressiveness of the treatment. The adjustments (degree of compression and distance of the spring tines from the cultivated plants) of torsion weeders must be done in the fields, taking into account crop stage and resistance to uprooting, weed sensitivity and soil conditions. Torsion weeders are often used in combination with precision cultivators equipped with a guidance system in order to perform a post-emergence, selective, gentle and very precise weeding treatment on the whole surface in only one passage. Torsion weeders were tested both in Europe and in North America on many herbaceous and horticultural crops with very good results in terms of intra-row weed control; consistently reducing the time required to finish weed removal in the rows by hand-pulling and/or manual hoeing (Ascard and Bellinder, 1996). This tool was also used with very good results in poorly rooted vegetable crops such as carrot, reducing weed density in the rows by 60–80% (Peruzzi *et al.*, 2005b).

Other flexible tools for intra row cultivation

There are other flexible tools (spring-hoe weeders or flats) that can be used in a similar way to the torsion weeders in that they work within the crop rows, exploiting the vibration around their vertical axis to selectively control weeds at the white-thread stage. Selectivity is also determined by the difference in anchorage existing between crop plants and weeds. Again, as for the torsion weeders, the adjustment of the flexible tools must be determined in the fields, taking into account crop growth stage and resistance to uprooting, weed sensitivity and soil conditions. The available assortment of different flexible implements, including torsion weeders, ensures that some tool will be available for use with different crop and weed types and anchorage, density, growth stages and development, aggressiveness, soil type and field conditions (Bowman, 1997; Mohler, 2001; Melander, 2004; Peruzzi *et al.*, 2005b,c).

Spring-hoe weeders are set up similarly to the torsion weeders except that they are in paired sets and the weeding section is a long flat metal blade that is driven along the crop row. The blade is held vertically instead of horizontally. The weeding effect is obtained by the uprooting of weeds when the blades move slightly below the soil surface. Being less flexible than the torsion weeders, they are more aggressive and can be used in well-established crops (Bowman, 1997).

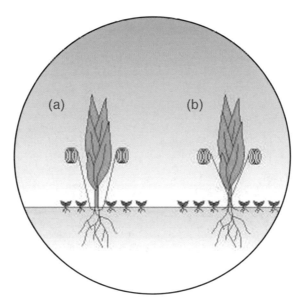

Fig. 8.11. Schematic diagram of torsion weeder action with different adjustments: (a) tines close to crop plants; (b) tines crossed on crop plants. (Drawn by Andrea Peruzzi.)

An implement has recently been developed that uses vibrating teeth made up of steel spring tines bent downward horizontally and then bent toward the soil surface with a first angle of 135° and a second angle of 45° in order to have two segments 5–10 cm long to work (vibrating around their axis) within the rows and very close to crop plants (Fig. 8.12). The tension of the vibrating teeth and distance from the cultivated plants must be adjusted, taking field conditions into account. The action of these tools is similar to the torsion weeders but less aggressive. Consequently, it is possible to perform intra-row weeding without damaging the crop, even if it is poorly anchored, provided that it is better rooted than the weeds (Peruzzi et al., 2005b,c).

Another type of flexible tool, recently built and developed in Italy, is made up of three pairs of vertical spring tines of 6 mm diameter, mounted on a horizontal bar. This set-up makes the two internal tines of the two external pairs vibrate and work in concert with the central pair within the rows and very close to crop plants (Fig. 8.13). Tool adjustment for the intensity of pressure against weeds is simple, while their distance from the cultivated plants must be adjusted by changing their position on the horizontal bar (see also Fig. 8.3). These tools were created to be mounted to rolling harrows used as precision cultivators in order to perform considerable intra-row weeding action. The results obtained on transplanted vegetable crops, such as fennel and chicory, are very encouraging and on the same level as those obtained with torsion weeders (reduction of weed density by about 80%) (Peruzzi et al., 2005a).

Pneumat

The Pneumat weeder is an inter-row cultivator that also controls weeds on the crop row by using compressed air to blow small weeds out of the row (Fig. 8.14). Nozzles are placed on each side of the crop row, slightly staggered so as not to blow at each other, cancelling their action. Some results indicate that there could be additional advantages of using the Pneumat in crops with widely spaced rows, such as tulip, or in situations with more developed weeds that have several leaves. The crop must pass precisely through the Pneumat's nozzles. The intensity of the cultivation increases with decreasing distance between the nozzles and increasing air pressure. The crop damage decreases with increasing tractor speed. A pressure of as much as 1 MPa can be used with crops that are extremely well rooted. In that case, even weeds with several true leaves can be controlled. The best effect is obtained when cultivator working depth, air pressure and tractor speed are adjusted according to crop growth stage and field conditions. The tractor's power might become limiting because the air compressor requires a lot of power, in particular when several crop rows are treated at the same time.

Ridging the crop

Another way of controlling weeds on the crop row is by ridging or hilling the crop. This consists in throwing soil on the crop row by using discs or specialized blades such as wings or ridgers (Fig. 8.15). This technique is being used extensively in various crops such as maize,

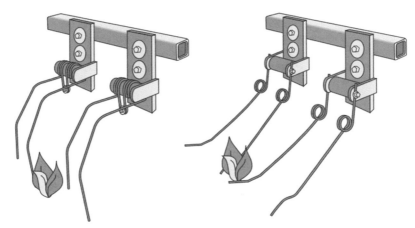

Fig. 8.12. Drawing of the vibrating teeth and torsion weeders. (Drawn by Andrea Peruzzi.)

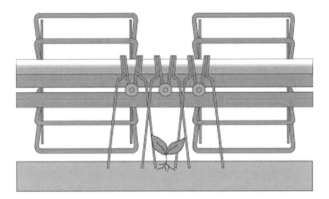

Fig. 8.13. Drawing of the three pairs of flexible tine tools in action. (Drawn by Andrea Peruzzi.)

Fig. 8.14. Blowing away small weeds in seeded onion with the Pneumat. (Photo by Wageningen UR Applied Plant Research.)

Fig. 8.15. Ridging in maize utilizing special wings mounted on an inter-row cultivator. (Photo by Daniel Cloutier.)

sorghum, soybean, potato and leek, among others. It must be done at a crop growth stage that can tolerate partial burial. It is particularly effective when the weeds are completely covered by soil. This technique is used extensively in field crops in North America.

Intelligent weeders

For selective control of well-developed weeds without damaging the crop plants, intelligent weeders are needed. One of the first commercially available new intelligent weeders, the Sarl Radis from France, has a simple crop detection system based on light interception and moves a hoe in and out of the crop row around the crop plants (Fig. 8.16). The machine was designed and built for transplanted crops such as lettuce. It is very effective, but only when weeds are smaller than the crop plants. The working speed is limited to 3 km/h.

Currently several small companies, in cooperation with researchers, are developing other intelligent weeders that use computer vision to recognize crop plants for them to guide weeding tools in and out of the crop row. Denmark (Melander, 2004) and Germany (Gerhards and Christensen, 2003) are focusing on developing sensors or cameras to distinguish between crop and weed plants. The University College of

Halmstad in Sweden has a prototype working in sugarbeet (see http://www2.hh.se/ staff/bjorn/ mech-weed). In the Netherlands, Wageningen University is developing and/or testing different intelligent intra-row weeders together with Danish, German and Dutch companies.

Cropping systems with no soil inversion

Cropping systems with annual soil inversions can have detrimental effects on the environment through erosion (Håkansson, 2003). Cropping systems with no soil inversion require less energy to operate, and conserve soil and water (Wiese, 1985). Another advantage of non-inversion tillage is that the organic matter is kept within the top of the soil profile where it can be most useful to the crop. Traditionally, an increased reliance on herbicides for weed control has been associated with cropping systems with no soil inversion (Wiese, 1985). However, there are some non-inversion cropping systems where weed management can be done mechanically. Examples of these systems are field crop production on permanent ridges, semiarid farming on the Great Plains of North America, vegetable production on permanent raised bed systems, and the Kemink exact soil management system, among others.

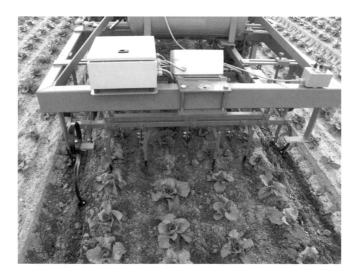

Fig. 8.16. Intelligent weeder manufactured by Sarl Radis, France. (Photo by Wageningen UR Applied Plant Research.)

Ridge tillage is common in North America with maize and soybean. Other field crops can also be produced on ridges but they require wide inter-row spacing. Ridge tillage can increase yields and economic returns from decreased costs, increased soil warming, drainage and aeration, decreased soil compaction, and decreased nutrient leaching (Mohler, 2001; Henriksen *et al.*, 2006). This cropping system consists of planting a crop in a ridge that was built during the previous growing season. At planting, the top of the ridge is cut or scraped approximately 5 cm down, a furrow is opened, seeds are dropped, and the furrow is closed by a press wheel, all in a single operation. When the top of the ridge is cut, the soil, weed seeds and plant debris are thrown in the inter-ridge. The plant material is usually rapidly broken down in this inter-ridge space. Weed control is initially done by using a rotary hoe adapted for high-residue conditions. The rotary hoe will normally only touch and cultivate the top of the ridges. The rotary hoe can be used a few times and, when weeds become taller, an inter-row cultivation can be done while the crop row is protected by shields. Once the crop is sufficiently developed to tolerate partial covering by soil, the ridge is rebuilt by one or several passages of an inter-row cultivator equipped with a special ridging tool. The inter-row cultivation will usually destroy all the weeds present in the inter-ridge space, while the soil projected on the crop row will partially or totally cover the weeds. After harvest the ridges are left intact with the crop stubble until the next growing season.

Non-inversion tillage is extensively used in the semiarid production areas of the Great Plains of North America. A number of non-inversion tillage tools were developed primarily for the purpose of summer fallowing; the practice of tilling the land for an entire growing season to control weeds, conserve moisture and stabilize yields. These tillage implements were designed to maintain crop residues on the soil surface and leave the soil with a higher percentage of large aggregates. The wide-blade sweep plough, developed in southern Alberta, employs V-blades that are approximately 1.5–3.0 m in width and are operated at a depth of 7–14 cm (Bowman, 1997). The sharp blades cut the weeds and the crop plants below ground while leaving approximately 85–95% of the

original plant residues on the soil surface. The stubble mulch blade plough is often used for the first tillage after crop harvest, and has V-shaped blades which work at a depth of 7–4 cm. The blades are 1–1.5 m wide and this plough leaves 75–95% of surface residue (Bowman, 1997). The rod weeder is particularly well adapted to work under dry conditions. It is used in final seedbed preparation or for pre-crop emergence cultivation when the crop seeds are deep enough for the rod to pass over them. It is passed 4–6 cm below the surface and its rotating action pulls and uproots weeds, depositing them on the soil surface. The rods can be round, square or hexagonal, vary in width from 2 to 4 m, and are powered by ground-driven systems, PTO or hydraulics (Bowman, 1997).

Raised beds are widely used in horticulture where soil inversion is used as primary tillage. Recently new techniques have been developed using permanent raised beds where no soil inversion occurs (Schonbeck, 2004). The raised bed is planted with a cover crop which is subsequently cut and/or mechanically destroyed by heavy crimper/roller tools, leaving a mulch of plant debris into which a crop is planted, normally by a conventional no-till planter, a subsurface tiller/transplanter, or possibly by a dibber drill (Rasmussen, 2003; Schonbeck, 2004). The raised bed can be rebuilt after crop harvest and/or before seeding a cover crop.

Another non-inversion system that was developed at the end of the 20th century is called the Kemink exact soil management system. Developed in Germany, it is a non-inversion soil management system based on subsoiling, ridging and controlling traffic where weeds are controlled mechanically (Henriksen *et al.*, 2005).

Systems with inversion of only the top 5–10 cm of the soil have been developed to provide a suitable seedbed while keeping the organic matter at the top of the soil profile. In these cases, weed control can be done mechanically, as in conventional systems with deep soil inversion. The types of plough that can be used are the 'spot plough' (Shoji, 2004), two-layer plough (Lazauskas and Pilipavicius, 2004), shallow ploughs, and shallow ploughs combined with a shank that goes deeper into the soil and to which various types of points, sweeps or blades are attached.

8.3 Cutting and mowing

Cutting and mowing are weed management methods commonly used in turf, in rights of way, in vineyards, in orchards, in pastures and in forage crops. These techniques are used to promote crop establishment, to control weed size and seed production, and to minimize competition with the crop (Schreiber, 1973; Kempen and Greil, 1985; Ross and Lembi, 1985; Lampkin, 1990; Smith, 1995; Frick, 2005; Donald, 2006).

Mowing and cutting are rarely sufficient to totally control weeds but, combined with other management techniques, they can favour the crop to the detriment of weed development. Cutting weeds reduces their leaf area, slows their growth and decreases or prevents seed production. Mowing is most effective against annual weeds and it will also affect stationary, but not creeping, perennial weeds. Dicotyledonous weeds are more vulnerable than monocotyledonous weeds to mowing and cutting.

New tools have been developed to mow weeds in crop inter-rows. For example, Donald et al. (2001) have developed a specialized mower to cut weeds between soybean and maize rows. Some of the cutting of weeds on the crop row could be done by lasers (Heisel et al., 2002). Laser cutting has been described also as a potential energy-efficient alternative to non-chemical weed control that delays the growth of weeds, decreases their competitive ability, and that eventually kills them. It is concluded that lasers have potential for reducing seed production in certain weed species and may be cost-effective on a field scale, although it is noted that further development is necessary. Regrowth appeared after laser cutting when plant stems were cut above their meristems, indicating that is important to cut close to the soil surface to obtain a significant effect (Heisel et al., 2001). Water-jet cutting using water at very high pressure (2000–3000 bar) using 5–25 l/min could also be an efficient way to cut weeds (Fogelberg, 2004).

8.4 Pulling

Several mechanical weed pullers have been developed to remove weeds that grow taller than the crop (Anonymous, 1979; Wicks et al., 1995). Most rely on rubber tyres rolling together in opposite directions. The stems of the weeds are pulled by the tyres and are either uprooted or broken where they are in contact with the tyres (Fig. 8.17). A moist soil will facilitate the uprooting of the weeds. Although this technique is faster than removing weeds by hand, it should not be used as a primary weed control technique but rather as a last attempt at removing weeds that have escaped other weed control efforts in

Fig. 8.17. Weed puller developed to pull wild mustard from soybean fields in south-western Quebec. (Photo by Daniel Cloutier.)

the course of the growing season. Since it must be used in the field when weeds are growing above the crop, this involves a very late machinery passage, with the risk of the tractor damaging the crop.

8.5 Economics of mechanical weed management

Economically, mechanical weed management can be profitable when compared with conventional production systems involving herbicides (Wicks *et al.*, 1995; Leblanc and Cloutier, 2001b; Mohler, 2001; Peruzzi *et al.*, 2005a,b). The reverse can be reported in some circumstance, but it is often in a context where there is no added value to the crop. Also, the use of herbicides, compared with mechanical weed management, would probably be uneconomical if the cost of decontaminating the environment from herbicides or their metabolites was to be considered when comparing production and weed control costs. This is not normally done as, in most studies, costs start and finish at the farm gate. Also, government commodity programmes might be unfavourable to the use of mechanical weed control instead of herbicides by decreasing crop insurance or support programme funds when non-conventional weed control methods are used (DeVuyst *et al.*, 2006).

8.6 Summary

Tillage remains the most important technique for mechanical weed management. However, tillage alone should not be relied on as a sole weed control technique but instead it should be part of an overall cropping management strategy. Individual weed control techniques will rarely be sufficient to provide season-long weed control. Rather it is a combination of weed control techniques with cropping management systems that will provide acceptable levels of weed control during a growing season.

Most tillage tools are constantly being improved. One of the current trends in mechanical weed management is the development of automatic guidance systems that improve equipment precision in the field. Eventually, this approach will result in self-guided, self-propelled and autonomous machines that will cultivate crops with minimal operator intervention. To this end, real-time image acquisition and analysis will be performed using one or several video cameras, while GPS-based positioning systems will be used to map fields and crop and weed locations. Another trend is the development of cultivators that will be able to selectively control weeds within the crop row while being assisted by sensors that will enable the machine to differentiate crop plants from weeds in order to be able to selectively destroy the latter.

8.7 References

Anonymous (1979) Bourquin weed puller. *Weeds Today* 10, 11.

ASAE (2004) Terminology and definitions for agricultural tillage implements. *ASAE Standards: ASAE S414 FEB04*, pp. 270–282.

ASAE (2005) Terminology and definitions for soil tillage and soil–tool relationships. *ASAE Standards: ASAE EP291.3 FEB05*, pp. 129–132.

Ascard, J. and Bellinder, R.B. (1996) Mechanical in-row cultivation in row crops. In: *Proceedings of the 2nd International Weed Control Congress*, Copenhagen, Denmark, pp. 1121–1126.

Balsari, P., Berruto, R. and Ferrero, A. (1994) Flame weed control in lettuce crop. *Acta Horticulturae* 372, 213–222.

Bàrberi, P. (2002) Weed management in organic agriculture: are we addressing the right issues? *Weed Research* 42, 176–193.

Barthelemy, P., Boisgontier, D. and Lajoux, P. (1987) *Choisir les outils du travail du sol*. ITCF, Paris, France.

Blake, G.R. and Aldrich, R.J. (1955) Effect of cultivation on some soil physical properties and on potato and corn yields. *Soil Science Society Proceedings* 19, 400–403.

Bleeker, P., van der Weide, R. and Kurstjens, D. (2002) Experiences and experiments with new intra-row weeders. In: *Proceedings of the 5th EWRS Workshop on Physical and Cultural Weed Control*, Pisa, 11–13 March, pp. 97–100.

Bowman, G. (ed.) (1997) *Steel in the Field: A Farmer's Guide to Weed Management Tools*. Handbook Series No. 2, Sustainable Agriculture Network, Beltsville, MD, USA.

Buckingham, F. (1984) *Tillage*, 2nd edn. Deere and Company, Moline, IL, USA.

Buhler, D.D., Doll, J.D., Proost, R.T. and Visocky, M.R. (1995) Integrating mechanical weeding with reduced herbicide uses in conservation tillage corn production systems. *Agronomy Journal* 87, 507–512.

Cavers, P.B. and Kane, M. (1990) Response of proso millet (*Panicum miliaceum*) seedlings to mechanical damage and/or drought treatments. *Weed Technology* 4, 425–432.

Cloutier, D. and Leblanc, M.L. (2001) Mechanical weed control in agriculture. In: Vincent, C., Panneton, B. and Fleurat-Lessard, F. (eds) *Physical Control in Plant Protection*. Springer-Verlag, Berlin, Germany, and INRA, Paris, France, pp. 191–204.

Cloutier, D.C., Leblanc, M.L., Benoit, D.L., Assémat, L., Légère, A. and Lemieux, C. (1996) Evaluation of a field sampling technique to predict weed emergence. Xième Colloque International sur la Biologie des Mauvaises Herbes à Dijon, 11–13 September 1996. *Annales de l'Association Nationale pour la Protection des Plantes* 10, 3–6.

Dal Re, L. (2003) Il controllo non chimico delle infestanti: strategie collaudate [Weed management without chemicals: tested strategies]. *Il Divulgatore* 10, 32–67.

DeVuyst, E.A., Foissey, T. and Kegode, G.O. (2006) An economic comparison of alternative and traditional cropping systems in the northern Great Plains, USA. *Renewable Agriculture and Food Systems* 21, 68–73.

Donald, W.W. (2006) Mowing for weed management. In: Singh, H.P., Batish, D.R. and Kohli, R.K. (eds) *Handbook of Sustainable Weed Management*. Food Products Press, New York, pp. 329–372.

Donald, W.W., Kitchen, N.R. and Sudduth, K.A. (2001) Between-row mowing banded herbicide to control annual weeds and reduce herbicide use in no-till soybean (*Glycine max*) and corn (*Zea mays*). *Weed Technology* 15, 576–584.

Fogelberg, F. (2004) Water-jet cutting of potato tops: some experiences from Sweden. In: Cloutier, D. and Ascard, J. (eds) *6th EWRS Workshop on Physical and Cultural Weed Control*, Lillehammer, Norway, 8–10 March 2004, p. 111.

Fogelberg, F. and Kritz, G. (1999) Intra-row weeding with brushes on vertical axis: factors influencing intra-row soil height. *Soil and Tillage Research* 50, 149–157

Frick, B. (2005) Weed control in organic systems. In: Ivany, J.A. (ed.) *Weed Management in Transition: Topics in Canadian Weed Science, Volume 2*. Canadian Weed Science Society, Sainte-Anne-de-Bellevue, Quebec, Canada, pp. 3–22.

Gerhards, R. and Christensen, S. (2003) Real-time weed detection, decision making and patch spraying in maize, sugarbeet, winter wheat and winter barley. *Weed Research* 43, 385–392.

Gunsolus, J.L. (1990) Mechanical and cultural weed control in corn and soybeans. *American Journal of Sustainable Agriculture* 5, 114–119.

Håkansson, S. (2003) Weeds and weed management on arable land: an ecological approach. CABI Publications, Wallingford, UK.

Hartmann, K.M. and Nezadal, W. (1990) Photocontrol of weeds without herbicides. *Naturwissenschaften* 77, 158–163.

Heisel, T., Schou, J., Christensen, S. and Andreasen, C. (2001) Cutting weeds with a CO_2 laser. *Weed Research* 41, 19–29.

Heisel, T., Schou, J., Andreasen, C. and Christensen, S. (2002) Using laser to measure stem thickness and cut weed stems. *Weed Research* 42, 242–248

Henriksen, C.B., Rasmussen, J. and Søgaard, C. (2005) Kemink subsoiling before and after planting. *Soil and Tillage Research* 80, 59–68.

Henriksen, C.B., Rasmussen, J. and Søgaard, C. (2006) Ridging in autumn as an alternative to mouldboard ploughing in a humid-temperate region. *Soil and Tillage Research* 85, 27–37.

Juroszek, P. and Gerhards, R. (2004) Photocontrol of weeds. *Journal of Agronomy and Crop Science* 190, 402–415.

Kempen, H.M. and Greil, J. (1985) Mechanical control methods. In: Kurtz, E.A. and Colbert, F.O. (eds) *Principles of Weed Control in California*. Thomson Publications, Fresno, CA, USA, pp. 51–62.

Kouwenhoven, J.K. (2000) Mouldboard ploughing for weed control. In: Cloutier, D. (ed.) *4th EWRS Workshop on Physical and Cultural Weed Control*, Elspeet, The Netherlands, pp. 19–22.

Kurstjens, D.A.G. and Kropff, M.J. (2001) The impact of uprooting and soil-covering on the effectiveness of weed harrowing. *Weed Research* 41, 211–228.

Kurstjens, D.A.G. and Perdok, U.D. (2000) The selective soil covering mechanism of weed harrows on sandy soil. *Soil and Tillage Research* 55, 193–206.

Kurstjens, D.A.G., Perdok, U.D. and Goense, D. (2000) Selective uprooting by weed harrowing on sandy soils. *Weed Research* 40, 431–447.

Lampkin, N. (1990) *Organic Farming*. Farming Press Books, Ipswich, UK.

Lazauskas, P. and Pilipavicius, V. (2004) Weed control in organic agriculture by the two-layer plough. *Zeitschrift fur Pflanzenkrankheiten und Pflanzenschutz* 19 (Special Issue), 573–580.

Leblanc, M.L. and Cloutier, D.C. (1996) Effet de la technique du faux semis sur la levée des adventices annuelles [Effect of the stale seedbed technique on emergence of annual weeds]. Xième Colloque International sur la Biologie des Mauvaises Herbes à Dijon, 11–13 September 1996. *Annales de l'Association Nationale pour la Protection des Plantes* 10, 29–34.

Leblanc, M.L. and Cloutier, D.C. (2001a) Susceptibility of row-planted soybean (*Glycine max*) to the rotary hoe. *Journal of Sustainable Agriculture* 18, 53–61.

Leblanc, M.L. and Cloutier, D.C. (2001b) Mechanical weed control in corn (*Zea mays* L.). In: Vincent, C., Panneton, B. and Fleurat-Lessard, F. (eds) *Physical Control in Plant Protection*. Springer-Verlag, Berlin, Germany, and INRA Paris, France, pp. 205–214.

Leblanc, M.L., Cloutier, D.C., Leroux, G.D. and Hamel, C. (1998) Facteurs impliqués dans la levée des mauvaises herbes au champ [Factors involved in emergence of weeds in the field]. *Phytoprotection* 79, 111–127.

Melander, B. (2004) Nonchemical weed control: new directions. In: *Encyclopedia of Plant and Crop Science*. Marcel Dekker [doi: 10.1081/E-EPCS-120020279, pp. 1–3, accessed 26 May 2006].

Melander, B., Rasmussen, I.A. and Bàrberi, P. (2005) Integrating physical and cultural methods of weed control: examples from European research. *Weed Science* 53, 369–381.

Mohler, C.L. (2001) Mechanical management of weeds. In: Liebman, M., Mohler, C.L. and Staver, C.P. (eds) *Ecological Management of Agricultural Weeds*. Cambridge University Press, Cambridge, UK, pp. 139–209.

Peruzzi, A. and Sartori, L. (1997) *Guida alla scelta ed all'impiego delle attrezzature per la lavorazione del terreno* [Guide to the choice and the proper use of tillage operative machines]. Edagricole, Bologna, Italy, 255 pp.

Peruzzi, A., Ginanni, M., Raffaelli, M. and Di Ciolo, S. (2005a) The rolling harrow: a new implement for physical pre- and post-emergence weed control. In: *Proceedings of 13th EWRS Symposium*, Bari, 19–23 June.

Peruzzi, A., Ginanni, M., Raffaelli, M. and Fontanelli, M. (2005b) Physical weed control in organic carrots in the Fucino Valley, Italy. In: *Proceedings of 13th EWRS Symposium*, Bari, 19–23 June.

Peruzzi, A., Ginanni, M., Raffaelli, M. and Fontanelli, M. (2005c) Physical weed control in organic spinach in the Serchio Valley, Italy. In: *Proceedings of 13th EWRS Symposium*, Bari, 19–23 June.

Rasmussen, I.A. (2004) The effect of sowing date, stale seedbed, row width and mechanical weed control on weeds and yields of organic winter wheat. *Weed Research* 44, 12–20.

Rasmussen, J. (1991) A model for prediction of yield response in weed harrowing. *Weed Research* 31, 401–408.

Rasmussen, J. (1992) Testing harrows for mechanical control of annual weeds in agricultural crops. *Weed Research* 32, 267–274.

Rasmussen, J. (1993) Yield response models for mechanical weed control by harrowing at early crop stages in peas (*Pisum sativum* L.). *Weed Research* 33, 231–240.

Rasmussen, J. (2003) Punch planting, flame weeding and stale seedbed for weed control in row crops. *Weed Research* 43, 393–403.

Riemens, M.M., van der Weide, R.Y., Bleeker, P.O. and Lotz, L.A.P. (2006) Effect of stale seedbed preparations and subsequent weed control in lettuce (cv. Iceboll) on weed densities. *Weed Research* 47, 149–156.

Ross, M.A. and Lembi, C.A. (1985) *Applied Weed Science*. Burgess Publishing, Minneapolis, MN, USA, 340 pp.

Schonbeck, M. (2004) Organic no-till for vegetable production? www.newfarm.org/features/0104/no-till/index.shtml [accessed 26 May 2006].

Schreiber, M.M. (1973) Weed control in forages. In: Hearth, M.E., Metcalfe, D.S. and Barnes, R.F. (eds) *Forages: The Science of Grassland Agriculture*, 3rd edn. Iowa State University Press, Ames, IA, USA, pp. 396–402.

Shoji, K. (2004) Forces on a model 'spot plough'. *Biosystems Engineering* 87, 39–45

Smith, A.E. (ed.) (1995) *Handbook of Weed Management Systems*. Marcel Dekker, New York, 741 pp.

Souty, N. and Rode, C. (1994) La levée des plantules au champ: un problème mécanique? [Seedling emergence in the field: a mechanical problem?]. *Sécheresse* 5, 13–22.

Steinmann, H.-H. (2002) Impact of harrowing on the nitrogen dynamics of plants and soil. *Soil and Tillage Research* 65, 53–59.

Toukura, Y., Devee, E. and Hongo, A. (2006) Uprooting and shearing resistances in the seedlings of four weedy species. *Weed Biology and Management* 6, 35–43.

van der Schans, D., Bleeker, P., Molendijk, L., Plentinger, M., van der Weide, R., Lotz, B., Bauermeister, R., Total, R. and Baumann, D.T. (2006) *Practical Weed Control in Arable Farming and Outdoor Vegetable Cultivation without Chemicals*. PPO Publication 532, Applied Plant Research, Wageningen University, Lelystad, The Netherlands, 77 pp.

van der Weide, R. and Bleeker, P. (1998) Effects of sowing time, false seedbed and pre-emergence harrowing in silage maize. *3rd EWRS Workshop on Physical Weed Control*, Wye, p. 1 [available at www.ewrs.org/pwc].

Weber, H. and Meyer, J. (1993) Mechanical weed control with a brush hoe. In: Thomas, J.-M. (ed.) *Maîtrise des adventices par voie non chimique: 4th International Conference of the International Federation of the Organic Agriculture Movement*. ÉNITA, Quétigny, France, pp. 89–92.

Wicks, G.A., Burnside, O.C. and Warwick, L.F. (1995) Mechanical weed management. In: Smith, A.E. (ed.) *Handbook of Weed Management Systems*. Marcel Dekker, New York, pp. 51–99.

Wiese, A.F. (ed.) (1985) *Weed Control in Limited-Tillage Systems*. Weed Science Society of America, Champaign, IL, USA.

9 Use of Non-living Mulches for Weed Control

A.C. Grundy[1] and B. Bond[2]

[1]Weed Ecology and Management, Warwick HRI, University of Warwick, Wellesbourne CV35 9EF, UK; [2]Henry Doubleday Research Association, Coventry, CV8 3LG, UK

9.1 Introduction

The use of mulches has been practised since antiquity in raised-field agriculture such as the *chinampa* system still in operation in the Valley of Mexico (Yih and Vandermeer, 1988). In these closed ecosystems, soil fertility, structure and weed control are maintained by applications of canal muck and mulches of grasses and aquatic plants. In such a low-fertility system, the nutritional benefits of the mulch often mask the weed control effects. In modern agriculture, where weed control may be the primary objective, it is known that mulching will help to prevent soil erosion, reduce pest problems, aid moisture retention, and limit nitrate loss from the soil.

Non-living mulch is applied over the soil surface to suppress weed seed germination by the exclusion of light and to act as a barrier that will physically prevent weed emergence. Mulches may be composed of natural materials of organic or inorganic origin, or synthetic materials that have been manufactured specifically for this purpose or which are recycled products. They may take the form of flat sheets that are laid by hand or machine, or loose particles that are spread out to form a continuous layer. The availability and cost limit the application of some materials to amenity and landscaping uses, where appearance and novelty may be the important factors. Elsewhere, the physical characteristics of the mulch will often determine how and where it is used. Different types of mulch can be used in combination where this is advantageous. In certain crops, film mulch is laid over the planted area and particle mulch is spread along the paths and wheelings. Where landscape fabric mulch is used in an amenity area, it is often covered with particle mulch to improve the appearance and durability.

The period that mulch is in place can vary considerably. Mulch may be expected to remain intact for just one growing season or may be intended to persist for many years. Under perennial crops or in an amenity situation greater durability is an advantage, and so woven polypropylene mulch has a life expectancy of several years. However, mulch used as an alternative to cultivation to clear vegetation such as pasture before cropping is usually left in place for only 12–18 months. In annual crops, sheeted mulch that does not degrade will require lifting and disposal unless it can be re-used *in situ*. Mulches can also be used for very short periods. For example, black polythene can be laid over freshly prepared seedbeds for just a few weeks to disrupt weed seedling emergence, then lifted before crop planting.

9.2 Advantages and Limitations of Non-living Mulches

Non-living mulches provide a number of bene-fits. These include retention of soil moisture, prevention of leaching, improved soil structure, disease and pest control, improved crop quality and, in many crops, extended growing season which reaps financial rewards. However, the primary advantages are associated with weed control.

Weed suppression by both particle and sheet mulches can achieve significant long-term savings in labour and the need for herbicides. It is essential to choose the appropriate mulch for the particular circumstances. The efficacy of the mulch will vary depending on the prevailing weed problems and environment. For example, particle mulches are generally ineffective against established perennial weeds, although there may be some initial suppression.

When particle mulches are used, weed control usually improves as the thickness of the organic mulch increases (Ozores-Hampton, 1998). A 3 cm layer of composted waste is needed to prevent the emergence of annual weeds (Ligneau and Watt, 1995). However, the perennial grass, common couch (*Elytrigia repens*) is able to grow through a 10 cm deep layer of bark. Coarse materials such as bark are considered to be more effective for weed control than fine materials because they provide a more hostile environment for seed germination (Pickering, 2003). However, coarse mulch needs to be deeper than one consisting of fine particles. Mulch on sandy soils needs to be deeper than that on heavy or wet soils.

With sheeted mulches, managing weeds at the edges of mulched strips is difficult and there are practical problems with covering large areas for long periods. Weeds often establish in the hole cut for crop planting and weeds emerge in decayed leaf litter that accumulates on top of non-living mulch (Benoit et al., 2006). Creeping weeds may emerge around the edges of the covers or may penetrate thin or damaged parts of the mulch. Tall vegetation should also be cut down before covering, but regrowth may lift insecure sheets.

Susceptibility to damage from weathering remains perhaps one of the greatest limitations to the use of non-living mulches. Loose particles of light mulch materials such as straw and hay are susceptible to wind blowing, which may expose bare areas of soil. Sheets of plastic, paper and other mulch material are also suscep-tible to storm or physical damage. In orchards, despite burying the edges of sheeted mulches, using landscape staples and weighting down the mulch with stones, there can be problems with the wind blowing and tearing of mulches in the winter (Merwin et al., 1995). Some materials are more susceptible to ripping than others but this can vary with conditions. Paper is more easily damaged when it is wet; while heavy cellulose sheets deteriorate when they become brittle. Stretching and contracting following wetting and drying can cause a paper mulch to tear. Similarly, plastic sheeting that is stretched too much during laying may split on a hot day or pull free as it contracts when the temperature drops at night. Pressure from being walked on or punctures caused at or after laying reduce longevity. Mowers and other machinery being used in adjacent areas can snag the mulch; hence some sheeted mulches have lines of rein-forcement to prevent tears spreading

Both the advantages and limitations of using non-living mulches are extremely site- and product-specific. Other limitations, even disad-vantages, associated with the use of mulches are increased runoff into wheelings and surrounding land during periods of heavy rain, introduction of new weed problems, fire risk, crop contamination with phytotoxins and spores, and labour and disposal costs. Examples of the characteristics and potential problems associated with different mulching materials, including their economics and impact on the environment, soil, crop quality, pests and diseases, are described in more detail later in the chapter.

9.3 Types of Non-living Mulches

While a mulching material is usually brought in from elsewhere, surface soil made weed-free *in situ* by steam sterilization or direct heating will act as a non-living mulch layer and limit weed emergence. In 'no-till' agriculture, although not solely for weed control purposes, the crop residues that build up as a mulch layer over the soil surface will reduce weed emergence, aided in part by the limited soil disturbance. The result has been compared with the effect of leaf litter that builds up on a woodland floor.

In general though, non-living mulch takes the form of: (i) loose particles of organic or inorganic matter that is spread as a layer over the soil; or (ii) sheets of artificial or natural materials that are laid on top of the soil. The range of materials that can and have been used for mulching is vast. Details of some of the applications and characteristics for a selection of sheeted and particle mulches are given here, but this is by no means a complete list.

Sheeted mulches

Black polythene sheeting

This remains the most widely used mulch for weed control in organic and conventional systems. It can be planted through for growing vegetables or soft fruit, laid around trees in orchards or used under particle mulch. It can also be laid for just a short period to disrupt early weed emergence in freshly prepared beds or for several months over established vegetation to clear it prior to crop planting. The edges are usually secured by burial. Black plastic mulch absorbs most ultraviolet, visible and infrared wavelengths and this is then reflected as heat or long wavelength infrared radiation and is lost. Only if the mulch is in direct contact with the soil is the heat conducted into the soil, otherwise an air gap acts as insulation. In the daytime the soil beneath the mulch is around 2°C higher than that of bare soil. Black plastic lasts for 1–3 years; thicker films last longer than thin. It is impermeable and becomes brittle with age.

Clear polythene sheeting

This is generally better than black polythene for warming the soil but does not control the weeds, except in very sunny climates where weeds can be killed by solarization. Transparent plastic degrades faster than black when exposed to the sun's ultraviolet (UV) radiation. The degradation is slowed by the addition of stabilizers. Plastic mulches have been developed that selectively filter out the photosynthetically active radiation (PAR) but let through infrared light to warm the soil. Infrared transmitting (IRT) mulches have been shown to be more effective in controlling perennial weeds.

Coloured polythene sheeting

A number of coloured products have undergone field testing (Horowitz, 1993). White and green coverings had little effect on the weeds, while brown, black, blue, and white-on-black (double colour) films prevented weeds from emerging.

Geotextiles

Whether these are woven or spun-bonded, they tend to allow air and water to pass through and are less likely to scorch crops when temperatures are high. Certain weaving techniques allow faster passage of water than others to prevent puddles from forming when they are used where container plants are stood out. These materials can be UV-stabilized to give a life expectancy of approximately 5 years. They are held down with cover pegs or staples and can withstand pedestrian and light vehicular traffic. Some fabrics are also treated with copper hydroxide to prevent container plants rooting into and through the textile. This reduces damage to the fabric when the containers are lifted. Disinfectant can be applied between batches of pots or growing seasons to reduce sources of *Pythium* inoculum, a major cause of plant losses. The sheeting can be lifted after cleaning, rolled and then re-laid elsewhere. These materials are often used over the soil surface to suppress weeds inside polytunnels where tomatoes are grown in hanging gutters or strawberries are grown in a 'table-top system'. The surface can be swept clear of debris and if necessary disinfected between crops, and requires only occasional hand-weeding around the edges of the material. However, while annual weeds are controlled, perennials may penetrate these woven fabrics.

In addition to agricultural and horticultural applications, woven groundcover can be used in low-maintenance landscaping schemes for weed suppression, soil moisture conservation and to provide clean, hardwearing surfaces. Ground-cover fabrics are also used under decking to prevent weed growth. They often look more attractive and may last longer if covered with a layer of loose mulch material. Geotextiles may last 15 years if covered in this way.

Needle-punched fabrics

These mulches are made from natural fibres such as hemp, jute and flax and can be used as mulch mats, tree mats, pot tops or can be seeded for erosion control purposes (Drury, 2003). The fabric decomposes over a period of 12–18 months and can be incorporated into the soil.

Paper mulches

Paper mulch can be made from recycled fibres, is permeable, and can be laid by machine and planted into by hand or mechanically. However, rolls of paper tend to be heavier and bulkier that plastic. There are no disposal costs, it is biodegradable and improves the organic matter content of the soil. The majority of paper mulches begin to degrade part-way through the growing season, but soil and temperature conditions will affect this (Stewart *et al.*, 1995). Brown paper with or without a thin film coating of a vegetable oil to reduce paper degradation has been evaluated as an alternative to black plastic in transplanted basil (Miles *et al.*, 2005). The paper mulch maintained its integrity and gave good weed control throughout, whether treated with oil or not.

Paper mulches have compared favourably with black polythene in trials with transplanted lettuce, Chinese cabbage and calabrese in the UK (Runham and Town, 1995). Tearing and wind blowing can be a problem but correct laying of the paper and rapid crop establishment are the key to success (Runham, 1998a). In the Netherlands, brown and black paper mulches have been tested with salad and flower crops. Both gave good weed control, but stretching and contracting following wetting and drying caused the brown paper mulch to tear. The black paper mulch was creped to overcome this problem and did not tear (Wilson, 1990). Crimped paper mulches have also proved more resilient than flat paper in trials with transplanted brassicas (Davies *et al.*, 1993).

Newspapers and carpet

Newspapers are applied several layers deep, wetted and held down by stones or covered with grass clippings. The newspapers, eight pages thick, will gradually decompose but will last a season. Newspaper may acidify the soil as it breaks down. Colour-printed magazines should be avoided. Cardboard from flattened boxes makes a good weed control membrane that lasts for a growing season. It is held down with bricks and can be planted through. It has been laid over established vegetation to clear it prior to crop planting (Lennartsson, 1990). Cardboard can damage crop plants if it is poorly secured and moves in the wind.

Old carpet can be used for small areas including paths and alley strips. It can also be laid over established vegetation for 12–18 months to clear it prior to crop planting (Lennartsson, 1990). There is some concern about the breakdown products from carpet as it degrades.

Particle mulches

Many new inorganic and organic materials, as well as recycled products, are being considered as potential particle mulches. For a particle mulch to be effective it should be able to reduce weed growth, be of a consistent texture, be resistant to compaction, be resistant to erosion by wind and water, be fire resistant, be slow to decompose, and be non-phytotoxic and free of weed and crop seeds and vegetative propagules. Prior to the application of a particle mulch, it is advisable to incorporate any old mulch to prevent a build-up and to remove any emerged weeds and moisten and aerate the soil.

Shredded and chipped bark or wood mulch

For this to be effective it needs to be 5–7.5 cm deep to control most annual weeds. It decomposes slowly over a period of around 2 years and will need raking over and topping up periodically. Weeds that develop from wind-dispersed seeds are easily uprooted. The mulch is not usually blown around by the wind. It may initially lock up nitrogen in the surface soil, but this is released again as the mulch decomposes. Wood chip and bark mulch is available in various size grades. The bark may be from conifers or broadleaved trees. Even the choice of species used will impact on weed control efficacy and seedbank size (Kamara *et al.*, 2000). Wood chip can be made from soft or hard woods and may consist of recycled materials. Certain woods, such as walnut, may release chemicals that inhibit plant growth. Others, such as cedar, may have insect-repellent properties. Wood that has been chemically treated or is from diseased trees should be avoided. Wood chip is less attractive than bark mulch but can be coloured to improve its appearance if desired.

Finer particles of wood

An example of this is sawdust, which is readily available as waste from sawmills. The dense nature of sawdust and its slow decomposition, even in a tropical environment, aid weed control both before and after canopy closure. In Nigeria, the growth and yield of plantains was significantly improved by the addition of sawdust mulch (Obiefuna, 1986). There are additional environmental benefits where the previous disposal method for the sawdust was by burning. Aged or partially rotted sawdust makes a satisfactory mulch if laid 5 cm deep, but is prone to caking. Another example is that of highbush blueberry production where a 0.9 to 1.2 m-wide strip under the plants with a deep mulch of up to 15 cm provides organic growers with good weed control. At the same time, the breakdown products of sawdust and woodchip mulches has the added benefit of buffering the soil pH for these acid-loving plants (NCAT, 2004).

Crushed rock or gravel mulch

This product forms a relatively permanent feature. Different size grades of stone and gravel can be used depending on the appearance required and whether it will be travelled over. Coarser particles provide a firmer base for travelling over than smooth ones. The colour of the stones will affect heat absorption by the mulch. Dirt and debris will settle on the mulch, allowing blown-in seeds to germinate and grow, but these pull out easily. A tough weed fabric or landscape fabric is usually laid under the stone

mulch to provide a good base and suppress perennial weeds. It should be of a type that is water-permeable to prevent puddles forming. If the stone mulch needs to be removed for any reason, a strong fabric below it can facilitate this. Seashells, broken pieces of slate, and crushed glass with the edges ground away for safety can be used in a similar way but provide some variation in colour and texture (Pickering, 2003).

Straw and hay

Straw and hay of cereal, ryegrass and lucerne are often used as a mulch to give winter protection. They are more effective if used over a membrane. Hay has also been laid over established pasture for 12–18 months to clear it prior to crop planting (Lennartsson, 1990).

Grass clippings

Organic mulch, such as grass clippings, can be laid directly over the soil or on top of a thick layer of wet newspaper. This should be allowed to air-dry before applying as mulch and should not be used if the grass area has recently been treated with herbicide. Leaf mould mulch consists of decomposed leaves, is a rich source of nutrients and has good microbial activity. Fresh leaves are best cut or shredded before application as a mulch, otherwise they form a soggy mat and take a long time to break down. Pine needles are less prone to becoming a soggy mass than other leaves. Peat moss, as well as a needle mulch, are beneficial around acid-loving plants. However, peat moss is expensive and a current trend is to limit the use of peat in any form because of the environmental cost. Spent mushroom compost, a waste material of the mushroom industry with a pH on the alkaline side, is weed-free and will aid soil fertility.

Crop wastes

Flax shrives are the hard woody residue left after processing harvested flax into fibre for paper manufacturing. It compares favourably with wood chips as mulch, does not break down readily, and has been used during the establishment of shelter belts. Ground corn cobs makes a good mulch although it is criti-cized for its light colour. There are other uses for this material so it is not readily available.

The waste materials from the processing of various nuts have been used as mulch. Walnut hulls and almond shells without hulls have been used as mulch in orange and in prune orchards in the USA (Heath and Krueger, 2000). Partially decomposed walnut hulls in a layer 7.5 cm deep were somewhat less effective than almond shells as a weed control mulch but contain 5% nitrogen, which may act as a fertilizer. The hulls may also contain juglone, which is known to be a plant inhibitor. Untreated groundnut shells at a depth of 7.5–10 cm will stop annual weeds emerging. Cocoa shells make a wind-resistant, relatively non-combustible mulch that contains 3% nitrogen and is high in potash. These nutrients are released as the shells decompose, but certain plants may be harmed by the high potash levels. Coffee grounds are fine particles that cake badly when used as mulch and should only be applied in a 2.5 cm layer. Coffee grounds are naturally acidic, with a pH of 4.5–5.2, and are also likely to lower the soil pH. Availability of this and other miscellaneous materials such as tea leaves is too limited for commercial use as mulch.

Industrial waste materials

A single 3 cm deep application of rubber mulch made from recycled tyres will suppress annual weeds, retain soil moisture and prevent soil compaction, but lasts much longer than organic mulch. It does not rot and can withstand pedes-trian traffic. For garden and amenity use, it can be produced in a range of colours. The rubber can be shredded to give the appearance of chipped bark but is slower to decompose.

9.4 Effects on Soil Components and Soil Conservation

Regulating soil temperature

A mulch can buffer the soil temperature against rapid fluctuations. There is great potential for mulches to increase soil temperature and hence provide added benefits during crop establishment and growth, but this is complex and product-specific.

Generally, sheeted plastic mulches can warm up the soil beneath and accelerate plant growth, leading to earlier harvests. The soil temperature under some plastics is higher both during the day and at night (Fig. 9.1) and this can increase mineralization of nitrogen from organic residues (Runham, 1998b).

In contrast, under a ground cover of particu-late mulches such as dry straw, shredded news-paper or leaves, the summer soil temperature is lower during the day and higher at night than soil without a covering (Munn, 1992). This can be advantageous to certain crops.

Plastic mulches are now available in a greater array of colours, and the colour influ-ences the energy-radiating behaviour of the mulch. It determines the surface temperature of the mulch and the underlying soil temperature. Under a white, white-on-black or silver mulch, the soil temperature may be lower by a degree or so because much of the incoming radiation is reflected back into the crop canopy. The lower soil temperature may be beneficial in hot weather. Wavelength-selective mulches absorb photosynthetically active radiation and transmit infrared radiation. This raises the soil tempera-ture but maintains good weed control.

It is difficult to make generalizations about the relative merits of different types of poly-thene products and colours. This is because mulches of the same colour from different manufacturers have been found to vary in light reflectance wavelengths, colourfastness and longevity (Orzolek and Otjen, 2005). In addi-tion, site-to-site variability makes it hard to achieve consistent results. However, some general observations can be made. White poly-thene film reflected a maximum of 85% of light uniformly over the visible range (Lieten, 1991). Black film warmed the soil but did not reflect any light. Blue film reflected light in the blue spectrum to a maximum of 33%. Red film reflected up to 15% of light in the violet/blue and up to 65% in the red to far-red range. In spring, the daytime soil temperature under the blue film was 2–3°C more than under white film and 4°C more than under white-on-black film. At night, the temperature under the blue poly-thene remained half a degree higher than under white and 1°C higher than under the white-on-black film. The white-on-black film temperature was 3°C less than under black plastic.

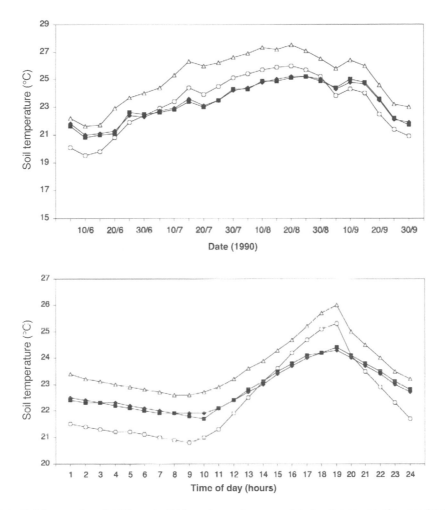

Fig. 9.1. Soil temperature (at 20 cm depth) in a vineyard measured during the course of a year (top) and during the course of a single day on 18 August 1990 (bottom) in the grapevine row. The soil temperature is recorded under a control (white circles) and under a range of mulches: polythene mulch (white triangles), biosolid/bark compost mulch (black diamonds) and MSW compost mulch (black squares). From Pinamonti (1998).

Soil structure and nitrate concentration

Soil structure is likely to benefit from the use of organic mulches (Feldman *et al.*, 2000) and the soil beneath a mulch tends to remain loose with a good crumb structure. Organic mulch has the potential to improve soil structure and fertility once broken down but, as it decomposes, soil nitrogen may be locked-up, at least initially. This normally affects only the surface layer of soil but if the mulch is incorporated a greater depth of soil will be affected. An application of

nitrogen fertilizer can help to overcome the problem.

Nitrate concentrations have been recorded as being higher under black polythene mulches than uncovered soil (Davies *et al.*, 1993). Soil N, and sometimes N uptake detected in leaves, under black polythene mulch has also been shown to be higher in studies with cucumber crops (Cohen *et al.*, 2000). Arora and Yaduraju (1998) also observed significant increases in soil N under clear plastic mulch compared with plots that were not covered.

Oxygen content and microbial activity

Currently, little is known about the effect of mulches on factors related to soil microbial activity. Mulches based on composted waste and recycled ground wood pallets have been shown to have a significant effect on soil respiration, microbial biomass, organic matter content and nutrient content. In particular, the composted waste material strongly and positively influenced the structure of the microbial rhizosphere community (Tiquia *et al.*, 2002). Carbon dioxide builds up under plastic film due to root and microbial activity. It escapes around the planting holes made for the crop, where it can become concentrated, enhancing photosynthesis by the crop plants

A mulch usually aids soil aeration but some materials and some conditions can limit it. A 7.5 cm layer of particle mulch is a sufficient depth for weed suppression, but over-mulching can reduce oxygen reaching the soil. This is a particular problem in wet or waterlogged conditions. On heavy, wet, clay soils, plastic mulches restrict soil microbial activity, leading to anaerobic soil conditions. The toxic by-products resulting from anaerobic decomposition can scorch, defoliate or kill plants growing in the mulch. Souring can also occur prior to application of the mulch if the material is left in large, saturated heaps that become depleted of oxygen (Gleason and Iles, 1998). There is a greater risk of air exclusion from fine-textured materials such as sawdust.

Soil moisture conservation

All types of mulch will retain moisture in the upper layers of soil and reduce water loss. The high degree of impermeability of plastic films to water vapour prevents the evaporation of soil moisture and achieves substantial water economy. For example, planting raspberries through black polythene mulch is beneficial in retaining soil moisture during establishment. In Greece, the control of weeds in olive and citrus orchards with black polythene has been promoted. Nevertheless, in some crops, impermeable mulch may necessitate the use of drip or trickle irrigation installed prior to mulching to moisten the soil beneath it. In extreme cases, the moisture-insulating nature of mulches has the potential to create problems with winter drought in perennial crops such as tree plantations.

Reduced leaching

The use of mulches tends to reduce leaching of nitrates from the soil and so nutrients are retained in the root zone, allowing more efficient uptake by the crop. However, fertilizer applications or other soil treatments need to be made before the mulch is laid. In perennial crops like raspberries that are grown through plastic mulch, annual fertilizer applications will need to be applied with trickle irrigation. This is sometimes referred to as fertigation.

Soil conservation

A mulch can reduce water erosion by allowing rain to slowly infiltrate the soil rather than hit the soil and run off. If the soil is dry when an organic mulch is laid, or dries out under drought conditions, the mulch will prevent the water from light showers reaching the soil. The mulches also prevent a crust forming on the soil surface. Stubble mulching serves to protect the soil against wind erosion and also provides some protection against water erosion.

9.5 Economics of Non-living Mulches

Transport and labour cost considerations

Loose materials such as straw, bark and composted municipal green waste provide effective annual weed control, but the depth of mulch needed to suppress weed emergence is likely to make transport costs prohibitive unless the material is produced on site (Merwin *et al.*, 1995). Particle mulches require considerable labour to lay them, although equipment has been developed for spreading some materials. Bark can be spread using a blower, a blowing chute or discharge hoses. Mulch tenders can carry the mulch material and discharge it directly to verges, traffic islands and other amenity areas.

Sheeted materials are relatively expensive and also require labour and often additional equipment to lay them efficiently (Runham, 1998b). There may also be a requirement for other equipment for initial bed shaping and, after laying, for cutting or punching holes for planting or seeding into the mulch. Bed-making machines have been designed and built that raise the soil and lay the plastic mulch, securing it at the sides in one operation. Beds need to be uniformly smooth to ensure good contact of mulch with the soil in order to ensure heat conductance and prevent lifting in the wind.

Improved economics of production

The high costs associated with mulching often limit its use to high-value horticultural crops (Runham and Town, 1995) unless there are other reasons for its use. In these situations, while the mulches may be expensive, the weeding costs are reduced in the long term. There are several examples where non-living mulches have significantly improved the economics of a production system. In some cases the use of mulches has actually exceeded the yields obtained by conventional chemical methods. For example, in the USA, the increased crop value from mulched apple orchards justified the greater costs of mulching with various films and fabrics (Merwin *et al.*, 1995). Cut ryegrass mulch spread between planted rows of tomatoes and peppers was more expensive than herbicide or cultivation treatments, but the higher financial returns from the mulched crops made it the most profitable system (Edwards *et al.*, 1995).

Mulching can also have significant impacts on local economies by increasing the practicability of growing certain commercial crops where, for example, soil moisture and temperature were previously limiting. This has been shown for chive production in Finland, where previously the majority of chives have been imported (Suojala, 2003).

The economic benefits of using mulches versus the costs and practicality of mulching will often be site- and situation-specific. Low weed pressure may mean that the cost of the mulch does not match the savings in weed control and gain in crop yield. Sites with optimal soil conditions may also not see significant benefits that

outweigh the additional costs of the mulching. It is often, therefore, the marginal sites, or those with high weed pressure, where the greatest economic gains from mulches can be seen (Green *et al.*, 2002).

However, increased crop yield is not the only economic consideration. In orchards and other long-term crops, economic assessments need to take into account factors such as laying cost, longevity/durability and any maintenance costs, as well as the benefits in terms of improved weed control and crop yield (Merwin *et al.*, 1995). Thus, the economics are very complex and theoretical studies have shown that leaf mulches produced from three different species gave dramatically different outcomes in terms of their economic feasibility (Bohringer, 1991). Therefore, the economic variability involved in the case-by-case appraisal of mulches has resulted in the economics of very few systems being reported, making generalization difficult.

9.6 Disposal Problems and Environmental Impacts

Disposal

After cropping, lifting and disposal may be a problem with plastic and other durable mulches. Recycling of dirty plastic is not economic (Runham and Town, 1995) and even the degradable plastics may break into fragments that litter the soil. On-site burning or burial is environmentally unacceptable.

Between 1991 and 1999 there was at least a 50% increase in the area mulched with plastics, but precise figures are difficult to determine (Jouët, 2001). In 1999 it has been estimated that 30 million acres worldwide were covered with plastic mulch, all of which would have needed to be disposed of (Miles *et al.*, 2005). In the USA, most mulch has been put in landfills, but evaluation studies have been made of field incineration of the mulch during the lifting process. A mulch collector was modified with the addition of four LPG gas burners to the rear of the machine. The plastic mulch tended to melt into globules rather than being completely incinerated and these dropped onto the soil, where they were likely to persist. Sheeting made from paper and other natural fibres has the advantage of breaking down naturally, and can be incorporated into the soil after use.

Biodegradable plastic products

There have been advances in the development of products that will degrade naturally with weathering and which also have the potential to be used in the field for more than one season, and hence more than one crop. There has been increasing demand for biodegradable mulches, and work on photodegradable polyolefin polymer and poly-ethylene co-polymer and biodegradable starch-based films has been ongoing since the 1960s. These tend to be more expensive than polythene films. Photodegradable films are designed to break down and become brittle after set periods of exposure to sunlight. The rate of breakdown of films made from biodegradable polymers will depend on soil and weather conditions.

However, the results of some studies have been highly variable. Different-coloured plastic mulches have been tested and there appears to be a significant trade-off between the thermal properties of the mulch to provide adequate soil warming in the second season and the ability of the mulch to degrade. Those that provide the most efficient warming also tend to be those that degrade slowly and therefore may still pose significant disposal issues if the intention is to gain two seasons' effective use (Ngouajio and Ernest, 2005). Generally, black plastic film breaks down more slowly than clear plastic film (Fig. 9.2). The buried edge of the sheeting is the slowest to degrade.

Biodegradable non-plastic products

Mulching film made from maize starch is bio-degradable and does not need removal and disposal after cropping. It has a life expectancy of 1.5–4.5 months depending on the season. It has the same resistance, elasticity and mulching characteristics as plastic film and can be laid by machine and planted into. A biocompostable polymer known as P2 breaks down into non-toxic residues under aerobic and non-aerobic conditions. It can be readily degraded in well-managed composting systems and is water-soluble above 60°C. It can be formulated to

break down in UV light and become increasingly water-soluble throughout the growing season.

Biodegradable polymerized vegetable-oil-coated paper mulches offer another alternative to traditional polythene mulches in that they can be used to effectively stop weed growth but will degrade naturally at the end of the season (Shogren and Rousseau, 2005). Foam mulches offer another alternative and, because they can be incorporated into the soil, disposal is not an issue (Masiunas *et al.*, 2003)

With all these biodegradable products, there is a constant trade-off between achieving sufficient weed suppression and the rate of degradation (Greer and Dole, 2003). An example of this has been observed in strawberry production using biodegradable polymer (clear or black)-covered paper, compared with uncoated paper mulches. Generally the polymer-coated papers were effective for weed suppression but degraded too slowly to allow essential runner rooting to take place, while the paper mulches degraded too quickly, allowing premature tearing and therefore poor weed control in the second season (Weber, 2003).

Increased water runoff

Where black plastic is laid over raised beds and the wheelings between are left uncovered, there

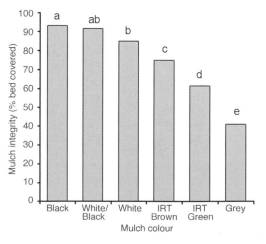

Fig. 9.2. Integrity of coloured plastic mulches measured as percentage of bed covered by the mulches after 1 year of field exposure. Means with the same letter are not significantly different (LSD 0.05). From Ngouajio and Ernest (2005).

is increased runoff of rainfall from the imperme-able mulch into the wheelings. The result may be greater soil erosion and, if chemical treatments have been applied to the crops, residues washed from the foliage will accumulate in the runoff water (Durham, 2003). This has the potential to cause an environmental hazard through runoff into watercourses (Sikora and Szmidt, 2001). In the USA, residues of copper-based fungicides have been found in runoff water, often attached to soil particles. The residues could harm fish and other aquatic life (Raloff, 2001). Sowing cereal rye in the wheeling may help to reduce the problem.

Introduction of new weeds

Compost and other soil-like materials remain moist and are the perfect environment for the germination of wind-blown seeds, but the result-ing seedlings are easily pulled out. Weed seeds in the mulch itself can be a problem with straw and other harvested materials used directly from the field. In straw mulch, shed cereal grains and even whole ears remaining in the straw after crop harvest make volunteer cereal seedlings a par-ticular problem. Weed-free mulch that has been stored in the open before use can soon become contaminated with wind-blown weed seeds. Composted material that has not been processed properly may contain seeds or vegetative propagules that are still viable. This in itself may have an environmental impact by introducing new and potentially invasive or exotic weed species or extending the geographical range of existing weed species to a new location.

Introduction of contaminants, phytotoxins and spores

With organic mulch made from pesticide-treated crop material, there is a potential risk that residues may linger and damage future crops. Some herbicides persist longer on organic matter than they do in the soil. Natural chemicals released from the decomposing organic mulch can also reduce the stand and vigour of desirable plants, particularly direct-sown, small-seeded crops (Table 9.1). These phytotoxins can be of benefit by reducing weed seed germination and

growth in tolerant crops (Ozores-Hampton, 1998). Cocoa shell mulch has a high potash content that can harm some plants when it is released during decomposition. The shells also contain theobromine, a chemical found in chocolate that is poisonous if ingested by dogs (Hansen *et al.*, 2003). With any organic mulch there is a health risk to operators from the fungal spores that are produced by decaying or partially composted materials. Clouds of these can be released when compost is turned or laid as mulch, and can cause potential allergic reactions if inhaled by a sensitive person. Human health pathogens are also a consideration when using composted waste material (Pinamonti and Sicher, 2001) and preventative measures should be taken to reduce any health risks to humans or animals (Gleason and Iles, 1998). Unfortunately, composts used as mulches may have a detrimen-tal build-up of some nutrients and metals in the soil (Roe, 2001) and, in hot dry weather, the mulch can also become a fire risk if it consists of combustible material.

Contaminants within the mulches, particularly composted waste material, are also a potential hazard (Ozores-Hampton *et al.*, 2001). Some contaminants, such as glass, may be potentially dangerous in the soil, to field workers or indeed to the final product if it is to be consumed.

9.7　Effect on Crop Quality

The impact of mulches on improved crop yield and quality are well documented. For example, most mulches will prevent rain splashing soil onto crop plants and spoiling their appearance. There are also several pest- and disease-reduc-ing effects of mulches that ultimately improve crop quality, as discussed later in this chapter.

However, there are also some less well-known quality benefits associated with the use of mulches. For example, straw mulches were shown to raise alpha-acid levels compared with a conventional residual herbicide in hops (Blackman *et al.*, 1996). The light spectrums and certain wavelengths reflected by different mulch colours can be beneficial to particular crops. Studies in strawberries have shown that, apart from keeping the fruit clean, which is important for the marketability of the crop, mulch colour can also have quality benefits. Kasperbauer *et al.*

Table 9.1. The phytotoxicity of several compounds found in compost (from Ozores-Hampton, 1998).

Phytotoxic compound	Compost type/age	Species affected	References
Acetic acid	Wheat straw, 4 weeks	Barley (*Hordeum vulgare* L.)	Lynch (1978)
Acetic acid	Municipal solid waste (MSW), immature	Cabbage (*Brassica oleracea* L. Capita group) Cauliflower (*Brassica oleracea* L. Botrytis group) Cress (*Lepidium sativum* L.) Lettuce (*Lactuca sativa* L.) Onion (*Allium cepa* L.) Tomato (*Lycopersicon esculentum* Mill.)	Keeling *et al.* (1994)
Ammonia	Biosolids	Brassica *campestris* L.	Hirai *et al.* (1986)
Ammonia and copper	Spent pig litter, <24 weeks	Lettuce Snap peas (*Phaseolus vulgaris* L.) Tomato	Tam and Tiquia (1994)
Ammonia, ethylene oxide	MSW, <16 weeks	*Brassica parachinensis* L.	Wong (1985)
Organic acid	Cow manure, 12 weeks	Tomato	Hadar *et al.* (1985)
Organic acid	MSW, <4 weeks	*Brassica campestris* L.	Hirai *et al.* (1986)
Organic acids and other compounds	Yard trimming waste, <17 weeks	Australian pine (*Casuarina equisetifolia* J. R & G. Frost) Bahiagrass (*Paspalum notatum* Flugge.) Brazilian pepper (*Schinus terebinthifolius* Raddi.) Ear tree (*Enterolobium cyclocarpum* Jacq.) Punk tree (*Melaleuca leucadendron* L.) Ragweed (*Ambrosia artemisiifolia* L.) Tomato Yellow nutsedge (*Cyperus esculentus* L.)	Schiralipour *et al.* (1991)
Phenolic acids	Pig slurries <24 weeks	Barley Wheat (*Triticum aestivum* L.)	Maureen *et al.* (1982)

(2001) demonstrated that by using red as opposed to the more normal black plastic, strawberries were larger, had greater sugar content and emitted higher concentrations of flavour/aroma compounds. They hypothesized that the far-red and the far-red to red ratio of light reflected from the red mulch modified gene expression through the natural phytochrome system. The spectrum of light reflected by a mulch can therefore have significant and largely un-researched impacts on subsequent crop quality. Other studies have shown significant increases in biomass and essential oil production of summer savory (*Satureja hortensis*) when using white mulch, with blue producing the least effect (Walker *et al.*, 2004).

Not all of the impacts of mulches on crop quality are beneficial. Weed suppression with composted or non-composted organic mulches is due to the physical presence of the materials on the soil surface and the action of phytotoxic chemicals generated by microorganisms in the composting process (Ozores-Hampton, 1998). The type and severity of plant injury is related to compost maturity (Ozores-Hampton *et al.*, 1998). Chemicals such as acetic, propionic and butyric acids and ammonia are found in partially decomposed green waste. Crop injury has been linked to the use of immature compost.

9.8 Effects on Disease and Pest Problems

Reduction of fungal disease problems

Bark mulches, such as pine bark, have been investigated for their ability not only to suppress weeds but also to reduce the incidence of fungal diseases. In a study of the effect of pine bark mulch on ginseng production, improved yield and root shape were observed when using pine bark compared with conventional straw mulch. However the added benefit was a suppression of the incidence of damping-off (*Rhizoctonia solani*). The leachates from the pine bark itself were not found to be inhibitory, but a bacterium and yeast isolated from the mulch did appear to be effective in reducing the incidence of *R. solani* (Reeleder *et al.*, 2004). Polythene mulches have also been seen to control verticillium wilt in tomatoes and reduce nematode populations through soil solarization (Morgan *et al.*, 1991; Coates-Beckford *et al.*, 1998). In potatoes, the use of the trickle irrigation, which is often required when using mulches, has the benefit of reducing scab and blight problems that are associated with foliar irrigation.

Many diseases require a vector to carry them to a new host, and if these are not present in or not attracted to the mulch material, the risk of disease spread is low. It has been shown that the nematode that causes pine wilt, *Bursaphelenchus xylophilus*, can be transmitted from infested wood chips to young pine trees, but only if the mulch is made from fresh wood (Gleason and Iles, 1998). The risk of transmission becomes negligible if the wood chips are composted for a few weeks. The same treatment appears to prevent transmission of the fungus that causes verticillium wilt (*Verticillium dahliae*) to seedlings growing in potting mixtures containing wood chips from infested trees. Recontamination of roses by rose blackspot spores splashing up from the soil is reduced by a surface dressing of organic mulch. The presence of the mulch may also encourage faster breakdown of disease spores. Those mildews that are aggravated by dry soil conditions may be less severe when a mulch is in place.

Increased risk of fungal diseases

Greater activity of *Rhizoctonia solani* has been found in soil under plastic mulch planted through with strawberries. Blackspot, an increasing problem in strawberries, is carried by many weed species that do not exhibit any symptoms, and is spread by rain splash. Plastic mulch along the crop row and straw mulch between the rows eliminated the weeds and prevented water splashing the soil onto the crop. However, on the plastic mulch itself, water splash could spread any disease already present on the strawberries. Damping off is also encouraged by the warm, damp conditions created by a mulch.

Reduction of insect pests

Composted mulches have several advantages over polythene mulches in terms of pest suppression; however, they can cause harvesting problems. There has been considerable work looking at the impact of plastic mulches on insect populations and the viruses they vector. Light reflectance may also affect the behaviour of certain insects. Aluminium foil and aluminium-painted mulches have been shown to be effective at repelling insect pests; the brightness and contrast with the soil is thought to be important (Greer and Dole, 2003). However, these types of mulch have been shown to have a differential effect on insect groups and favoured many beneficial insects. There is the additional benefit of reducing the spread of virus diseases if insect vectors are deterred from landing, but this effect diminishes as the crop cover increases.

Encouraging invertebrate and vertebrate pests

A mulch can encourage slugs and wireworms. In the USA, wood mulches hold moisture and attract termites and other insects. A stone mulch is sometimes used under trees but can lead to mouse damage. A hay mulch can encourage rodents, slugs and insect pests, while a straw mulch may give rise to problems with thrips. Wood mulch encourages rodents, ants and woodlice. Impermeable mulches create a favourable habitat for slugs and snails, but coarse-textured particle mulches can discourage slugs.

In orchard situations, wood mulch should not be laid too close to tree trunks, as it can encourage pest and disease problems. Nor should wood and bark mulches be laid close to wooden structures, as they may act as vectors for termites. Organic mulches may also serve as overwintering sites for hibernating pests. In apple orchards in the USA, where voles are a serious pest, there were more voles, and hence tree damage, on mulched plots than on bare soil (Merwin *et al.*, 1995). In soft fruit plantations they are an ideal habitat for vine weevil, a notorious pest.

9.9 Promoting Beneficial Invertebrates and Vertebrates

There is considerable evidence for the effects of mulches on increasing populations of natural enemies and non-pest alternative prey (Halaj and Wise, 2002). The high number of invertebrates associated with organic mulches can also have the added advantage of attracting predatory birds and other beneficial animals to feed on them. For example, Schmidt *et al.* (2004) demonstrated that plots mulched with straw as opposed to bare soil had lower cereal aphid densities. This was attributed to the enhanced densities of ground predators, such as spiders, attracted by the environmentally rich soil surface litter. Mulches also promote earthworm populations and are particularly important in hot, dry conditions, which would normally drive them deep into the soil.

9.10 Effects on Weed Seed Germination and Seedbank Dynamics

The weed seedbank and its management are vital to the long-term success of any weed control strategy (Grundy and Jones, 2002; Swanton and Booth, 2004). Mulches act directly on stages in the life cycle of weeds by either promoting or preventing seed germination and also reducing the likelihood of new weeds entering the soil seedbank beneath.

It is not just the physical impedance by the mulch that can reduce weed emergence. Leaching of phytotoxic fatty acids, such as acetic acid, may inhibit weed seed germination (Ozores-Hampton *et al.*, 2001). This could, however, be problematic in also inhibiting crop seed germination. Allelopathic effects have been reported for many crop residues both on weeds and on other crops. The residues of winter crops of oat, rye, forage rape and rape, and spring-sown sun-

flowers and buckwheat chopped and spread over the soil after crop harvest have been found to reduce weed seedling emergence (Gawronski *et al.*, 2002; Golisz *et al.*, 2002). The intensity of the allelopathic effect has been correlated with the amount of mulch material and its rate of decomposition. Varietal differences in allelopathic intensity of sunflowers, as well as differences between parts of the plant, have been found (Gawronski *et al.*, 2002). Composted green waste can reduce weed seed germination in soil but, conversely, has the potential disadvantage of introducing additional weed seeds to the seedbank.

Clear plastic mulches, as opposed to black plastic, will tend to increase weed growth as well as crop growth by raising the temperature of the soil beneath, but not excluding light; hence promoting weed seed germination. Such mulches can create as many weed problems as they solve. Coates-Beckford *et al.* (1998) observed that clear plastic mulch actively promoted the germination and growth of grass weeds. Bond and Burch (1989) observed similarly increased weed numbers and biomass under clear plastic mulches compared with uncovered plots. However, these clear polythene mulches can play an important role in managing the weed seedbank by providing a tool for actively depleting it of viable weed seeds. For solarization to be effective, clear daytime skies are required to heat the soil sufficiently to kill weed seeds (Standifer *et al.*, 1984). In contrast pre-planting black polythene has been shown to cause a delay in weed establishment and subsequent weed growth (Davies *et al.*, 1993; Grundy *et al.*, 1996). The manipulation of the soil temperature may have an impact on seed dormancy, thus creating potential opportunities for seedbank manipulation. However, not all species respond in the same way and reproducibility depends on soil moisture prior to laying the mulch.

Observations of some weed species being depleted, while others were not affected, led Arora and Yaduraju (1998) to suggest a solarization reduction index (SRI) for each species. However success and reproducibility would depend on a better understanding of dormancy cycles. There may be some value in such an index in helping to make an objective judgement regarding whether the species composition of a given weed flora is such that there would be an economic benefit from solarization.

Black polythene is probably the most frequently used and most effective mulch for suppressing weed seed germination; however, other materials such as cover-crop residues and straw can also be very effective in reducing weed emergence from the seedbank.

In an attempt to quantify the potential impact of mulches on subsequent weed emergence from the seedbank, Teasdale and Mohler (2000) have proposed the use of a mulch area index: a two-parameter model relating weed emergence to mulch area index and the solid faction of the mulch. The model was shown to give reasonable predictions of weed emergence through a range of mulches (Fig. 9.3).

9.11 Future Scope and Technology

Some mulch materials are not currently available in sufficient quantities to be used commercially, and this restricts their use at present. New commercial avenues for materials presently used as mulches may limit their future availability. In the UK, for example, the availability of bark and other forest residues has been estimated at more than 6 million m^3 by 2020, but bark is increasingly being used in potting mixtures as a peat substitute. However, new technologies may help create new products with promising mulching properties.

Light transmission

The control of light transmission will improve as plastics chemistry develops. For example, infrared-transmitting brown and green plastic mulches show good correlations between weed infestation and light transmission. Light quality is very important. The high light transmission with these mulches makes them valuable in situations where good weed control as well as soil warming are essential (Ngouajio and Ernest, 2005; see also Cohen and Rubin, Chapter 11, this volume). As to the question of whether light transmission can be used as an indicator of weed population; in fact, good correlations have been identified between average light transmission and weed infestation (Ngouajio and Ernest, 2005)

The optical properties of paper mulches can change as the mulch ages, while those of polythene mulches have been shown to remain stable for longer. The colour of plastic mulches and foams appears to be important in crop ripening, and further work to explore these opportunities is needed.

Impregnated mulches

In non-organic systems, bark mulches impregnated with herbicides have been found to offer 'a promising alternative technology for weed control' (Mathers, 2003). It is thought that the lignin in the bark gives a slow release of the herbicide over time, and an improvement compared with the conventional herbicide used alone. Such technology may allow possible reductions in herbicide application rates under some circumstances, which carries environmental benefits, but the economic gains would need to be proved for it to be attractive to growers.

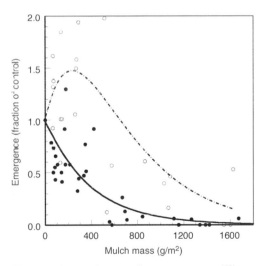

Fig. 9.3. *Amaranthus retroflexus* emergence (*E*) as a function of mulch mass (*M*) in 1996 (white circles, dotted line) and 1997 (black circles, solid line). Models were $E = (1 + 0.00778 \cdot M) \cdot \exp(-0.00277 \cdot M)$ for 1996 ($R^2 = 0.43$) and $E = \exp(-0.00255 \cdot M)$ for 1997 ($R^2 = 0.59$). The coefficients in the 1996 model were significantly different from 0, indicating that the stimulation of emergence at low mulch rates was significant. Data for *Trifolium incarnatum* and *Vicia villosa* mulches were pooled for presentation because responses to mulches separately were not significantly different according to 95% confidence intervals. From Teasdale and Mohler (2000).

Foam spray-on substrates

Foam spray-on substrates may offer sufficient integrity and weed control during the season by completely blocking out light, but they may also need to overcome problems with disposal (Masiunas *et al.*, 2003). Foams have been developed to form a thin film of cellulose fibres or latex over the soil surface to bond the particles together (Stout, 1985). These are used mainly to stabilize loose soil while vegetation becomes established, or to prevent wind-blow on open sites. Those for hydro-seeding motorway embankments consist of fibres that protect the grass seed during germination and form a carrier for water and nutrients. Once sprayed, the fibres mesh together around the seeds, forming a water-holding layer on the soil that resists wind, rain and erosion. A *papier mache*-type mulch that can be sprayed onto clean soil to form a solid barrier against weeds is a recent development. It can be coloured to absorb or reflect more light if required.

Seedmat systems

A seedmat system has been developed for forestry crops and baby-leaf salad based on pre-sown paper rolls that are designed to be laid mechanically on the soil surface and secured at the edges. Seeds are sandwiched between an absorbent paper base and a perforated biodegradable film overlay through which the germinating seedlings emerge. The film acts as a mulch, providing early weed control and retaining soil moisture. The seedmat can be incorporated into the soil after crop harvest. A starch-based version of the film may be available in future.

Exploring biocontrol benefits of mulches

Studies have shown that composts can be used to encourage a greater abundance of biocontrol agents in arable crops. However, the possibilities for enhanced pest suppression are only just beginning to be explored in horticultural systems and could have attractive benefits to growers (Sunderland, 2003).

9.12 Summary

Non-living mulches can provide an effective means of weed suppression in agricultural, horticultural and amenity areas. A diverse range of mulch materials are available, with widely different characteristics. These features may prove beneficial or detrimental depending on the use to which they are put. Improvement of the beneficial effects of non-living mulch on crop growth and quality is an area attracting increasing interest. In the future, it is likely that there will be wider range of new products available and the exploitation of novel mulches from recycled materials.

It is essential to give careful thought to the choice of mulch to ensure that it is appropriate for the intended purpose. The application costs associated with some materials and any limitations due to local availability are also important considerations when choosing the type of mulch for a given situation.

Current research is aimed at ensuring that environmental concerns regarding the persistence and disposal of certain synthetic mulches are met. However, there are also environmental benefits to be gained from using mulches, including moisture conservation in drier climates and the prevention of leaching and soil erosion in areas where precipitation may be an issue.

9.13 References

Arora, A. and Yaduraju, N.T. (1998) High-temperature effects on germination and viability of weed seeds in soil. *Zeitschrift fuer Acker und Pflanzenbau* 181, 35–43.

Benoit, D.L., Vincent, C. and Chouinard, G. (2006) Management of weeds, apple sawfly (*Hoplocampa testudinea* Klug) and plum curculio (*Conotrachelus nenuphar* Herbst) with cellulose sheeting. *Crop Protection* 25, 331–337.

Blackman, J.D., Rees, L. and Glendinning, P.J. (1996) The effects of alternatives to soil residual herbicides on weed control, yield and quality of hops. *Journal of Horticultural Science* 71, 629–638.

Bohringer, A. (1991) The potential of alleycropping as a labour-efficient management option to control weeds: a hypothetical case. *Tropenlandwirt* 92, 3–12.

Bond, W. and Burch, P.J. (1989) Weed control in carrots and salad onions under low-level polyethylene covers. In: *Brighton Crop Protection Conference – Weeds 1989*, pp. 1021–1026.

Coates-Beckford, P.L., Cohen, J.E., Ogle, L.R., Prendergast, C.H. and Riley, D.M. (1998) Effects of plastic mulches on growth and yield of cucumber (*Cucumis sativus* L.) and on nematode and microbial population densities in the soil. *Nematropica* 27, 191–207.

Cohen, J.E., Ogle, L.R. and Coates-Beckford, P.L. (2000) Effects of plastic mulches on the level of N, P and K in soil and leaves of cucumber (*Cucumis sativus* L.). *Tropical Agriculture* 77, 207–212.

Davies, D.H.K., Drysdale, A., McKinlay, R.J. and Dent, J.B. (1993) Novel approaches to mulches for weed control in vegetables. In: *Proceedings: Crop Protection in Northern Britain*, Dundee, UK, pp. 271–276.

Drury, S. (2003) Taking cover. *Horticulture Week* (13 March), 21–24.

Durham, S. (2003) Plastic mulch, harmful or helpful? *Agricultural Research* (July), 14–16.

Edwards, C.A., Shuster, W.D., Huelsman, M.F., Yardim, E.N. (1995) An economic comparison of chemical and lower-chemical input techniques for weed control in vegetables. In: *Proceedings of the Brighton Crop Protection Conference – Weeds*, Brighton, UK, pp. 919–924.

Feldman, R.S., Holmes, C.E., Blomgren, T.A. (2000) Use of fabric and compost mulches for vegetable production in low tillage, permanent bed system: effects on crop yield and labour. *American Journal of Alternative Agriculture* 15, 146–153.

Gawronski, S.W., Bernat, W. and Gawronska, H. (2002) Allelopathic potential of sunflower mulch in weed control. In: *Third World Congress on Allelopathy, 26–30 August, 2002, Tsukuba, Japan*. Sato Printing Co., Tsukuba, Japan, p. 160.

Gleason, M.L. and Iles, J.K. (1998) Mulch matters. *American Nurseryman* 187, 24–31.

Golisz, A., Ciarka, D. and Gawronski, S.W. (2002) Allelopathic activity of buckwheat – *Fagopyrum esculentum* Moench. In: *Third World Congress on Allelopathy, 26–30 August, 2002, Tsukuba, Japan*. Sato Printing Co., Tsukuba, Japan, p. 161.

Green, D.S., Kruger, E.L. and Stantosz, G.R. (2002) Effects of polyethylene mulch in a short-term, poplar plantation vary with weed-control strategies, site quality and close. *Forest Ecology and Management* 173, 251–260.

Greer, L. and Dole, J.M. (2003) Aluminum foil, aluminum-painted, plastic and degradable mulches increase yields and decrease insect-vectored viral diseases of vegetables. *HortTechnology* 13, 276–284.

Grundy, A.C., Bond, W. and Burston, S. (1996) The effect of short term covering of the soil surface with black polyethylene mulch on subsequent patterns of weed seedling emergence and the extent of seed loss from the weed seedbank. In: *Xème Colloque International sur la Biologie des Mauvaises Herbes*, Dijon, France, pp. 395–401.

Grundy, A.C. and Jones, N.E. (2002) Seedbanks. In: Naylor, R.E.L. (ed.) *Weed Management Handbook*, 9th edn. Blackwell Science for the BCPC.

Hadar, Y., Inbar, Y. and Chen, Y. (1985) Effect of compost maturity on tomato seedling growth. *Scientia Horticulturae* 27, 199–208.

Halaj, J. and Wise, D.H. (2002) Impact of a detrital subsidy on trophic cascades in a terrestrial grazing food web. *Ecology* 83, 3141–3151.

Hansen, S., Trammel, H., Dunayer, E., Gwaltney, S., Farbman, D. and Khan, S. (2003) Cocoa bean mulch as a cause of methylxanthine toxicosis in dogs. *Journal of Toxicology: Clinical Toxicology* 41, 720.

Heath, Z. and Krueger, B. (2000) *Use of Walnut Shells for Weed Control*. Research Report, Organic Farming Research Foundation Project 99-78, 7 pp.

Hirai, M.F., Katayama, A. and Kubota, H. (1996) Effect of compost maturity on plant growth. *BioCycle* 24(6), 54–56.

Horowitz, M. (1993) Soil cover for weed management. In: *Communications of the 4th International Conference I.F.O.A.M. Non Chemical Weed Control*, Dijon, France, pp. 149–154.

Jouët, J.-P. (2001) Plastics in the world. *Plasticulture* 120, 108–119.

Kamara, A.Y., Akobundu, I.O., Chikoye, D. and Jutzi, S.C. (2000) Selective control of weeds in an arable crop by mulches from some multipurpose trees in Southwest Nigeria. *Agroforestry Systems* 50, 17–26.

Kasperbauer, M.J., Loughrin, J.H. and Wang, S.Y. (2001) Light reflected from red mulch to ripening strawberries affects aroma, sugar and organic acid concentrations. *Photochemistry and Photobiology* 74, 103–107.

Keeling, A.A., Paton, I.K. and Mullet, J.A. (1994) Germination and growth of plants in media containing unstable refuse-derived compost. *Soil Biology and Biochemistry* 26(6), 767–772.

Lennartsson, M.E.K. (1990) The use of surface mulches to clear grass pasture and control weeds in organic horticultural systems. In: *Organic and Low Input Agriculture*. BCPC Monograph No. 45, BCPC, Farnham, UK, pp. 187–192.

Lieten, P. (1991) Multi-coloured crop production. *Grower* 116, 9–10.

Ligneau, L.A.M. and Watt, T.A. (1995) The effects of domestic compost upon the germination and emergence of barley and six arable weeds. *Annals of Applied Biology* 126, 153–162.

Lynch, J.M. (1987) Production and phytotoxicity of acetic acid in anaerobic soils containing plant residues. *Soil Biology and Biochemistry* 10, 131–135.

Masiunas, J., Wahle, E., Barmore, L. and Morgan, A. (2003) A foam mulching system to control weeds in tomatoes and sweet basil. *HortTechnology*, 13, 324–327.

Mathers, H.M. (2003) Novel methods of weed control in containers. *HortTechnology* 13, 28–34.

Maureen, A., Ramirez, E. and Garraway, J.L. (1982) Plant growth inhibitory activity of extracts of raw and treated pig slurry. *Journal of the Science of Food and Agriculture* 33, 1189–1196.

Merwin, I.A., Rosenberger, D.A ., Engle, C.A., Rist, D.L. and Fargione, M. (1995) Comparing mulches, herbicides and cultivation as orchard groundcover management systems. *HortTechnology* 5, 151–158.

Miles, C., Garth, L., Sonde, M. and Nicholson, M. (2005) Searching for alternatives to plastic mulch [see http://agsyst.wsu.edu].

Morgan, D.P., Liebman, J.A., Epstein, L. and Jimenez, M.J. (1991) Solarizing soil planted with cherry tomatoes vs solarizing fallow ground for control of verticillium wilt. *Plant Disease* 75, 148–151.

Munn, D.A. (1992) Comparisons of shredded newspaper and wheat straw as crop mulches. *HortTechnology* 2, 361–366.

NCAT [National Centre for Appropriate Technology] (2004) *Blueberries: Organic Production*. Horticultural Production Guide [see http://attra.ncat.org/attra-pub/PDF/blueberry.pdf].

Ngouajio, M. and Ernest, J. (2005) Changes in the physical, optical and thermal properties of polyethylene mulches during double cropping. *HortScience* 40, 94–97.

Obiefuna, J.C. (1986) The effect of sawdust mulch and increasing levels of nitrogen on the weed growth and yield of false horn plantains (*Musa* ABB). *Biological Agriculture and Horticulture* 3, 353–359.

Orzolek, M.D. and Otjen, L. (2005) Is there a difference in red mulch? [see http://plasticulture.cas.psu.edu/RedMulch.htm].

Ozores-Hampton, M. (1998) Compost as an alternative weed control method: municipal waste compost production and utilization for horticultural crops. *HortScience* 33, 938–940.

Ozores-Hampton, M., Obreza, T.A. and Hochmuth, G. (1998) Using composted wastes on Florida vegetable crops. *HortTechnology* 8, 130–137.

Ozores-Hampton, M., Obreza, T.A. and Stoffella, P.J. (2001) Mulching with composted MSW for biological control of weeds in vegetable crops. *Compost Science and Utilization* 9, 352–360.

Pickering, J. (2003) Laying it on thick. *The Garden* 128, 266–269.

Pinamonti, F. (1998) Compost mulch effects on soil fertility, nutritional status and performance of grapevine. *Nutrient Cycling in Agroecosystems* 51, 239–248.

Pinamonti, F. and Sicher, L. (2001) Compost utilization in fruit production systems. In: Stoffella, P.J. and Kahn, B.A. (eds) *Compost Utilization in Horticultural Cropping Systems*. Lewis Publishers, Boca Raton, FL, USA, pp. 177–200.

Raloff, J. (2001) Toxic runoff from plastic mulch. *Science News* 159(16), 16.

Reeleder, R.D., Capell, B.B., Roy, R.C., Grohs, R. and Zilkey, B. (2004) Suppressive effect of bark mulch on weeds and fungal diseases in ginseng (*Panax quinquefolius* L.). *Allelopathy Journal* 13, 211–231.

Roe, N.E. (2001) Compost effects on crop growth and yield in commercial vegetable cropping systems. In: Stoffella, P.J. and Kahn, B.A. (eds) *Compost Utilization in Horticultural Cropping Systems*. Lewis Publishers, Boca Raton, FL, USA, pp. 123–134.

Runham, S.R. (1998a) Clear edge for paper mulch. *Grower* 129, 21–22.

Runham, S.R. (1998b) Mulch it. *Organic Farming* 60, 15–17.

Runham, S.R. and Town, S.J. (1995) An economic assessment of mulches in field scale vegetable crops. In: *Brighton Crop Protection Conference – Weeds*, Brighton, UK, pp. 925–930.

Schmidt, M.H., Thewes, U., Thies, C. and Tscharntke, T. (2004) Aphid suppression by natural enemies in mulches cereals. *Entomologia Experimentalis et Applicata* 113, 87–93.

Shiralipour, A., McConnell, D.B. and Smith, W.H. (1991) Efffects of compost heat and phytotoxins on germination of certain Florida weed seeds. *Soil & Crop Science Society of Florida* 50, 154–157.

Shogren, R.L. and Rousseau, R.J. (2005) Field testing of paper/polymerized vegetable oil mulches for enhancing growth of eastern cottonwood trees for pulp. *Forest Ecology and Management* 208, 115–122.

Sikora, L.J. and Szmidt, R.A.K. (2001) Nitrogen sources, mineralization rates, and nitrogen nutrition benefits to plants from composts. In: Stoffella, P.J. and Kahn, B.A. (eds) *Compost Utilization in Horticultural Cropping Systems*. Lewis Publishers, Boca Raton, FL, USA, pp. 287–305.

Standifer, L.C., Wilson, P.W. and Porche-Rorbet, R. (1984) Effects of solarization on soil weed seed populations. *Weed Science* 32, 569–573.

Stewart, K.A., Jenni, S. and Martin, K.A. (1995) Accelerated test of paper mulch degradation. *HortScience* 30, 883.

Stout, G.J. (1985) 'Spray on' mulch demo. *American Vegetable Grower* (November) 82.

Sunderland, K. (2003) The relationship between weeds and invertebrates in horticulture. In: Grundy, A.C. (ed.) *The Impact of Herbicides on Weed Abundance and Biodiversity in Horticulture*. Department of Environment Food and Rural Affairs Final Report Project HH3403sx [see http://www2.warwick.ac.uk/fac/sci/hri2/research/weedecologyandmanagement/hh3403/].

Suojala, T. (2003) Yield potential of chive: effects of cultivar, plastic mulch and fertilisation. *Agricultural and Food Science Finland* 12, 95–105.

Swanton, C.J. and Booth, B.D. (2004) Management of weed seedbanks in the context of populations and communities. *Weed Technology* 18, 1496–1502

Tam, N.F.Y. and Tiquia, S. (1994) Assessing toxicity of spent pig litter using a seed germination technique. *Resources, Conservation and Recycling* 11, 261–274.

Teasdale, J.R. and Mohler, C.L. (2000) The quantitative relationship between weed emergence and the physical properties of mulches. *Weed Science* 48, 385–392.

Tiquia, S.M., Lloyd, J., Herms, D.A., Hoitink, H.A.J. and Michel, F.C. (2002) Effects of mulching and fertilization on soil nutrients, microbial activity and rhizosphere bacterial community structure determined by analysis of TRFLPs of PCR-amplified 16S rRNA genes. *Applied Soil Ecology* 21, 31–48.

Walker, K.L., Svoboda, K., Booth, E.J. and Walker, K.C. (2004) Coloured mulch technology as a weed control strategy for summer savory (*Satureja hortensis*). In: *Proceedings: Crop Protection in Northern Britain 2004*, pp. 81–86.

Weber, C.A. (2003) Biodegradable mulch films for weed suppression in the establishment year of matted-row strawberries. *HortTechnology* 13, 665–668.

Wilson, J. (1990) Black mulches go green. *Grower* 115, 12–15.

Wong, M.H. and Chu, L.M. (1985) Changes in properties of a fresh refuse compost in relation to root growth of *Brassica chinensis*. *Agricultural Wastes* 14, 115–125.

Yih, W.K. and Vandermeer, J.H. (1988) Plant mulches in a reconstructed chinampa in Tropical Mexico: effects on cowpea (*Vigna unguiculata*) and potential for weed control. *Biological Agriculture and Horticulture* 5, 365–374.

10 Thermal Weed Control

J. Ascard,[1] P.E. Hatcher,[2] B. Melander[3] and M.K. Upadhyaya[4]

[1]Swedish Board of Agriculture, Alnarp, Sweden; [2]School of Biological Sciences, The University of Reading, Reading, UK; [3]Aarhus University, Faculty of Agricultural Sciences, Department of Integrated Pest Management, Research Centre Flakkebjerg, Slagelse, Denmark; [4]Faculty of Land and Food Systems, University of British Columbia, Vancouver, BC, Canada

10.1 Introduction

Because of increasing public concern for health and environment, several non-chemical weed control options causing thermal injury to plant tissues have been developed. These include use of fire, flaming, hot water, steam and freezing. Radiation within the microwave, infrared, ultraviolet and laser wavebands, and electrocution have also been investigated for weed control. Some of these methods have also been exploited to control harmful insects and some soil-borne pathogens.

Thermal heating methods are attractive because they provide rapid weed control without leaving chemical residues in the soil and water. Unlike cultivation, they do not bring buried weed seeds to the soil surface, generally leave dead plant biomass on the soil surface (which protects the soil from erosion), and may kill some insect pests and disinfect plant residues and surface soil. However, they are often costly and slow, and do not provide residual weed control. Although these methods do not leave chemical residues in the soil and water, many thermal approaches use large amounts of energy per unit area, mainly from fossil fuels, which causes air pollution.

This chapter describes various options that use energy to cause thermal injury to plants for weed control. Some of these options, e.g. flame weeding, are commercially viable, but some are impractical at present, and others need further research and development before they can be used under field conditions.

10.2 Effects of High Temperatures on Plants

Heat injury involves denaturation and aggregation of cellular proteins and protoplast expansion and rupture, which results in plant desiccation (Ellwanger et al., 1973a,b). Depending on the exposure time, protein denaturation may start at 45°C (Sutcliffe, 1977; Levitt, 1980).

The effects of heat treatments on plants are influenced by several factors including temperature, exposure time and energy input. Since many of these methods kill only shoots, the affected plants (especially perennial weeds) may regenerate and repeated treatments may be necessary.

Temperatures in the range of 55–95°C have been reported to be lethal for leaves and stems (Daniell et al., 1969; Porterfield et al., 1971; Hoffmann, 1989). Exposure to flame for 0.065–0.130 s is enough to kill leaf tissue (Thomas, 1964; Daniell et al., 1969).

Several studies indicate that higher temperatures are more effective at causing plant damage. Daniell et al. (1969) found that the structural changes in cells were more pronounced when cellular temperature changed more rapidly (as in a flaming) compared to when the changes were gradual (as in a hot-water treatment). In general, lethal temperature varies inversely with exposure time, and a negative exponential relationship between lethal temperature and exposure time has been reported (Sutcliffe, 1977; Levitt, 1980). It has also been shown that when higher temperatures were used for weed control, the temperature sum (heat units) required to kill plants was lower (Storeheier, 1994). This is explained by the higher rate of heat transfer to plants at higher temperatures. Moreover, when the gas velocity and the energy density of a flame are increased, the forced convection to the plants also increases (Bertram, 1994). In practice, this is used to increase the driving speed of a flamer by using more powerful burners that can maintain high temperatures in operation (Ascard, 1995b).

10.3 Use of Fire and Prescription Burning for Vegetation Management

Controlled or prescription burning has been practised in parts of Africa, Australasia and North America for a number of land management objectives. These include site preparation for natural regeneration and replanting, improving growth and yield of grasslands and forests, insect and fungal pest management, improved forest accessibility, establishment of fire breaks in timber plantations and reduction of high-intensity wildfire hazards. In some areas, controlled burning has been an important woodland or grassland management tool. Fire has also been used in some arable cropping systems, particularly sugarcane and temperate cereals. Here, crop straw and stubble are burnt soon after harvest for disease control, ground clearance, avoidance of soil acidification and reduction of tillage for seedbed preparation. This practice, which declined during the 1990s due to concerns about pollution, soil erosion, and to a lesser extent for effects on non-target species, has become illegal in some European countries.

Prescription burning may lead to increased productivity and flowering of surviving plants, increased seed germination and improved seedling establishment (Whelan, 1995). Thus, managed fire regimes may lead to weed encroachment; for example, in a survey of experimental burnings to reduce invasive plants,

D'Antonio (2000) lists only five studies that unequivocally succeeded, but 14 that failed. These failed experimental burns had no effect, increased the abundance of the target species, or controlled some of the target species but allowed other invasive species to increase. Thus, prescription burning is used mainly for reasons other than weed control, and may even enhance weed problems.

Effects of fire on soil and soil organisms

The effect of fire on soil is a function of severity of the heat treatment, which is influenced by factors including the amount and type of fuel used, exposure duration, soil moisture and soil characteristics. Soil is a very good insulator and can absorb a great deal of heat with little increase in temperature (Whelan, 1995). For example, while high-intensity fires may reach 500–700°C on the soil surface, soil temperatures at 5 cm depth rarely exceed 150°C and no heating occurs at 20–30 cm depth (Certini, 2005). Low-intensity burning, as in the case of controlled burning, can increase available plant nutrients in the soil, whereas severe fires could cause adverse changes in the soil, including increased volatilization of nutrients, altered mineralization rates and C:N ratios, and nutrient loss through accelerated erosion, leaching and denitrification. Alteration of soil physical properties (e.g. structure, porosity and water-holding capacity) may change soil hydrology. Soils subjected to high-intensity fires may develop a discrete layer of soil aggregates coated with hydrophobic organic compounds, which may prevent soil wetting. Following a fire, water and wind erosions, landslides and soil creep may also increase (Neary et al., 1999; Certini, 2005; Shakesby and Doerr, 2006).

A considerable amount of work has been done on the effects of managed and unmanaged fires on insect and soil microbial populations. Many insect groups have been shown to decline immediately after a fire, with the extent of decline depending on the severity of the fire and insect mobility. The longer-term (up to a year) effects of burning on insects are quite varied. Some insect populations are reduced, while others (e.g. Coleoptera) become more abundant (Warren et al., 1987; Swengel, 2001).

The impact of fire on soil microbes is also variable. High-intensity fires have been shown to influence soil-inhabiting microbes. A decline in heterotrophic microbes and frequently an increase in autotrophic microbes, especially those involved in nitrogen cycling, have been reported. This decline can lead to a loss of N from the system by denitrification or leaching (Neary *et al.*, 1999; Hart *et al.*, 2005). Compared with fungi, bacteria are more tolerant to fire and recolonize a burned area faster (Guerrero *et al.*, 2005). The physical and chemical changes in soil due to fire and their effects on the plant communities could influence soil microbe and invertebrate communities (Neary *et al.*, 1999; Certini, 2005; Hart *et al.*, 2005).

10.4 Flame Weeding

Flame weeding is by far the most widely used thermal weed control method. Flaming heats plant tissues rapidly to rupture cells but not to burn them. It is currently widely used for weed control in organic farming in western Europe. Usually flaming is applied as a single application for non-selective weed control prior to crop emergence in carrots and other slow-emerging row crops (Dierauer and Stoppler-Zimmer, 1994; Rasmussen and Ascard, 1995). Flaming before crop emergence followed by post-emergence mechanical inter- and intra-row weeding has been particularly useful (Melander and Rasmussen, 2001). Selective post-emergence flaming is used, albeit less frequently, in crop rows in some heat-tolerant crops, e.g. maize and onions (Ascard, 1989), and as an emergency treatment between rows when the soil is too wet for mechanical cultivation. Occasionally, it is also used to control weeds on hard surfaces in urban areas and for desiccation of potato haulms prior to harvest (Lagüe *et al.*, 2001).

Flaming has also been investigated for insect control in crops. Thermal control using propane burners has been shown to cause 92% mortality of Colorado potato beetle (*Leptinotarsa decemlineata*) in commercial potato fields, and is efficient at top-killing without any loss of harvest quality (Duchesne *et al.*, 2001). The integration of flaming with other methods is further described by Hatcher and Melander (2003).

History and development of flame weeding

The large-scale agricultural application of flaming began in the early 1940s for selective weed control in cotton in the USA. Liquid fuels such as kerosene and oils were used initially but were gradually replaced by LPG (liquefied petroleum gas, mainly propane and butane). Between 1940 and the mid-1960s, flame weeding was widely used in the USA in e.g. cotton, maize, soybeans, beans, lucerne, potatoes, onions, grapes, blueberries and strawberries (Kepner *et al.*, 1978; Lagüe *et al.*, 2001). Open burners without covers were used for selective post-emergence flaming in crop rows (Fig. 10.1). Flaming for non-selective weeding prior to crop

Fig. 10.1. Selective flame weeding in maize, with one burner on each side of the row. The flames hit the small weed seedlings in the crop row and the lower, heat-tolerant part of the crop plant.

emergence was not common at that time. Covered flamers were developed in the 1960s in the USA for non-selective weed control between crop rows, for insect control in lucerne, and for potato haulm desiccation. Other applications included thermal defoliation of cotton plants, disease and insect control in various crops, and weed control on drainage ditch banks. With the greater availability of effective herbicides and increases in petroleum costs, flaming had become unpopular by the late 1970s (Kepner *et al.*, 1978). Because of increasing environmental concerns about herbicides, there is now a renewed interest for flaming in the USA and Canada (Laguë *et al.*, 2001; Leroux *et al.*, 2001).

Unlike the USA, except for some scattered reports on the use of flaming for weed control in nurseries and vineyards, flaming was not widely used in Europe in the 1960s. Its use prior to crop emergence in sugarbeet, and pre- and post-emergence weed control as well as pre-harvest haulm desiccation in potatoes, was investigated by the petroleum industry in various countries in western Europe. Organic farmers started using flame weeding in Germany and Switzerland in the early 1970s (Hoffmann, 1989).

Effect of flaming on weeds

Flaming kills plants mainly by rupturing cells, which leads to tissue desiccation. Young seedlings are more sensitive to high temperatures; with stem or hypocotyl close to the soil surface being the most vulnerable (Sutcliffe, 1977). Shoot apices of young plants are more susceptible to heat damage. In some species, the location of sensitive plant parts to heat varies with seedling development. In older plants, the shoot apex may be protected by leaves. Regrowth of old plants following flaming may be reduced or eliminated when flames penetrate the canopy enough to kill axillary buds at lower nodes, which may be protected by surrounding leaves, leaf sheaths and petioles. Moderate flaming may only partially damage plants, and their ability to regrow depends on their energy reserves, environmental conditions such as soil moisture, and competition from neighbouring plants.

S-shaped logistic models, of the same type as those used in herbicide bioassays, can be used to describe plant response to flaming treatments in terms of fuel input in kg/ha. In this context, LD_{95} is used to define the lethal dose needed to reduce the plant numbers by 95%, whereas ED_{95} represents the effective dose needed to reduce the plant weight by 95%. Plant size at the time of treatment has a greater influence on the rate of flaming required, than plant density (Ascard, 1994). When a mix of sensitive weed species in the 1–4 leaf stage were treated, propane doses of approximately 40 kg/ha (1840 MJ/ha) were needed to reduce weed numbers by 95%. Plants with 6–12 leaves required twofold to fourfold higher rates for control than those at the 1–4 leaf stage (Ascard, 1995a) (Fig. 10.2). The treatment dose must therefore be adjusted considering the weed species present and their growth stage. The LPG rates given here are calculated for broadcast application. In practice, banded flame weeding in the crop rows is often used and the rates per hectare are then considerably reduced.

The susceptibility of plants to flaming depends on their ability to avoid heating and their heat tolerance. The extent to which flame heat penetrates crop and weed stands, and therefore the efficacy of flame weeding, depends on flaming technique, soil structure and the presence of moisture on the leaf surface. Tolerance to

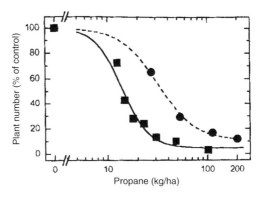

Fig. 10.2. The effect of flame weeding on total annual weed number at early treatment (1–4 leaves) (■——■) and late treatment (6–12 leaves) (●– –●). Propane rates of 40 kg/ha killed 90% of the small weeds, whereas rates of 200 kg/ha killed less than 90% of the larger weeds. From Ascard (1995a).

heat injury also depends on the protection offered by layers of hair and wax, lignification, external and internal water status of the plant, and the species regrowth potential (Ascard, 1995a; Laguë *et al.*, 2001; Leroux *et al.*, 2001).

Weed species can be divided into four groups on the basis of their susceptibility to flaming (Ascard, 1995a):

1. The first group consists of species with unprotected growing points and thin leaves (e.g. *Chenopodium album*, *Stellaria media* and *Urtica urens*). These species can be killed at early seedling stages (0–4 true leaves) using 20–50 kg propane/ha, depending on the species and treatment conditions. At later stages (4–12 leaves), higher rates, 50–200 kg propane/ha, are needed to kill these plants.
2. The second group, moderately sensitive weeds, contains species with relatively heat-tolerant leaves or protected growing points (e.g. *Polygonum aviculare*, *Polygonum persicaria* and *Senecio vulgaris*), which can be completely killed with a single flame treatment both at early and late developmental stages. Compared with the first group, control of these species requires a higher dose of fuel.
3. The third group consists of weeds with more protected growing points (e.g. *Capsella bursa-pastoris* and *Chamomilla suaveolens*), which allow the weed to regrow after one flame application. While these weeds can be completely killed at early growth stages (<4 true leaf stage), repeated treatments are needed at later stages due to their ability to regrow.

4. The weeds of the fourth group are very tolerant to flaming because of their creeping growth habit and protected growing points (e.g. *Poa annua* and several other grasses). Perennial weeds with large underground parts also belong to this very tolerant group. Following a complete shoot kill, they regrow from their below-ground meristems. Repeated flamings are needed to control these weeds (Ascard, 1995a).

Selective intra-row flame weeding

Selective post-emergence intra-row flaming can be done on relatively small weed seedlings in taller and heat-tolerant crops such as cotton, soybean, maize, brassica crops and onions (Fig. 10.3). The flames hit the lower part of the stem of the crop plant, but it can withstand this treatment. The success of the selective treatment depends on the smoothness of the soil surface and on directing the flame towards the weeds, avoiding damage to the crop (Ascard, 1989; Laguë *et al.*, 2001).

Flaming technology

Commercial flame weeders use LPG (propane-butane mixture) as fuel. While propane flames generate temperatures up to 1900°C, air temperatures measured by thermocouples are

Fig. 10.3. Burner settings in selective flame weeding in onions. (Photo by Johan Ascard.)

usually considerably lower (1200–1350°C); and even lower temperatures have been recorded when the flame weeder is operating under field conditions (Ascard, 1995b, 1998).

Several types of burners have been used for flaming (Kepner *et al.*, 1978; Hoffmann, 1989). They are commonly grouped according to the shape of the burner and the flame (flat or tubular) and the presence of a vapour chamber (liquid- or gas-phase burner). Both covered (Fig. 10.4) and open burners have been used for flame weed control. Flamers with an insulated cover over the burners are often used for non-selective weed control prior to crop emergence (Fig. 10.5), for weed control in urban areas, and for haulm destruction in potatoes.

While most covered flamers are intended for broadcast weed control, covered burners for band flaming over the crop rows are also available. Open burners (no cover) are used in small hand-held flamers and for selective flaming in the crop row.

Burners must be set at an appropriate angle and height for optimum weed control. If set too close to the ground, the flame deflects upwards, especially when the burner angle is steep and the fuel pressure high. If set too high, the hottest part of the flame does not reach the weeds. For a standard open burner, a 45° or 67° angle to the horizontal directed backwards with a height of 10–12 cm yields good results with flaming prior to crop emergence (Hoffmann, 1989; Ascard, 1995b).

The fuel consumption, and thereby the optimum travel speed of a flamer, can be increased by increasing the number of burners per working

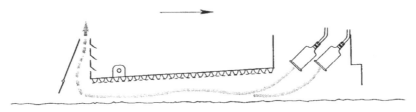

Fig. 10.4. Covered flame weeder with burners directed backwards underneath an insulated cover. This type of flame weeder works mainly by heat convection from the flames and to some extent indirectly by infrared radiation from the insulated cover, heated by the flames.

Fig. 10.5. Non-selective flame weeding pre-emergence of the sugarbeet crop using a Swedish flamer with one burner per row and one common insulated cover. (Photo by Johan Ascard.)

width, raising fuel pressure and using larger nozzles. The relation between fuel consumption, operating pressure and travel speed are further discussed by Ascard (1995b) and Laguë *et al.* (2001).

Improving the design of flamers

European research has improved the design of flame weeders in terms of heat transfer, energy efficiency and operating speed (Bertram, 1992; Storeheier, 1994; Ascard, 1995b). Shielding of burners has made retention of heat close to the ground for longer periods possible.

Thermodynamic modelling by Bertram (1992) suggests that with a standard open flamer using 50 kg propane/ha, only about 15% of the heat was actually transferred to the plants. A standard covered flamer improved heat transfer to plants to 30%, and a burner with an improved design of the cover to 60%. The performance of various open and covered flamers differs greatly depending on burner type, burner position and the burner cover design (Bertram, 1992; Storeheier, 1994).

For these reasons, flamers with covered burners are generally more effective than open flamers, especially on larger plants and tolerant species. Ascard (1995b) showed that, on average, an open flamer required 40% more fuel than a covered flamer to achieve good weed control.

Although covered flamers offer several advantages, inappropriately designed covers may cause oxygen deficiency and influence propane combustion efficiency (Laguë *et al.*, 2001). Therefore the shape, height and angle of the cover must be carefully evaluated and related to burner type and positioning (Storeheier, 1994).

Increasing the operating speed

The travel speed of flaming equipment determines the duration of exposure of weeds to heat. Exposure to heat can be increased by lowering the travel speed or by increasing the fuel input. An increase in fuel input may allow a flamer to be operated at higher speed (Ascard, 1995b). For example, a covered flamer, with a propane consumption of 34 kg/h (430 kW) per metre

working width, achieved 95% weed control with an effective ground speed of 7.9 km/h, whereas for a flamer with a more typical burner power of 12 kg/h (154 kW) per metre working width, the effective ground speed was 2.6 km/h. This increase in capacity was accomplished without increasing the propane consumption per hectare (Ascard, 1995b). Heat losses, however, may be greater at higher temperatures and energy inputs (Bertram, 1994). Therefore, it is important to adjust and optimize the design of the flamer when the burners are modified.

Advantages and disadvantages

Flaming is an attractive weed control option because it leaves no chemical residue in the crop, soil and water, it can control herbicide-tolerant or resistant weeds, and it can be used in crops where few or no herbicides are registered. There are also restrictions for herbicide use in several groundwater areas, which may increase the interest in flaming and other non-chemical weed control methods. In organic farming, flame weeding is a profitable method in several vegetable crops to reduce the need for hand-weeding.

In comparison with cultivation, flame weeding can be carried out on wet soils, does not bring buried weed seeds to the soil surface, and may kill some insect pests and plant pathogens. Flame weeding usually provides better weed control than cultivation before crop emergence in small-seeded crops.

The disadvantages of flame weeding include the high cost of labour, fuel and equipment compared with herbicide application, low selectivity, and lack of residual weed control, making repeated flaming treatments necessary. Flame weeders may have about the same capacity as mechanical weed control but are usually slower than chemical weed control. The working environment, involving gas and flames, can be uncomfortable for some operators. From a resource and environmental point of view, the high energy requirement and the release of carbon emissions could be seen as disadvantages. However, compared with other fossil fuels, propane combustion is relatively clean.

In North America, the cost of propane for two flamings is less than the cost of herbicides in maize, but flame weeding requires more time for application than herbicides do (Leroux, *et al.*, 2001). The total cost of flaming is often greater than that of chemical weed control, mainly due to high machinery costs and slow speed of field flaming (Nemming, 1994). When flaming is used in vegetable production, such as carrots, labour costs for supplementary hand-weeding might make up a large part of the total weed control costs as opposed to a strategy based on herbicides, where little or no hand-weeding is required. However, in heat-tolerant crops such as maize, selective flame weeding is possible, which eliminates the need for hand-weeding.

Effects of flaming on soil organisms

Studies on effects of flame weeding on benefi-cial insects and fungi are scarce. Flame weeding could be detrimental to some airborne as well as soil-surface-inhabiting organisms. However, soil is a very good insulator and can absorb a significant amount of heat with little increase in temperature (Reeder, 1971; Whelan, 1995). Furthermore, because in flame weeding the thermal treatment is brief, only the uppermost few millimetres of the soil are heated. For example, Rahkonen *et al.* (1999) found that extreme LPG flame weeding raised soil temperature by 4°C at 5 mm depth, and only 1.2°C at 10 mm depth. Therefore, a significant damage to the soil microflora or fauna is not expected during a normal flame weed control operation. Rahkonen *et al.* (1999) found that although flame weeding using 100 kg/ha LPG led to a 19% reduction in soil microbial biomass at 0–5 mm depth, it had little effect at

5–10 mm depth. In an attempt to control insect pests in the soil, Reeder (1971), using a fourfold greater flame intensity (370 kg/ha) and fivefold longer flame duration (5 s) than normally used in flame weeding, achieved a 11°C temperature rise at 6 mm depth and 4°C at 12 mm depth. The insulating properties of the soil are such that Reeder (1971) only killed a total of 75% pink bollworm (*Pectinophora gossypiella*) larvae in the soil even with an extremely high dose of 8800 kg/ha LPG and 5 min flame duration.

In a study of the effect of flame weeding on carabid beetles, a beneficial insect group, Dierauer and Pfiffner (1993) found no effect of flaming. In nine trials, flame-weeded plots had a slightly higher carabid activity than the control. However, in a maize experiment, cara-bid activity was lower in flame-weeded plots than in the control – this was the only trial in which flaming was carried out on a ground substantially covered with weeds. Since cara-bids have a higher activity under ground cover, compared with bare soil, the beetles might have been at a greater risk there from the flaming.

10.5 Infrared Radiation

Infrared (IR) radiation, produced by heating ceramic or metal surfaces, is used to induce ther-mal injury to weed tissues. IR radiators, driven by LPG, operate at red brightness temperatures of about 900°C with essentially no visible flame on the combustion surface (Fig. 10.6). This type of true IR-radiator should not be confused with flame weeders using ordinary gas flame burners covered by an insulated shield, sometimes marketed as 'infrared' weeders. This type of

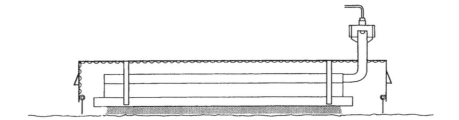

Fig. 10.6. True infrared radiant gas burner with no visible flame on the combustion surface. This type of IR-burner heats ceramic or metal surfaces which then radiate heat towards the target.

flame weeder is no different from the other types of covered flamers, which also work mainly by heat convection from the flames and to some extent indirectly by infrared radiation from the insulated cover, heated by the flames (Ascard, 1995b).

IR radiation was tested for thermal defoliation of cotton in the USA during the 1960s (Reifschneider and Nunn, 1965). It was used for weed control to some extent in Europe during the 1980s, but was replaced by flame weeders using covered flame burners. The use of IR radiators for weed control has not become popular because of their high cost compared to flame weeders. Furthermore, IR radiators produce lower temperatures and therefore transfer less heat to plants compared with most flame burners (Parish, 1989).

Ascard (1998) found that both an IR radiator and a covered flame weeder using 60 kg/ha of propane and travel speeds of ~1 km/h yielded a 95% reduction in *Sinapis alba* (white mustard) seedlings. Parish (1989) found that while *Sinapis alba* control with electrical infrared emitters and propane flamers at similar energy inputs was similar, the energy requirement to control *Lolium italicum* (Italian ryegrass) was considerably higher for IR radiators than for flame weeders. Thus, the relative performance of IR and flame weeders may differ between species.

10.6 Hot Water

Technologies to control weeds by application of hot water or steam have been developed. Unlike non-specific burning and flaming, they pose little danger of starting uncontrolled fires. The leaves of the treated plants change colour within a few minutes and the shoots desiccate in a couple of days. Many of the affected weeds may regenerate since the roots are not sufficiently damaged, making repeated applications necessary. Hot water may also be a useful option for weed control in orchards (Kurfess and Kleisinger, 2000).

Weed control can be achieved by applying hot water either as a foliar spray, on the soil surface and/or by injection into the soil followed immediately by cultivation. In the case of soil application, soil type, depth to be treated

and soil temperature are important in determining the amount of heat necessary to achieve the desired results.

In a study of *Sinapis alba* control using hot water as a foliar spray, the energy used to obtain a 90% reduction in plant number at the two-leaf stage was 3970 MJ/ha (corresponding to 110 kg/ha of diesel fuel) and a water use of approximately 10,000 l/ha (Hansson and Ascard, 2002). The developmental stage of the weed at the time of treatment significantly influenced the dose–response relationship and the energy-use efficiency (Fig. 10.7). For example, the energy requirement for 90% reduction following hot water application at the two-leaf stage was only one-third of that at the six-leaf stage. Similarly, travel speeds needed for 90% fresh weight reduction at the two- and six-leaf stages were 1.26 and 0.44 km/h, respectively.

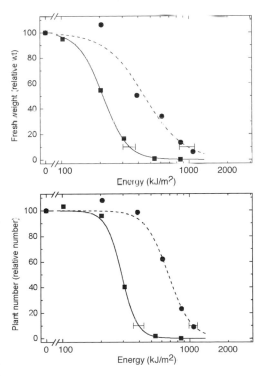

Fig. 10.7. Effect of hot water on *Sinapis alba* fresh weight (top) and plant number (bottom) in relation to the untreated controls at two stages of development: 6-leaf stage (●---●) and 2-leaf stage (■—■). Horizontal bars represent 95% confidence interval at ED_{90}. From Hansson and Ascard (2002).

Thus, application of hot water at an earlier stage of weed development, or when weed infestations are low, may reduce energy use and increase travel speed, thereby lowering the cost of weed control.

A biodegradable foam formulated from corn and coconut sugar extracts has been used to reduce heat dissipation during hot water application (Fig. 10.8) (Quarles, 2001). The addition of a surfactant to the water has been shown to allow higher travel speeds (up to 8 km/h) of hot water application equipment for effective weed control (Kurfess and Kleisinger, 2000). By using a long and narrow hot water applicator, the effective travel speed could be increased to 8 km/h (Hansson and Ascard, 2002).

Use of hot water, however, requires large amounts of water and energy, which is costly. Furthermore, the large amounts of water needed may not be available in drier regions. Transporting large volumes of water is also inconvenient and expensive. Hot water that does not come into contact with the weed foliage may represent energy losses. This option therefore may not be practical on a very large scale. It may, however, be a method of choice for small, environmentally sensitive areas, for spot treatments, for weed control around poles, near fences, in cracks in concrete and asphalt, and on gravel.

The use of hot water may also kill some desirable soil-borne microorganisms and insects. On the positive side, it does not add any harmful chemicals to the environment and may control some soil-borne pathogens and nematodes. Compared with flaming it has better canopy penetration and does not pose a fire hazard (Hansson and Ascard, 2002).

10.7 Steam

While steam has been extensively used for soil disinfection for over a century (Sonneveld, 1979; Runia, 1983) and its potential for weed control was demonstrated in the early 1990s (Upadhyaya et al., 1993), application of this option for weed control has not been fully realized. However, it has been used for weed control on rights-of-way (Fig. 10.9) in forestry (Norberg et al., 1997) and in cropping systems (Kolberg and Wiles, 2002).

The use of steam for weed control offers some advantages over hot water in that it uses less water and may provide better leaf canopy penetration. Steam can be superheated to very high temperatures to increase its effectiveness and shorten the required exposure time. In a study using superheated steam (400°C), Upadhyaya et al. (1993) found that 2–4 s

Fig. 10.8. A weed control system using hot water and a biodegradable foam formulated from corn and coconut sugar extracts. Photo from Quarles (2001).

Fig. 10.9. Weed control following steam treatment on railroad right-of-way. (Photo by D. Polster.)

exposures to steam could kill seedlings of a variety of weeds. The extent of injury depended on weed species, steam temperature, duration of exposure and plant size. Exposures as short as 0.1–0.2 s damaged foliar epicuticular wax and cell membrane integrity. Weeds, particularly perennial species, regenerated, making repeated exposures necessary. Seed production of *Chenopodium album* plants that survived was significantly reduced. Short exposure to superheated steam also killed weed seeds, with imbibed seeds being generally more susceptible. Seed coat and other coverings were found to offer protection from steam exposure in some species.

Mobile soil steaming

Soil steaming has the potential for reducing laborious intra-row hand-weeding in row crop systems where herbicides are not used. Mobile soil steaming is commercially used on raised beds, especially in short-term field salad crops with a strong need to control soil-borne pathogens (Pinel *et al.*, 1999). Steam is applied to the whole bed area and down 50–100 mm in the soil, depending on the steaming time. Steaming causes high mortality of weed seeds,

which could lead to effective and long-term weed control.

It should be noted that the current soil steaming technology has two major disadvantages. The consumption of fossil energy is extremely high, with diesel fuel use ranging from 3500 to 5000 l/ha, and secondly, it is time-consuming, requiring 70–100 h to treat 1 ha. This has led to the idea of band-steaming, where only a limited soil volume is steamed, enough to control weed seedlings that would otherwise emerge in the rows. The width of the treated intra-row band depends on how close to the crop plants inter-row cultivation is carried out. Inter-row hoeing in single-line-sown onion, leek and carrot can leave only a 50 mm wide untilled strip in the row, with no negative impact on crop growth and yield (Melander and Rasmussen, 2001). Steaming down to a moderate soil depth of 50–60 mm appears to be sufficient, considering that most weed seeds in the seedbank are small and will predominantly emerge from the top 20 mm of the soil profile.

Laboratory experiments have shown that the rise in soil temperature following steaming strongly affects subsequent weed seedling emergence (Fig. 10.10). Seedling emergence from natural seedbanks was reduced by 99% when the maximum soil temperature reached

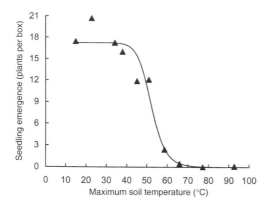

Fig. 10.10. The relationship between maximum soil temperature (°C) obtained by steaming the soil at different time intervals and weed seedling emergence from steamed soil. From Melander and Jørgensen (2005).

70°C (Melander and Jørgensen, 2005). In field trials using a Danish prototype band steamer, a 80–90°C soil temperature yielded the same level of weed control under field conditions as 70°C in the laboratory. To reach 80–90°C, the band steamer used 350–400 l/ha of diesel fuel when steaming a band width of 100 mm and down to 50 mm soil depth (M.H. Jørgensen and B. Melander, unpublished results).

In a laboratory study where a crop was seeded immediately after steaming had ended, and while the soil temperature was still close to its maximum, sugarbeet, maize, leek, onion and sometimes carrot seeds were surprisingly tolerant to the heat (M.H. Jørgensen and B. Melander, unpublished results). As shown by Melander and Jørgensen (2005), the soil temperature falls very slowly after steaming has ended.

Band steaming has been used commercially and evaluated in on-farm experiments on sandy soils in Sweden. A nine-row band steamer with a 700 kW diesel-driven steam generator treated 105 mm wide bands, 50 mm deep, before sowing sugarbeet and parsnips (Fig. 10.11). The travel speed was about 0.25 km/h, resulting in a treatment time of 8 h/ha, and a water consumption of 8000 l/ha. Field experiments showed that a maximum temperature of 86°C in the middle of the steamed band at 40 mm depth was needed to give a 90% reduction in annual weed numbers, with an energy use of 570 l/ha of diesel fuel. At this rate, the labour requirement for hand-weeding was 49 h/ha compared with 132 h/ha in the non-steamed plots. At a higher energy use of 650 l/ha of diesel fuel, the hand-weeding time was further reduced to 32 h/ha (Hansson and Svensson, 2007).

Effects of steaming on soil organisms

A major concern about steaming is its lethal effect on soil organisms other than weed seeds, and a slow soil recovery from these effects. Soil bacteria responsible for the oxidation of ammonium-N were significantly inhibited by steam treatment

Fig. 10.11. Band steaming for weed control and partial soil sterilization before sowing of sugarbeets. This nine-row band steamer, built by a Swedish farmer, has a capacity of 1 ha per 8 h and requires 690 l/ha of diesel fuel to obtain a 95% reduction in weed emergence. (Photo by Johan Ascard.)

and the population had not recovered after 90 days. While fungal and enzyme activities were reduced significantly in response to steam treatment, soil water and nitrate contents, pH, water-soluble carbon and *in situ* respiration were not affected (Melander *et al.*, 2004). The impact of these effects on crop performance has not been investigated.

Improving hot water and steaming technology

When designing hot water and steam application equipment, consideration must be given to water and energy-use efficiency, the spectrum of weed control, canopy penetration, travel speed, the extent of damage to the weed plant (which determines its possible recovery and the need for repeat applications), effects of ambient temperature, moisture status and wind velocity, use of barriers to prevent loss of energy by dissipation, inclusion of site-specific application technology and operator safety. Effects of weed density, stage of development and application pressure on canopy penetration and energy-use efficiency should also be determined. The possibility of mixing hot air exhausts from boilers with the steam or hot water must also be explored. Further research is needed to determine the optimum superheating of steam in order to improve the weed control and energy-use efficiencies, and to lower the cost of the control operation.

10.8 Electrical Energy

The concept of using electrical energy for weed control (Diprose and Benson, 1984; Diprose *et al.*, 1984; Vigneault and Benoît, 2001; Vigneault, 2002) was developed in the late 1800s, with the first US patent registered in 1895 (Scheible, 1895). The control equipment consists of a generator, a transformer, one or more electrodes, and rolling coulters (Fig. 10.12) (Vigneault, 2002). Because of the plant's resistance to electrical current, electrical energy is converted to heat, volatilizes cellular water and other volatiles, and ruptures cells, causing plant death. Electric current travels through the root system and is dissipated into the soil. Plants with large below-ground parts are damaged to a lesser extent, and the root damage is greater in drier soils. At higher weed density, sheltering may reduce the contact of some plants with the electrode, which may reduce the effectiveness of this method. While heating is considered the main cause of plant damage, other effects of electrical current cannot be ruled out. The extent of plant damage depends on the voltage used, contact duration, plant species, age, morphology, electrical resistance, soil properties and soil moisture content. This approach has also been used to kill sugarbeet plants in the reproductive phase ('bolters') (Diprose *et al.*, 1985).

Two systems have been developed for weed control using electrical energy. One system uses one or several short-duration, high-voltage pulses with two or more electrodes around the

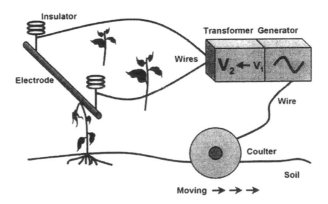

Fig. 10.12. Schematic diagram of a weed electrocution system showing thermal injury of a plant which is in contact with a horizontally placed electrode. From Vigneault (2002).

plant; and the other uses a continuous contact with electrode(s), with the current passing through the plant for the duration of the contact (Diprose and Benson, 1984).

Field equipment used in the past for electrocution typically had a current of 6–25 kV and a power of 50–200 kW, and used diesel fuel. In experiments with Lasco EDS equipment at North Dakota State University in the early 1980s, an average travel speed of 5 km/h was used with a diesel use of about 8 l/ha, out of which 4 l/ha was used to generate electricity and the rest to run the tractor (Kaufmann and Schaffner, 1982). The energy use of electrical methods is dependent on factors such as the type and size of weeds, and the soil moisture content. Vigneault and Benoît (2001) calculated that the energy use is very dependent on weed density, and the energy input for electrical weed control was between 418 and 16,500 MJ/ha for weed densities between 5 and 200 weeds/m^2, respectively. They concluded that electricity has advantages for controlling tall escaped weeds at low densities, but is not suitable as the primary method of weed control at densities of more than 200 weed-stems/m^2. Even at low weed densities of 15 plants/m^2, electrical weed control requires twice as much energy and takes five times longer than chemical control (Vigneault and Benoît, 2001).

While electrocution appears to be an interesting and attractive option, and may be useful on a small scale or in high-value crops (e.g. herbs), several factors limit its wide commercial use. These include high equipment cost, poor and inefficient control of small emerging weeds, and a concern for the operator's safety. Kaufmann and Schaffner (1982) stated that weed electrocution had to be used on 210 ha each year to obtain the same low cost per hectare as using chemical herbicides.

The available technology requires weeds to be taller than the crop in a mixed weed–crop population to allow selective electrode contact with the weeds. Also, a significant amount of energy is wasted when weeds of different heights are controlled (Diprose et al., 1984). The electrode has to be lowered to kill the smaller weeds, which may increase the duration of electrode contact with the taller plants. As a result, more energy is passed through the taller plants than is required to destroy them, resulting in energy wastage. The technology to assess the electrical resistance of individual plants and their lethal energy dose requirement, to monitor when a lethal energy dose has been provided, and to stop any additional flow of electric current to a plant, needs to be developed in order to reduce wastage.

10.9 Microwave Radiation

Microwaves are electromagnetic radiation in the 300 MHz to 300 GHz frequency range. The wavelength of microwave radiation ranges from 1 m to 1 mm. Microwaves are used mainly for telecommunications, but some frequencies are set aside for heating and other purposes (Pelletier and Colpitts, 2001). Absorption of microwaves causes water molecules within tissues to oscillate, thereby converting electromagnetic energy into heat. This dielectric heating has been exploited to kill weeds, seeds (Davis et al., 1971; Barker and Craker, 1991; Sartorato et al., 2006) and insects (Nelson, 1996; Pelletier and Colpitts, 2001).

Microwave action is based on energy absorption and internal heating of the target, as opposed to most other thermal methods based on heat transfer from outside the seed or plant. At least theoretically, microwaves could kill the seeds or pests with high efficiency if their humidity is high and soil humidity is low.

Exposure to microwaves has been demonstrated to kill weed seeds under laboratory conditions, with imbibed seeds being more susceptible than dry seeds. Barker and Craker (1991) mixed seeds with soil, exposed the mixture to microwave radiation and monitored seedling emergence. Treatments that raised and maintained soil temperature above 80°C for 30 s reduced seed germination. While the relationships between duration of microwave exposure, soil temperature, and seed kill suggests the potential for using this option to kill soil-borne weed seeds, the effectiveness of microwaves in killing soil-borne seeds under field conditions is limited. Microwave radiation does not penetrate the soil to substantial depths due to its rapid attenuation, particularly in moist soils. Furthermore, since dielectric heating is the mechanism by which soil-borne tissues are killed and selective heating of target tissues in soil is difficult to achieve, a substantial amount of energy is wasted in heating a large volume of soil in order to damage the intended targets. This also reduces the travel speed of the application equipment. Knowledge of relative specific heat, moisture content, and density of soil-borne targets and the surrounding soil are important to achieving selective heating of the targets intended to be damaged. Exposure to microwaves has also been shown to stimulate seed germination in some species (Nelson and Stetson, 1984).

Microwave application equipment to control weeds has been designed, but the factors discussed above and the concern for operator safety has limited its practical use. The energy use of microwave-based weed control in a field test ranged from 10,000 to 34,000 MJ/ha. Considering the low conversion efficiency from diesel fuel to microwave energy, these figures correspond to diesel fuel consumptions of between 1000 and 3400 kg/ha (Sartorato et al., 2006). While the use of this option under field conditions is far from a reality, it could be practical for killing weed seeds and other reproductive parts in small volumes of soil. The effects of this treatment on soil chemistry and biology need further investigation.

10.10 Ultraviolet Radiation

UV radiation is subdivided into three spectral bands: UV-A (320–400 nm), UV-B (280–320 nm) and UV-C (100–280 nm); where 100 nm corresponds to 3×10^{17} Hz. While UV-B radiation levels slightly above those found in solar radiation have been reported to influence weed and crop seedling growth and morphology, with species differing in their response (Furness and Upadhyaya, 2002; Furness et al., 2005), UV-C radiation is the most damaging to plants.

High levels of UV radiation (1–100 GJ/ha range) have been shown to control weeds (Fig. 10.13) (Andreasen et al., 1999). Weeds are damaged due to heating of the foliage following the absorption of UV radiation by plant tissues. Exposure to UV radiation (1–100 GJ/ha) under glasshouse conditions severely reduced fresh weight of *Capsella bursa-pastoris*, *Senecio vulgaris*, *Urtica urens* and *Poa annua* (Andreasen et al., 1999). About 10 GJ/ha was needed to obtain a 95% fresh weight reduction on a mixed weed flora, and even higher if the energy loss for converting fuel into UV radiation is considered.

The extent of UV-induced damage was influenced by weed species, stage of plant growth, and the height of the UV lamp above the canopy. Annual bluegrass buds protected by other tissue coverings escaped UV damage and the exposed plants produced new tillers.

This approach is being explored for its weed control potential under field conditions. Trial equipment for applying high doses of UV radiation has been developed (Fig. 10.14). However, before this option can be implemented, its effec-

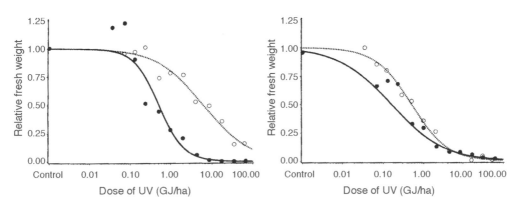

Fig. 10.13. UV radiation damage dose-response curves for *Poa annua*: left; 4 leaves (●) and 20 leaves (○); and *Capsella bursa-pastoris*: right; 12 leaves (●) and 16 leaves (○). From Andreasen et al. (1999).

Fig. 10.14. Experimental UV application equipment. Demonstration of the instantaneous effect on grass. Regrowth can be a problem (see http://www.kaj.dk/weed-by-uv.htm; Kaj Jensen's homepage).

tiveness under field conditions, energy require-ments, influence on soil properties, and any unwanted side-effects (e.g. mutation induction) must be investigated. Additionally, an economic feasibility and risk assessment (e.g. fire hazard) must be carried out.

10.11 Lasers

Lasers can be used to cut weed stems. Light absorption from CO_2 lasers by water molecules heats tissue contents and causes their explosive boiling (Langerholc, 1979). Heisel *et al.* (2001) showed that a far infrared regime laser (CO_2 laser, wavelength 10.6 μm) was more effective and energy-efficient than UV (355 nm) and IR (1064 nm) lasers. When stems were cut below the meristems, 0.9 and 2.3 J/mm of CO_2 laser energy was sufficient to cause 90% biomass

reduction in *Chenopodium album* and *Sinapis arvensis*, respectively. Regrowth appeared when stems of dicotyledonous plants were cut above meristems, showing the importance of cutting as close to the soil surface as possible to obtain significant effects. Heisel *et al.* (2002) showed that more energy was needed to cut thicker stems. To cut stems of *Solanum nigrum* seedlings with two true leaves, about 10 J/mm was needed to obtain a 95% reduction in the plant dry weight. However, with a realistic power conversion ratio of 5%, the authors concluded that approximately 150 J/mm is needed to cut the stems of *Solanum nigrum*. This corresponds to 2955 MJ/ha for a banded treatment of 10% of the surface of a sugarbeet field with 250 plants/m², and 29,550 MJ/ha to treat the whole surface.

10.12 Freezing

Exposure to low temperature has been exploited in order to control some aquatic weeds. Lowering of the water level in small ponds during the winter months in temperate regions kills weed shoots by exposing them to freezing temperatures (Gangstad, 1982). This approach has been used to control Eurasian water milfoil (*Myriophyllum spicatum* L.) and water lily (*Nymphaea* spp.). Alternate freezing and thawing following water drawdown could also contribute to freeze-killing of below-ground reproductive parts (e.g. rhizomes) of aquatic weeds.

Freezing treatments (liquid nitrogen, −196°C) and CO_2 snow (dry ice, −78°C) have been compared with flaming for weed control. Freezing affected the weeds in a similar manner to flame weeding, but freezing required 6000 and 12,000 MJ/ha, for liquid nitrogen and CO_2 snow, respectively. Freezing, therefore consumed about 3–6 times more energy to obtain the same level of weed control as flame weeding (Fergedal, 1993).

10.13 Energy Use and Environmental Impacts of Thermal Weed Control

Thermal weed control methods are often used as alternatives to chemical herbicides, and it is important that the environmental impacts of these methods are lower than those of the herbicides. An important advantage of using thermal weed control over herbicides is the reduction or the elimination of the risk of chemical residues in soil, water and in the crop. However, many thermal methods consume a relatively high amount of energy, obtained by burning a significant amount of fossil fuel, and their value in sustainable farming systems has been debated.

Energy use

While caution should be exercised in comparing the energy use efficiencies of thermal methods, by using data from different studies with different kinds of fuels and energy conversion factors, some rough comparisons can be made.

The thermal methods described in this chapter generally use more energy compared with mechanical and chemical weed control. For example, in potato production, a broadcast application of flaming required about ten times more energy than mechanical or chemical weed control (Fykse, 1985) or haulm killing (Jolliet, 1993, 1994). These data take into account the energy needed for herbicide synthesis and application, but not the energy required to eliminate herbicide residues from the water, soil and atmosphere.

For flame weeding, the energy use was based on a broadcast application using 50 kg/ha of propane, which corresponds to about 2300 MJ/ha, and the energy for transport and processing of propane, as well as the diesel used for application, which together will be about 2700 MJ/ha (Jolliet, 1993).

Weed control using IR radiation requires roughly similar amounts of energy as flame weeding, but the high equipment costs and low capacity of IR weeders have hindered their use (Ascard, 1998). Hot water treatment uses about twice as much energy (Hansson and Ascard, 2002) compared with flame weeding for equivalent weed control.

The energy input for electrical weed control can be higher or lower than for flame weeding, depending on the weed density (Vigneault and Benoît, 2001). However, equipment costs are high and, in addition, electricity is not suitable for weed control at high weed densities. Microwave weed control uses about 40 times more energy than flame weeding (Sartorato *et al.*, 2006). In addition, high machinery costs, low travel speed, unwanted side-effects, and concerns about operator safety have limited its practical use. UV radiation uses about four times as much energy (Andreasen *et al.*, 1999) as flame weeding for equivalent weed control, and even more if the energy efficiency is considered for converting fuel to UV radiation. The use of lasers to cut weed stems uses more than ten times, and freezing 3–6 times the energy compared with flame weeding for equivalent weed control (Fergedal, 1993; Heisel *et al.*, 2002).

All of the above energy figures are calculated for broadcast treatments, for the purposes of comparison, but in practice several of these methods can be justified as banded applications in order to reduce energy input and costs. For example, conventional broadcast soil steaming for deep soil sterilization uses about 50 times more fuel, but the new type of shallow band steaming (Melander and Jørgensen, 2005; Hansson and Svensson, 2007) uses 5–10 times more fuel than flame weeding, for equivalent weed control. However, soil steaming will result in a longer-lasting reduction of seedling emergence than is achieved by flaming.

Several of the energy requirements listed above were obtained by using prototype equipment, and improvements in their energy-use efficiencies are possible. However, these weed control methods also involve high machinery costs and relatively low travel speeds or working widths, which makes their use costly.

Flaming and other thermal methods cannot serve as the principal weed control options in most agricultural systems. In a crop rotation, flame weeding is used mainly in certain vegetable and row crops, where it is often used as a single application when mechanical methods are less effective and herbicides are not an option, e.g. in organic farming.

Although flaming requires approximately ten times more energy input than chemical or mechanical control, when considered at the farm gate, the energy inputs of weed control methods should be considered in relation to other farm inputs. The energy use for weed control is often a minor part of the total energy use in crop production. At the farm level, Pimentel (1992) found that the total energy used to produce a maize crop was 47,900 MJ/ha, and a major component of this, 13,400 MJ/ha, was nitrogen fertilizer.

Environmental impacts

The environmental impacts of thermal control (flaming) and chemical control in agriculture on soil, water, air and energy resources have been studied in Canada. The study showed that traffic-induced soil compaction and unwanted heating of the soil caused by the thermal treatments are not important. However, thermal control has greater negative impacts on the air than does chemical control. These impacts are directly related to the combustion by-products (CO, CO_2, nitrous and sulphur oxides), which are important pollutants related to global warming. These impacts are considered more important than those associated with volatiles and spray drift of pesticides. On the other hand, thermal control has no negative impacts on surface or underground water. However, the energy input into thermal weed control is usually much higher than that for chemical control, since thermal methods require great quantities of fossil fuels. In conclusion, thermal control methods have environmental benefits in terms of impacts on soil and water, but environmental costs in terms of air quality and energy use (Laguë *et al.*, 2001).

A similar study compared different combinations of chemical, mechanical and thermal methods for potato plant top-killing. It was suggested that mechanical top-killing had the least negative overall environmental impact on air and soil, chemical top-killing the most, with the thermal method being intermediate (Jolliet, 1993, 1994).

Many factors should be considered in the evaluation of weed control and farming methods. Pollution from chemical pesticides, and the connected health hazards and environmental costs, have to be weighed against other types of air pollution and the use of natural resources. The result of any environmental impact assessment or life cycle analysis, such as those mentioned above, will depend on the methods used for weighting and comparing different environmental impacts.

10.14 Conclusions

With increasing public concern regarding health and the environment, and increasing governmental and consumer pressure to regulate pesticides, many thermal weed control methods have been developed. These include the use of fire, flaming, infrared radiation, hot water, steam, electrical energy, microwave radiation, ultraviolet radiation, lasers and freezing temperatures. Of these, mainly flame weeding, and to some extent infrared radiation, steam and electrocution, have been used commercially. They are mainly used as an alternative to chemical herbicides, e.g. in organic farming, and when mechanical methods are not sufficient.

Thermal weed control options are attractive because they do not leave chemical residues in the crop, soil and water and can control herbicide-tolerant or resistant weeds, and provide rapid weed control. However, several thermal methods use much fossil energy and generally have high equipment costs, slow treatment speeds and do not give residual weed control. Some methods also involve risks of injury to the operator and of fire, which has hindered their application.

In comparison with cultivation, thermal methods can be carried out on wet soils, do not bring buried seeds to the soil surface, generally leave dead plant biomass on the soil surface, protecting the soil from erosion, and may kill some insect pests and pathogens. Thermal weed control usually provides better weed control than cultivation, pre-emergence to the crop.

The availability of inexpensive herbicides and their acceptability has hindered research on thermal weed control options. A thorough cost–benefit analysis in comparison with other options, technology development to lower the cost and raise energy efficiency, and their integration at the farm level is essential for the greater adoption of many of these options.

Thermal control methods have shown environmental benefits in terms of impacts on soil and water, but negative impacts on air quality and energy usage, compared with mechanical and chemical methods. More research is needed in order to develop effective and sustainable thermal methods for weed control.

10.15 References

Andreasen, C., Hansen, L. and Streibig, J.C. (1999) The effect of ultraviolet radiation on the fresh weight of some weeds and crops. *Weed Technology* 13, 554–560.

Ascard, J. (1989) Thermal weed control with flaming in onions. In: *30th Swedish Crop Protection Conference: Weeds and Weed Control. Vol. 2.* Swedish University of Agricultural Sciences, Uppsala, Sweden, pp. 35–50.

Ascard, J. (1994) Dose-response models for flame weeding in relation to plant size and density. *Weed Research* 34, 377–385.

Ascard, J. (1995a) Effects of flame weeding on weed species at different developmental stages. *Weed Research* 35, 397–411.

Ascard, J. (1995b) Thermal weed control by flaming: biological and technical aspects. Dissertation, Department of Agricultural Engineering, Swedish University of Agricultural Sciences, Alnarp, Sweden. Report 200.

Ascard, J. (1998) Comparison of flaming and infrared radiation techniques for thermal weed control. *Weed Research* 38, 69–76.

Barker, A.V. and Craker, L.E. (1991) Inhibition of weed seed germination by microwaves. *Agronomy Journal* 83, 302–305.

Bertram, A. (1992) Thermodynamische Grundlagen der Abflammtechnik. *Landtechnik* 7/8, 401–402.

Bertram, A. (1994) Wärmeübergang und Pflanzenschädigung bei der thermischen Unkrautbekämpfung [Heat transfer and plant damages in thermal weed control]. *Zeitschrift für Pflanzenkrankheiten und Pflanzenschutz* 14, 273–280.

Certini, G. (2005) Effects of fire on properties of forest soils: a review. *Oecologia* 143, 1–10.

Daniell, J.W., Chappell, W.E. and Couch, H.B. (1969) Effect of sublethal and lethal temperatures on plant cells. *Plant Physiology* 44, 1684–1689.

D'Antonio, C.M. (2000) Fire, plant invasions, and global changes. In: Mooney, H.A. and Hobbs, R.J. (eds) *Invasive Species in a Changing World.* Island Press, Washington, DC, USA, pp. 65–93.

Davis, F.S., Wayland, J.R. and Merkle, M.G. (1971) Ultra-high frequency electromagnetic field for weed control: phytotoxicity and selectivity. *Science* 173, 535–537.

Dierauer, H.U. and Pfiffner, L. (1993) Auswirkungen des Abflammens auf Laufkäfer [Effects of flame weeding on carabid beetles]. *Gesunde Pflanzen* 45, 226–229.

Dierauer, H.U. and Stöppler-Zimmer, H. (1994) *Unkrautregulierung ohne Chemie.* Ulmer, Stuttgart, Germany.

Diprose, M.F. and Benson, F.A. (1984) Electrical methods of killing plants. *Journal of Agricultural Engineering Research* 30, 197–209.

Diprose, M.F., Benson, F.A. and Willis, A.J. (1984) The effect of externally applied electrostatic fields, microwave radiation and electric current on plants and other organisms, with special reference to weed control. *Botanical Review* 50, 171–223.

Diprose, M.F., Fletcher, R., Longden, P.C. and Champion, M.J. (1985) Use of electricity to control bolters in sugar beet (*Beta vulgaris* L.): a comparison of the electrothermal with chemical and mechanical cutting methods. *Weed Research* 25, 53–60.

Duchesne, R.-M., Laguë, C., Khelifi, M. and Gill, J. (2001) Thermal control of Colorado potato beetle. In: Vincent, C., Panneton, B. and Fleurat-Lessard, F. (eds) *Physical Control Methods in Plant Protection.* Springer-Verlag, Berlin, Germany, pp. 61–73.

Ellwanger, T.C., Jr, Bingham, S.W. and Chappell, W.E. (1973a) Physiological effects of ultra-high temperatures on corn. *Weed Science* 21, 296–299.

Ellwanger, T.C., Jr, Bingham, S.W., Chappell, W.E. and Tolin, S.A. (1973b) Cytological effects of ultra-high temperatures on corn. *Weed Science* 21, 299–303.

Fergedal, S. (1993) Weed control by freezing with liquid nitrogen and carbon dioxide snow: a comparison between flaming and freezing. In: *Communications 4th International Conference IFOAM, Non-chemical Weed Control.* Dijon, France, pp. 163–166.

Furness, N.H., Jolliffe, P.A. and Upadhyaya, M.K. (2005) Competitive interactions in mixtures of broccoli and *Chenopodium album* grown at two UV-B radiation levels under glasshouse conditions. *Weed Research* 45, 449–459.

Furness, N.H. and Upadhyaya, M.K. (2002) Differential susceptibility of agricultural weeds to ultraviolet-B radiation. *Canadian Journal of Plant Science* 82, 789–796.

Fykse, H. (1985) Arbeids-, energiforbruk og kostnad ved olike metodar for ugrastyning [Labour and energy requirements and costs of different weed control methods]. *Aktuellt fra Statens Fagtjeneste for Landbruget: Informasjonsmöte Plantevern* 2, 71–78 (in Norwegian).

Gangstad, E.O. (ed.) (1982) *Weed Control Methods for Recreation Facilities Management.* CRC Press, Boca Raton, FL, USA.

Guerrero, C., Mataix-Solera, J., Gómez, I., García-Orenes, F. and Jordán, M.M. (2005) Microbial recolonization and chemical changes in a soil heated at different temperatures. *International Journal of Wildland Fire* 14, 385–400.

Hansson, D. and Ascard, J. (2002) Influence of developmental stage and time of assessment on hot water weed control. *Weed Research* 42, 307–316.

Hansson, D. and Svensson, S.-E. (2007) Steaming soil in narrow bands to control weeds in row crops. In: *Proceedings of the 7th EWRS (European Weed Research Society) Workshop on Physical and Cultural Weed Control*, Salem, Germany, 12–14 March 2007 [see http://www.ewrs.org/pwc/proceedings.htm].

Hart, S.C., DeLuca, T.H., Newman, G.S., MacKenzie, M.D. and Boyle, S.I. (2005) Post-fire vegetative dynamics as drivers of microbial community structure and function in forest soils. *Forest Ecology and Management* 220, 166–184.

Hatcher, P.E. and Melander, B. (2003) Combining physical, cultural and biological methods: prospects for integrated non-chemical weed management strategies. *Weed Research* 43, 303–322.

Heisel, T., Schou, J., Christensen, S. and Andreasen, C. (2001) Cutting weeds with a CO_2 laser. *Weed Research* 41, 19–29.

Heisel, T., Schou, J., Andreasen, C. and Christensen, S. (2002) Using laser to measure stem thickness and cut weed stems. *Weed Research* 42, 242–248.

Hoffmann, M. (1989) *Abflammtechnik.* KTBL-Schrift 331. Landwirtschaftsverlag, Münster-Hiltrup, Germany.

Jolliet, O. (1993) Life cycle analysis in agriculture: comparison of thermal, mechanical and chemical process to destroy potato haulm. In: Weidema, B.P. (ed.) *Life Cycle Assessments of Food Products: Proceedings of the 1st European Invitational Expert Seminar on Life Cycle Assessments of Food Products.* The Ecological Food Project, Interdisciplinary Centre, Technical University of Denmark, Lyngby, Denmark, pp. 27–42.

Jolliet, O. (1994) Bilan écologique de procédés thermique, méchanique, et chimique pour le defange des pommes de terre. *Revue Suisse d'Agriculture* 26, 83–90.

Kaufmann, K.R. and Schaffner, L.W. (1982) Energy and economics of electrical weed control. *Transactions of the ASAE* 25, 297–300.

Kepner, R.A., Bainer, R. and Barger, E.L. (1978) *Principles of Farm Machinery*, 3rd edn. AVI, Westport, CT, USA.

Kolberg, R.L. and Wiles, L.J. (2002) Effect of steam application on cropland weeds. *Weed Technology* 16, 43–49.

Kurfess, W. and Kleisinger, S. (2000) Effect of hot water on weeds. *Zeitschrift für Pflanzenkrankheiten und Pflanzenschutz* 17, 473–477.

Laguë, C., Gill, J. and Péloquin, G. (2001) Thermal control in plant protection. In: Vincent, C., Panneton, B. and Fleurat-Lessard, F. (eds), *Physical Control Methods in Plant Protection.* Springer-Verlag, Berlin, Germany, pp. 35–46.

Langerholc, J. (1979) Moving phase transitions in laser-irradiated biological tissue. *Applied Optics* 18, 2286–2293.

Leroux, G.D., Douhéret, J. and Lanouette, M. (2001) Flame weeding in corn. In: Vincent, C., Panneton, B. and Fleurat-Lessard, F. (eds) *Physical Control Methods in Plant Protection.* Springer-Verlag, Berlin, Germany, pp. 47–60.

Levitt, J. (1980) *Responses of Plants to Environmental Stresses. Vol. I: Chilling, Freezing, and High Temperature Stresses*, 2nd edn. Academic Press, New York.

Melander, B. and Jørgensen, M.H. (2005) Soil steaming to reduce intrarow weed seedling emergence. *Weed Research* 45, 202–211.

Melander, B. and Rasmussen, G. (2001) Effects of cultural methods and physical weed control on intrarow weed numbers, manual weeding and marketable yield in direct-sown leek and bulb onion. *Weed Research* 41, 491–508.

Melander, B., Jørgensen, M.H. and Elsgaard, L. (2004) Recent results in the development of band steaming for intra-row weed control. In: *Proceedings of the 6th EWRS (European Weed Research Society) Workshop on Physical and Cultural Weed Control*, Lillehammer, Norway, 8–10 March 2004, p. 154 [see http://www.ewrs.org/pwc/proceedings.htm].

Neary, D.G., Klopatek, C.C., DeBano, L.F. and Ffolliott, P.F. (1999) Fire effects on belowground sustainability: a review and synthesis. *Forest Ecology and Management* 122, 51–71.

Nelson, S.O. (1996) A review and assessment of microwave energy for soil treatment to control pests. *Transactions of the ASAE* 39, 281–289.

Nelson, S.O. and Stetson, L.E. (1984) Germination responses of selected plant species to RF electrical seed treatment. *Transactions of the ASAE* 28, 2051–2058.

Nemming, A. (1994) Costs of flame weeding. *Acta Horticulturae* 372, 205–212.

Norberg, G., Jaderlund, A., Zackrisson, O., Nordfjell, T., Wardle, D.A., Nilsson, M.-C. and Dolling, A. (1997) Vegetation control by steam treatment in boreal forests: a comparison with burning and soil scarification. *Canadian Journal of Forestry Research* 27, 2026–2033.

Parish, S. (1989) Investigations into thermal techniques for weed control. In: Dodd, V.A. and Grace, P.M. (eds) *Proceedings of the 11th International Congress on Agricultural Engineering, Dublin.* Balkema, Rotterdam, The Netherlands, pp. 2151–2156.

Pelletier, Y. and Colpitts, B.G. (2001) The use of microwaves for insect control. In: Vincent, C., Panneton, B. and Fleurat-Lessard, F. (eds) *Physical Control Methods in Plant Protection.* Springer-Verlag, Berlin, Germany, pp. 125–133.

Pimentel, D. (1992) Energy inputs in production agriculture. In: Fluck, R.C. (ed.) *Energy in World Agriculture 6, Energy in Farm Production.* Elsevier, Amsterdam, The Netherlands, pp. 13–29.

Pinel, M.P.C., Bond, W., White, J.G. and de Courcy Williams, M. (1999) *Field Vegetables: Assessment of the Potential for Mobile Soil Steaming Machinery to Control Diseases, Weeds and Mites of Field Salad and Related Crops.* Final Report on HDC Project FV229, Horticultural Development Council, East Malling, UK.

Porterfield, J.G., Batchelder, D.G., Bashford, L. and McLaughlin, G. (1971) Two stage thermal defoliation. In: *Proceedings of the Fifth Annual Symposium on Thermal Agriculture.* National LP-Gas Association and Natural Gas Processors Association, Dallas, TX, USA, pp. 32–34.

Quarles, W. (2001) Improved hot water weed control system. *IPM Practitioner* 23, 1–4.

Rahkonen, J., Pietikäinen, J. and Jokela, H. (1999) The effects of flame weeding on soil microbial biomass. *Biological Agriculture and Horticulture* 16, 363–368.

Rasmussen, J. and Ascard, J. (1995) Weed control in organic farming systems. In: Glen, D.M., Greaves, M.P. and Anderson, H.M. (eds) *Ecology and Integrated Farming Systems: Proceedings of the 13th Long Ashton International Symposium.* Wiley, Chichester, UK, pp. 49–67.

Reeder, R. (1971) Flaming to heat soil and control pink bollworms in cotton. In: *Proceedings of the Eighth Annual Symposium on Thermal Agriculture,* Natural Gas Processors Association and National LP-Gas Association, Dallas, TX, USA, pp. 26–31.

Reifschneider, D. and Nunn, R.R. (1965) Infrared cotton defoliation or desiccation. In: *Proceedings of the Second Annual Symposium: Use of Flame in Agriculture.* Natural Gas Processors Association and National LP-Gas Association, St Louis, MO, USA, pp. 25–29.

Runia, W.T. (1983) A recent development in steam sterilization. *Acta Horticulturae* 152, 195–200.

Sartorato, I., Zanin, G., Baldoin, C. and De Zanche, C. (2006) Observations on the potential of microwaves for weed control. *Weed Research* 46, 1–9.

Scheible, A. (1895) Apparatus for exterminating vegetation. US Patent No. 546,682, granted 24 Sept 1895.

Shakesby, R.A. and Doerr, S.H. (2006) Wildfire as a hydrological and geomorphological agent. *Earth Science Reviews* 74, 269–307.

Sonneveld, C. (1979) Changes in chemical properties of soil caused by steam sterilization. In: Mulder, D. (ed.) *Developments in Agricultural and Managed-Forest Ecology 6: Soil Disinfestation.* Elsevier, New York, pp. 39–49.

Storeheier, K. (1994) Basic investigations into flaming for weed control. *Acta Horticulturae* 372, 195–204.

Sutcliffe, J. (1977) *Plants and Temperature.* Edward Arnold, London.

Swengel, A.B. (2001) A literature review of insect responses to fire, compared to other conservation managements of open habitat. *Biodiversity and Conservation* 10, 1141–1169.

Thomas, C.H. (1964) Technical aspects of flame weeding in Louisiana. In: *Proceedings of the First Annual Symposium: Research on Flame Weed Control.* Natural Gas Processors Association, Memphis, TN, USA, pp. 28–33.

Upadhyaya, M.K., Polster, D.F. and Klassen, M.J. (1993) Weed control by superheated steam. *Weed Science Society of America Abstracts* 33, 115.

Vigneault, C. (2002) Weed electrocution. In: Pimentel, D. (ed.) *Encyclopedia of Pest Management.* Marcel Dekker, New York, pp. 896–898.

Vigneault, C. and Benoît, D.L. (2001) Electrical weed control: theory and applications. In: Vincent, C., Panneton, B. and Fleurat-Lessard, F. (eds) *Physical Control Methods in Plant Protection.* Springer-Verlag, Berlin, Germany, pp. 174–188.

Warren, S.D., Scifres, C.J. and Teel, P.D. (1987) Response of grassland arthropods to burning: a review. *Agriculture, Ecosystems and Environment* 19, 105–130.

Whelan, R.J. (1995) *The Ecology of Fire.* Cambridge University Press, Cambridge, UK.

11 Soil Solarization and Weed Management

O. Cohen[1] and B. Rubin[2]

[1]Department of Geography and Environmental Development, Ben Gurion University of the Negev, Beer Sheva 84105, Israel; [2]RH Smith Institute of Plant Sciences and Genetics in Agriculture, Faculty of Agricultural, Food and Environmental Quality Sciences, The Hebrew University of Jerusalem, Rehovot 76100, Israel

11.1 Introduction

Cropland area has increased over the past 300 years from 265 million ha in 1700 to 1470 million ha in 1990 (Goldewijk, 2001) with recent parallel increases in both pesticide use (Pimentel et al., 1992; Tilman et al., 2001) and soil fertilization (Vitousek et al., 1997). The increase in land use has had a detrimental effect on the environment and biodiversity (Chapin et al., 2000). Concern for the environment, an increase in awareness of issues such as pesticide effects on human health, and the depletion of the stratospheric ozone layer by methyl bromide have directed attention to alternatives for chemical pest control (Altieri, 1992; Ristaino and Thomas, 1997; Katan, 1999, 2000).

Soil solarization or 'solar heating' (SH) is a non-chemical disinfestation practice that has potential application as a component of sustainable integrated pest management (IPM) programmes. In addition, it also increases the availability of soil mineral nutrients and reduces crop fertilization requirements (Stapleton and DeVay, 1986). SH was originally developed to control soil-borne pathogens by Katan et al. (1976), but it was soon discovered to be an effective treatment against a wide range of soil-borne pests and weeds. The list of the pests controlled by SH includes more than 40 fungal plant pathogens, a few bacterial pathogens, 25 species of nematodes and numerous weeds (Stapleton, 1997).

SH utilizes solar energy to heat the soil. Soil is mulched with a transparent polythene film and temperatures in the upper layer of soil reach 40–55°C. There is a gradient of temperature from the upper to the lower soil layers (Fig. 11.1). The temperature elevation is facilitated by wetting the soil before and/or during mulching with the polythene sheet. The main factor involved in the pest control process is the physical mechanism of thermal killing. In addition, chemical and biological mechanisms may also be involved. The meaning of the term 'soil solarization' implies the process of heating the soil by solar radiation. However, there is nothing new under the sun in utilizing its rays for soil disinfestation, since exposing soil or plant material to solar heating has been known for centuries (Katan and DeVay, 1991). The innovation was in developing a new practical technique using a modern tool, namely plastic sheeting, for achieving this purpose.

In this chapter we present the principles of SH and its technological evolution, then we discuss the underlying mechanism(s) involved in weed control, focusing on the seedbank deterioration processes.

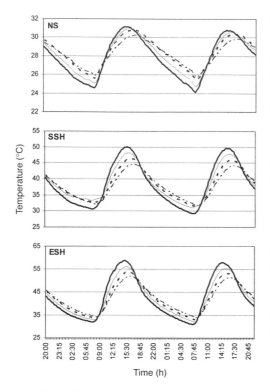

Fig. 11.1. Effect of soil solarization (SH) treatments on soil temperature at different soil depths: 3 (thick solid line), 6 (thin solid line), 9 (thick dotted line) and 12 cm (thin dotted and dashed line) at the Experimental Farm, Rehovot, Israel during July–August 2004. Plots were 3 × 4 m each. NS: non-solarized (control) – for details on the procedure see Gamliel et al. (2000). SSH: soil mulched with polythene. ESH: the soil surface was sprayed with a black polymer (Ecotex, Nir-Oz, Israel) before mulching with polythene sheet.

11.2 Implementation of Soil Solarization for Weed Control

Principles and technology

The basic principles

The basic principle of the SH technique is to produce an extreme environment, characterized by a high moist soil temperature (Fig. 11.1) which directly affects the viability of sensitive organisms. In addition, SH may also induce other environmental and biological changes in the soil that

indirectly affect the survival of soil-borne pests as well as beneficial organisms (Katan, 1981). Both the maximum soil temperature and the accumulated soil temperature (duration × temperature) determine thermal effects on soil-borne pests (Katan, 1987) and weed seeds (Stapleton et al., 2000a, b). Currently, the most common practice of SH is based on mulching moistened soil with transparent polythene. Commonly, the soil mulching duration required for successful disinfestation is 4–5 weeks, depending on weed population, soil characteristics, climatic conditions and the properties of the polythene (Katan, 1981, 1987; Rubin and Benjamin, 1984). Weed population and environmental conditions are unmanageable variables, while soil moisture and polythene properties can be modified as necessary. Soil pre-treatment and appropriate polythene technology may overcome unfavourable environmental conditions that prevail in some regions or in certain seasons, increasing weed control and shortening the duration of soil mulching (Stevens et al., 1991).

Soil preparation

Soil moisture improves temperature conductivity of soil, thus improving pest control in deeper soil layers as compared with dry soil (Mahrer et al., 1984). In addition, most weed seeds are more sensitive to 'wet heating' than 'dry heating', presumably because moist seeds are likely to have greater metabolic activity (Delouche and Baskin, 1973; Egley, 1990). This assumption is also true for other soil pests (Shlevin et al., 2004). Therefore, all soil pretreatments that improve water capacity, such as soil cultivation or drip irrigation during mulching may improve SH efficacy. Moreover, good soil preparation, which leads to a smooth soil surface, facilitates plastic mulching and prevents tearing.

The use of organic amendments, such as animal manure or incorporated cover crop, combined with SH may increase solarized soil temperature by an additional 1–3°C (Gamliel and Stapleton, 1993a, b; Gamliel et al., 2000; Lira-Saldivar et al., 2004). Gamliel et al. (2000) proposed that this elevation is a result of improved thermal conductivity in moist soil, exothermic microbial activity, or a combination of both. Furthermore, combining SH with organic amendments leads to the generation of biotoxic volatile compounds that accumulate under the plastic mulch and consequently enhance the vulnerability of soil organisms to SH (Gamliel *et al.*, 2000). The role of these volatile compounds in the weed control is questionable (Rubin and Benjamin, 1984).

Polythene properties and mulching techniques

Polythene technology offers a wide range of sheets differing in their chemical and physical properties such as thickness, colour and wavelength transmission, UV protection and durability. In general, all types of transparent polythene sheets commonly used in agriculture are appropriate for weed disinfestation purposes.

The effect of transparent polythene on the soil heating process resembles the greenhouse effect to some extent. Part of the solar radiation is transmitted through the transparent polythene, absorbed by the soil surface, and transformed to conserved heat (Fig. 11.1). The polythene largely prevents the escape of long-wave radiation and water evaporation from the soil to the atmosphere, consequently exerting the greenhouse effect. In addition, the water vapour accumulated on the inner surface of the polythene sheet further enhances the greenhouse effect, resulting in a higher soil temperature (Stevens *et al.*, 1991). Black polythene, however, absorbs most of the solar radiation and heats up, but does not transmit the radiation, resulting in lower soil temperature and poorer weed control (Horowitz *et al.*, 1983; Rubin and Benjamin, 1983; Standifer *et al.*, 1984; Mudalagiriyappa *et al.*, 1996; Abu-Irmaileh and Thahabi, 1997; Singh, 2006).

As mentioned above, transparent polythene, regardless of its thickness, is suitable for weed control; however, economic considerations such as cost and durability play a major role in the selection of polythene type. Thin polythene is cheaper and reflects less radiation than the thicker sheets, resulting in a slightly higher soil temperature. Unfortunately, thin polythene tends to get damaged more easily under field conditions than the thicker film.

Avissar *et al.* (1986a, b) reported that aged polythene (previously used) for SH is more efficient in soil temperature elevation than new polythene. The authors concluded that significant changes occurred in the photometric properties of polythene during the aging process, resulting in increased radiation influx at the soil surface.

The most common mulching technique is usually based on a single polythene sheet, but the double-tent technique in which the soil is mulched with two layers of polythene (3–7 cm apart) increases soil temperature by an additional 10°C compared with a single-layer solarization (Ben-Yaphet *et al.*, 1987). The double-tent method technique was found to be more effective than a single polythene layer (McGovern *et al.*, 2004), especially against weeds in nursery containers (Stapleton *et al.*, 2000b, 2002). It is obvious that the double-tent technique raises both the economic cost and the environmental hazard due to polythene pollution, and should be used only in special cases.

During the last decade, alternative technologies to polythene have been suggested, e.g. soil mulching with sprayable polymers (Gamliel and Becker, 1996), or the use of paraffin-wax emulsion as a mulching material (Al-Kayssi and Al-Karaghouli, 2002). However, their cost-effectiveness and efficacy have not been studied compared to the common polythene mulching. The effect of different soil mulching treatments on soil temperature elevation is presented in Fig. 11.1.

Implementation of SH for weed control: historical and global perspectives

Originally, SH was developed for controlling soil-borne pathogens, as indicated by Katan *et al.* (1976), but it was soon recognized as an effective treatment against weeds. The first significant findings on the high efficiency of SH against weeds were already known in the early 1980s from studies in Israel (Grinstein *et al.*, 1979; Horowitz, 1980; Jacobsohn *et al.*, 1980; Horowitz *et al.*, 1983; Rubin and Benjamin, 1983, 1984), California (Hejazi *et al.*, 1980), Mississippi (Egley, 1983) and Louisiana (Standifer *et al.*, 1984). In the following years, additional reports on effective weed control by SH were published, as indicated below.

Jacobsohn *et al.* (1980) reported that SH effectively controlled Egyptian broomrape (*Orobanche aegyptiaca*) and other weeds in the field (Table 11.1). Rubin and Benjamin (1983) demonstrated efficient control of a wide range of annual weeds and only partial control of perennials or hard-seeded weeds (Table 11.1). They also showed that SH results in significant changes in soil atmosphere while preserving soil moisture (Rubin and Benjamin, 1984). Similarly, Egley (1983) reported that SH treatment for 7–28 days in midsummer increased maximum temperature up to 69°C at 1.3 cm soil depth. The weed control efficacy was well correlated with the duration of SH and the type of seeds. The effect of SH on hard-seeded weeds such as spurred anoda (*Anoda cristata*) and morning glory (*Ipomoea* spp.) required longer exposure, and emergence reduction was not always significant compared with grasses.

Furthermore, purple nutsedge (*Cyperus rotundus*) was not controlled at all by SH; its sprouting even increased following the SH treatment. Standifer *et al.* (1984) indicated that SH efficacy is highly dependent on seed location in the vertical soil profile and the mulching duration. Surprisingly, in spite of different geographical locations, the maximum soil temperature recorded in Louisiana was the same as that previously recorded in Israel. It was also noted that frequent rain showers during mulching did not reduce the effectiveness of SH.

Based on the fact that the weed seedbank in soil consists of a wide range of species that differ in sensitivity to SH, Katan (1987) suggested utilizing the level of weed re-infestation following

Table 11.1. Response of weeds to solarization. The common name and botanical family are according to the terminology of the Weed Science Society of America (see http://www.wssa.net/CLNAMES/namesearch.asp). Plants were classified into four categories according to their response to SH treatment. **S** – susceptible, completely controlled; **MS** – moderately susceptible, partially controlled; **R** – resistant, poorly controlled; **St** – stimulated, weeds grew better after SH.

Scientific name	Common name	Family	Response to SH treatment
Abutilon theophrasti	Velvetleaf	Malvaceae	**S** (Egley, 1983) **MS** (Rubin and Benjamin, 1983)
Acrachne racemosa	Ohwi goosegrass	Poaceae	**S** (Kumar *et al.*, 1993; Singh, 2006)
Adonis aestivalis	Summer pheasant's eye	Ranunculaceae	**S** (Standifer *et al.*, 1984)
Alhagi maurorum	Camelthorn	Fabaceae	**S** (Rubin and Benjamin, 1983)
Amaranthus albus	Tumble pigweed	Amaranthaceae	**S** (Elmore, 1991; Stapleton, 2000b)
Amaranthus gracilis	Slender amaranth	Amaranthaceae	**S** (Linke, 1994)
Amaranthus hybridus	Slender pigweed	Amaranthaceae	**S** (Moya and Furukawa, 2000)
Amaranthus retroflexus	Redroot pigweed	Amaranthaceae	**S** (Rubin and Benjamin, 1983; Elmore, 1991; Linke, 1994; Caussanel *et al.*, 1997; Boz, 2004; Benlioglu *et al.*, 2005)
Amaranthus spp.	Pigweed	Amaranthaceae	**S** (Jacobsohn *et al.*, 1980; Egley, 1983; Horowitz *et al.*, 1983; Ioannou, 2000)
Amaranthus. blitoides	Prostrate pigweed	Amaranthaceae	**S** (Elmore, 1991; Al-Masoom *et al.*, 1993; Linke, 1994)
Anagallis coerulea	Blue pimpernel	Primulaceae	**S** (Horowitz *et al.*, 1983; Rubin and Benjamin, 1983)
Anchus aggregata	Large blue alkanet	Boraginaceae	**MS** (Rubin and Benjamin, 1983)
Anoda cristata	Spurrred anoda	Malvaceae	**S** (Egley, 1983; Elmore, 1991)
Aristolochia maurorum	Aristolochia	Aristolochiaceae	**R** (Elmore, 1991)
Arum italicum	Italian arum	Araceae	**S** (Elmore, 1991)
Asphodelus tenuipholius	Onionweed	Liliaceae	**R** (Arora and Yaduraju, 1998)
Astragalus boeticus	Milkvetch	Fabaceae	**R** (Rubin and Benjamin, 1983)
Astragalus spp.	Milkvetch	Fabaceae	**MS** (Jacobsohn *et al.*, 1980)
Avena fatua	Wild oat	Poaceae	**S** (Elmore, 1991; Arora and Yaduraju, 1998)
Avena sterilis	Sterile wild oat	Poaceae	**S** (Jacobsohn *et al.*, 1980)
Bellevalia sp.	Bellevalia	Liliacaceae	**R** (Linke, 1994)
Brassica nigra	Black mustard	Brassicaceae	**S** (Elmore, 1991)

Scientific name	Common name	Family	Response to SH treatment
Bunium elegans	Bunium	Apiaceae	**St** (Linke, 1994)
Calendula arvensis	Field marigold	Asteraceae	**S** (Ioannou, 2000)
Capsella bursa-pastoris	Shepherd's purse	Brassicaceae	**S** (Elmore, 1991; Haidar and Iskandarani, 1997; Tekin *et al.*, 1997)
Capsella rubella	Pink shepherd's purse	Brassicaceae	**S** (Elmore, 1991)
Carthamus flavescens		Asteraceae	**S** (Linke, 1994)
Carthamus syriacum		Asteraceae	**S** (Jacobsohn *et al.*, 1980)
Centaurea iberica	Iberian starthistle	Astraceae	**S** (Rubin and Benjamin, 1983)
Chenopodium album	Common lambsquarters	Chenopodiaceae	**S** (Rubin and Benjamin, 1983; Stevens *et al.*, 1989; Elmore, 1991; Caussanel *et al.*, 1997)
Chenopodium murale	Nettleleaf goosefoot	Chenopodiaceae	**S** (Jacobsohn *et al.*, 1980; Rubin and Benjamin, 1983)
Chenopodium pumila	Clammy goosefoot	Chenopodiaceae	**S** (Elmore, 1991)
Chenopodium spp.	Goosefoot		**S** (Ioannou, 2000)
Chloris gayana	Rhodesgrass	Poaceae	**S** (Elmore, 1991)
Chrozophora tinctoria	Dyer's litmus	Euphorbiaceae	**S** (Vizantinopoulos and Kataranis, 1993) **MS** (Rubin and Benjamin, 1983)
Chrysanthemum coronarium	Crown daisy	Asteraceae	**S** (Elmore, 1991)
Chrysanthemum spp.		Asteraceae	**S** (Ioannou, 2000)
Cichorium intybus	Chicory	Asteraceae	**S** (Linke, 1994)
Cirsium arvense	Canada thistle	Asteraceae	**S** (Vizantinopoulos and Kataranis, 1993)
Commelina communis	Common dayflower	Commelinaceae	**S** (Standifer *et al.*, 1984; Singh *et al.*, 2004)
Convolvulus althaeoiedes	Hollyhock bindweed	Convolvulaceae	**S** (Sauerborn *et al.*, 1989) **R** (Linke, 1994)
Convolvulus arvensis	Field bindweed	Convolvulaceae	**S** (Rubin and Benjamin, 1983) **MS** (Elmore, 1991; Elmore *et al.*, 1993) **R** (Linke, 1994; Ioannou, 2000)
Conyza bonarinsis	Hairy fleabane	Asteraceae	**S** (Elmore, 1991)
Conyza canadensis	Horseweed	Asteraceae	**R** (Horowitz *et al.*, 1983; Boz, 2004; Benlioglu *et al.*, 2005)
Coronilla scorpioides	Trailing crownvetch	Fabaceae	**St** (Sauerborn *et al.*, 1989; Linke, 1994)
Coronopus didymus	Lesser swinecress	Brassicaceae	**S** (Moya and Furukawa, 2000)
Cuscuta campestris	Field dodder	Cuscutaceae	**R** (Abu-Irmaileh and Thahabi, 1997)
Cuscuta monogyna	Dodder	Cuscutaceae	**R** (Abu-Irmaileh and Thahabi, 1997)
Cuscuta spp.	Dodder	Cuscutaceae	**S** (Haidar and Iskandarani, 1997)
Cynodon dactylon	Bermudagrass	Poaceae	**S** (Elmore *et al.*, 1993) **MS** (Rubin and Benjamin, 1983; Elmore, 1991)
Cyperus esculentus	Yellow nutsedge	Cyperaceae	**R** (Stapleton *et al.*, 2002)
Cyperus rotundus	Purple nutsedge	Cyperaceae	**S** (Chase *et al.*, 1999; Ricci *et al.*, 1999) **R** (Egley, 1983; Rubin and Benjamin, 1983; Chauhan *et al.*, 1988; Elmore, 1991; Kumar *et al.*, 1993; Tekin *et al.*, 1997; Ioannou, 2000; Stapleton *et al.*, 2002; Kumar and Sharma, 2005; Singh, 2006)

Continued

Table 11.1. – *Continued*

Scientific name	Common name	Family	Response to SH treatment
Cyperus spp.	Nutsedge	Cyperaceae	S (Standifer *et al.*, 1984)
Dactyloctenium aegyptium	Crowfootgrass	Poaceae	S (Kumar *et al.*, 1983; Singh, 2006)
Datura stramonium	Jimsonweed	Solanaceae	S (Elmore, 1991)
			MS (Rubin and Benjamin, 1983)
Daucus aureus	Wild carrot	Apiaceae	S (Rubin and Benjamin, 1983)
Daucus sp.		Apiaceae	St (Linke, 1994)
Descurainia sophia	Flixweed	Brassicaceae	S (Linke, 1994)
Digera arvensis	Kanjero	Amaranthaceae	S (Singh, 2006)
Digitaria sanguinalis	Large crabgrass	Poaceae	S (Rubin and Benjamin, 1983; Elmore, 1991; Vizantinopoulos and Kataranis, 1993)
			R (Elmore, 1991)
Diplotaxis erucoides	Rocket	Brassicaceae	S (Haidar and Iskandarani, 1997)
Echinochloa colona	Jungle rice	Poaceae	S (Singh *et al.*, 2004; Singh, 2006)
Echinochloa crus-galli	Barnyardgrass	Poaceae	S (Standifer *et al.*, 1984; Elmore, 1991; Benlioglu *et al.*, 2005)
Eleucine indica	Goosegrass	Poaceae	S (Rubin and Benjamin, 1983; Standifer *et al.*, 1984)
Emex spinosa	Spiny emex	Polygonaceae	S (Rubin and Benjamin, 1983)
Equisetum arvense	Field horsetail	Equisetaceae	S (Elmore, 1991)
Equisetum ramosissimum	Branched horsetail	Equisetaceae	S (Elmore, 1991)
Eragrostis magastachys	Stinkgrass	Poaceae	S (Elmore, 1991)
Erodium aegyptiacum	Filaree	Geraniaceae	S (Linke, 1994)
Erodium spp.	Filaree	Geraniaceae	S (Elmore, 1991)
Euphorbia aleppica	Spurge	Euphorbiaceae	S (Linke, 1994)
Euphorbia heliscopia	Sun spurge	Euphorbiaceae	S (Linke, 1994)
Euphorbia peplus	Petty spurge	Euphorbiaceae	S (Linke, 1994)
Falcaria vulgaris	Sickleweed	Apiaceae	MS (Linke, 1994)
Fumaria judaica	Fumitory	Fumariaceae	S (Jacobsohn *et al.*, 1980)
Fumaria mularis	Fumitory	Fumariaceae	S (Elmore, 1991)
Galinsoga parviflora	Small flower galinsoga	Asteraceae	S (Elmore, 1991)
Galium tricorne	Three-horned bedstraw	Rubiaceae	S (Linke, 1994)
Geranium tuberosum	Geranium	Geraniaceae	St (Linke, 1994)
Gladiolus aleppicus	Sword lily	Iridaceae	R (Linke, 1994)
Heliotropium kotsychyi	Heliotrope	Boraginaceae	S (Al-Masoom *et al.*, 1993)
Heliotropium sp.	Heliotrope	Boraginaceae	S (Linke, 1994)
Heliotropium suaveoleus	Heliotrope	Boraginaceae	S (Horowitz *et al.*, 1983; Rubin and Benjamin, 1983)
Hordeum leporinum	Mouse barley	Poaceae	S (Elmore, 1991)
Hypericum crispum	St John's wort	Hypericaceae	MS (Jacobsohn *et al.*, 1980)
Hypericum triquetrifolium	St John's wort	Hypericaceae	R (Linke, 1994)
Ipomoea lacunosa	Pitted morning glory	Convolvulaceae	S (Stevens *et al.*, 1989; Elmore, 1991)
			MS (Elmore, 1991)
Ipomoea spp.	Morning glory	Convolvulaceae	S (Egley, 1983; Stevens *et al.*, 1989)
Lactuca serriola	Prickly lettuce	Asteraceae	S (Linke, 1994)
Lactuca orientalis	Wild lettuce	Asteraceae	S (Linke, 1994)
Lactuca scariola	Prickly lettuce	Asteraceae	S (Jacobsohn *et al.*, 1980)
Lagonychium farctum[a]	Syrian mesquite	Fabaceae	MS (Linke, 1994)
Lamium amplexicaule	Henbit deadnettle	Lamiaceae	S (Jacobsohn *et al.*, 1980; Horowitz *et al.*, 1983; Rubin and Benjamin, 1983; Stevens *et al.*, 1989)
Lavatera cretica	Cornish mallow	Malvaceae	R (Rubin and Benjamin, 1983)

Scientific name	Common name	Family	Response to SH treatment
Leontice leontopetalum	Taqaiq	Berberidaceae	**R** (Horowitz *et al.*, 1983)
Linaria chalepensis	Toadflax	Scrophulariaceae	**S** (Linke, 1994)
Linum sp.	Flax	Linaceae	**S** (Linke, 1994)
Lolium rigidum	Rigid ryegrass	Poaceae	**S** (Ioannou, 2000)
Malva nicaeensis	Bull mallow	Malvaceae	**MS** (Jacobsohn *et al.*, 1980)
			R (Horowitz *et al.*, 1983; Rubin and Benjamin, 1983)
Malva parviflora	Little mallow	Malvaceae	**S** (Elmore, 1991)
			MS (Elmore, 1991)
Malva spp.	Mallow	Malvaceae	**S** (Ioannou, 2000)
Malva sylvestris	High mallow	Malvaceae	**S** (Elmore, 1991)
Matricaria chamomilla	Wild chamomile	Asteraceae	**S** (Boz, 2004)
Medicago arabica	Spotted burclover	Fabaceae	**S** (Moya and Furkawa, 2000)
Medicago polymorpha	California burclover	Fabaceae	**S** (Elmore, 1991)
Melilotus sulcatus	Sweetclover	Fabaceae	**R** (Rubin and Benjamin, 1983; Elmore, 1991)
Melucella laevis	Bells of Ireland	Lamiaceae	**S** (Linke, 1994)
Mercurialis annua	Annual mercury	Euphorbiaceae	**S** (Jacobsohn *et al.*, 1980; Rubin and Benjamin, 1983)
Montia perfoliata	Miner's lettuce	Portulacaceae	**S** (Rubin and Benjamin, 1983)
Muscari racemosum	Grape hyacinth	Liliaceae	**St** (Linke, 1994)
Myagrum perfoliatum	Bird's-eye cress	Brassicaceae	**S** (Linke, 1994)
Notobasis syriaca	Syrian thistle	Asteraceae	**S** (Rubin and Benjamin, 1983)
Ornithogalum narbonense	Star of Bethlehem	Liliaceae	**R** (Linke, 1994)
Orobanche cernua	Nodding broomrape	Orobanchaceae	**S** (Elmore, 1991; Abu-Irmaileh and Thahabi, 1997)
Orobanche ramosa	Hemp broomrape	Orobanchaceae	**MS** (Elmore, 1991)
			S (Abu-Irmaileh, 1991a; Abu-Irmaileh and Thahabi, 1997; Mauromicale *et al.*, 2005)
Orobanche spp.	Broomrape	Orobanchaceae	**S** (Tekin *et al.*, 1997)
Orobanche aegyptiaca	Egyptian broomrape	Orobanchaceae	**S** (Jacobsohn *et al.*, 1980; Sauerborn *et al.*, 1989; Elmore, 1991; Linke, 1994)
Orobanche crenata	Crenate broomrape	Orobanchaceae	**S** (Horowitz *et al.*, 1983; Linke, 1994; Abu-Irmaileh and Thahabi, 1997)
Oxalis corniculata	Creeping woodsorrel	Oxalidaceae	**S** (Elmore, 1991)
Papaver dubium	Field poppy	Papaveraceae	**S** (Elmore, 1991)
Papaver rhoeas	Corn poppy	Papaveraceae	**S** (Linke, 1994)
Phalaris brachystachys	Short-spike canarygrass	Poaceae	**S** (Jacobsohn *et al.*, 1980; Sauerborn *et al.*, 1989; Elmore, 1991; Linke, 1994)
Phalaris minor	Littleseed canarygrass	Poaceae	**S** (Arora and Yaduraju, 1998)
Phalaris paradoxa	Hood canarygrass	Poaceae	**S** (Rubin and Benjamin, 1983; Rubin and Benjamin, 1984)
Phyllanthus fraternus	Gulf leaf-flower	Euphorbiaceae	**MS** (Singh *et al.*, 2004)
Plantago spp.	Plantain	Plantaginaceae	**MS** (Jacobsohn *et al.*, 1980)
Poa annua	Annual bluegrass	Poaceae	**S** (Rubin and Benjamin, 1983; Standifer *et al.*, 1984; Elmore, 1991)
			MS (Peachey *et al.*, 2001; Boz, 2004; Benlioglu *et al.*, 2005)
Polygonum aviculare	Prostrate knotweed	Polygonaceae	**S** (Linke, 1994)

Continued

Table 11.1. – *Continued*

Scientific name	Common name	Family	Response to SH treatment
Polygonum equisetiforme	Horsetail knotgrass	Polygonaceae	**S** (Rubin and Benjamin, 1983)
Polygonum persicaria	Ladysthumb	Polygonaceae	**S** (Linke, 1994)
Polygonum polyspermum	Knotweed	Polygonaceae	**S** (Elmore, 1991)
Portulaca oleracea	Common purslane	Portulacaceae	**S** (Horowitz *et al.*, 1983; Rubin and Benjamin, 1983; Elmore, 1991; Al-Masoom et al., 1993; Vizantinopoulos and Kataranis, 1993; Stapleton, 2000b; Boz, 2004; Benlioglu *et al.*, 2005)
			MS (Tekin *et al.*, 1997)
Primula sp.	Primrose	Primulaceae	**S** (Stevens *et al.*, 1989)
Prosopis farcta	Syrian mesquite	Fabaceae	**S** (Rubin and Benjamin, 1983)
Ranunculus arvensis	Corn buttercup	Ranunculaceae	**S** (Linke, 1994)
Ranunculus repens	Creeping buttercup	Ranunculaceae	**S** (Linke, 1994)
Raphanus raphanistrum	Wild radish	Brassicaceae	**S** (Rubin and Benjamin, 1983; Elmore, 1991; Caussanel *et al.*, 1997; Haidar and Iskandarani, 1997; Boz, 2004)
Rhagadiolus stellatus	Endive daisy	Asteraceae	**MS** (Linke, 1994)
Roemeria hybrida	Violet horned poppy	Papaveraceae	**S** (Linke, 1994)
Rumex acetocella	Red sorrel	Polygonaceae	**S** (Linke, 1994)
Rumex crispus	Curly dock	Polygonaceae	**S** (Linke, 1994)
Scandix pecten-veneris	Venus comb	Apiaceae	**S** (Tekin *et al.*, 1997)
Scandix pecten-veneris	Shepherd's needle	Apiaceae	**S** (Linke, 1994)
Scorpiurus muricatus	Prickly scorpion's-tail	Fabaceae	**St** (Sauerborn *et al.*, 1989; Linke, 1994)
			R (Rubin and Benjamin, 1983)
Senecio vernalis	Spring groundsel	Asteraceae	**S** (Rubin and Benjamin, 1983)
Senecio vulgaris	Common groundsel	Asteraceae	**S** (Elmore, 1991)
Setaria glauca	Yellow foxtail	Poaceae	**S** (Elmore, 1991)
Sida spinosa	Prickly sida	Malvaceae	**S** (Egley, 1983)
Sinapis arvensis	Wild mustard	Brassicaceae	**S** (Rubin and Benjamin, 1983; Sauerborn *et al.*, 1989; Elmore, 1991; Linke, 1994; Caussanel *et al.*, 1997; Haidar and Iskandarani, 1997)
Sisymbrium spp.	London rocket	Brassicaceae	**S** (Jacobsohn *et al.*, 1980)
Solanum luteum	Nightshade	Solanaceae	**S** (Vizantinopoulos and Kataranis, 1993)
			MS (Rubin and Benjamin, 1983)
Solanum nigrum	Black nightshade	Solanaceae	**S** (Elmore, 1991; Stapleton, 2000b)
			MS (Rubin and Benjamin, 1984)
Sonchus oleraceus	Annual sowthistle	Asteraceae	**S** (Rubin and Benjamin, 1983; Elmore, 1991; Moya and Furukawa, 2000)
Sorghum halepense	Johnsongrass	Poaceae	**S** (Sauerborn *et al.*, 1989)
			MS (Rubin and Benjamin, 1983; Elmore, 1991; Linke, 1994)
Spergula fallax	Corn spurry	Caryophyllaceae	**S** (Linke, 1994)
Stellaria media	Common chickweed	Caryophyllaceae	**S** (Rubin and Benjamin, 1983; Elmore, 1991; Moya and Furukawa, 2000)
Striga asiatica	Witchweed	Scrophulariaceae	**R** (Osman *et al.*, 1991)
Striga hermonthica	Witchweed	Scrophulariaceae	**S** (Elmore, 1991)

Scientific name	Common name	Family	Response to SH treatment
Thesium humile	Lesser bastard toadflax	Santalaceae	**S** (Linke, 1994)
Thlaspi perfoliatum	Field pennycress	Brassicaceae	**S** (Linke, 1994)
Torilis leptophylla	Bristlefruit hedgeparsley	Apiaceae	**MS** (Linke, 1994)
Trianthema monogyna	Desert horse purslane	Aizoaceae	**S** (Elmore, 1991)
			R (Arora and Yaduraju, 1998)
Trianthema portulacastrum	Horse purslane	Aizoaceae	**S** (Egley, 1983; Singh, 2006)
Tribulus terrestris	Puncturevine	Zygophullaceae	**S** (Linke, 1994)
Urtica urens	Burning nettle	Urticaceae	**S** (Rubin and Benjamin, 1983; Tekin *et al.*, 1997; Ioannou, 2000)
Vaccaria pyramidata	Cowcockle	Caryophyllaceae	**S** (Linke, 1994)
Xanthium pensylvanicum	Canada cocklebur	Asteraceae	**S** (Egley, 1983)
Xanthium spinosum	Spiny cocklebur	Asteraceae	**S** (Rubin and Benjamin, 1983)
Xanthium strumarium	Common cocklebur	Asteraceae	**MS** (Rubin and Benjamin, 1984)

[a] See *Prosopis farcta*.

SH as a visual, easy and reliable tool for estimating the level of soil pathogen control. In spite of the different climate, Sauerborn *et al.* (1989) confirmed the previous observations (Jacobsohn *et al.*, 1980) that SH significantly controls broomrapes by more than 90% (Table 11.1), particularly when the treatment duration exceeded 10 days in the hot season. Abu-Irmaileh (1991a,b) indicated that SH with black or clear polythene in large tomato field trials completely eliminated both nodding broomrape (*Orobanche cernua*) and hemp broomrape (*Orobanche ramosa*) during the growing season. Similarly, SH effectively controlled *Orobanche crenata* in Egypt (Abdel-Rahim *et al.*, 1988).

Based on the response of numerous annual and perennial species, grasses and broadleaved plants from around the world, Elmore (1990, 1991; see also Table 11.1) suggested that winter annual weeds that germinate during short days and cool temperatures are effectively controlled by SH, due to their low thermotolerance. Egley (1990) indicated that SH is involved in the seed dormancy-breaking process and noted that moist heating is more effective than dry heating for reducing seed viability. Vizantinopoulos and Katranis (1993) emphasized that SH has a significant effect on weed infestation even when commenced at relatively low temperatures (maximum soil temperature measured was 53°C). Furthermore, results indicated that 7–28 SH days provided better weed control than the pre-emergence herbicide examined.

Under tropical Indian conditions, Kumar *et al.* (1993) have shown that SH for 32 days reduced the emergence of dominant weeds (including purple nutsedge) by over 90%. Al-Masoom *et al.* (1993) reported from the United Arab Emirates that 60 days of SH resulted in the complete control of various annual weeds for one season, but the effect faded in the subsequent crop. Under Mediterranean conditions, Linke (1994) showed that SH was effective against perennials such as bermudagrass (*Cynodon dactylon*), johnsongrass (*Sorghum halepense*) and partially effective against bindweed (*Convolvulus arvensis*). Intensive work has been done in Central American countries, where successful control of annual and perennial weeds has been demonstrated (FAO, 1995). A species-specific study conducted in Oregon, USA, found that SH effectively controlled annual bluegrass (*Poa annua*) in the upper 5 cm of soil, but did not reduce – or even improved – seed survival below 5 cm (Peachey *et al.*, 2001).

Solarization was found to be effective in a glasshouse soil for ornamental crops in Argentina (Moya and Furukawa, 2000), Brazil (Marenco and Lustosa, 2000), as well as in potting mixes in containers for nursery cultivation (Stapleton *et al.*, 2002). Successful weed control was demonstrated in a rainfed upland rice field ecosystem in India (Khan *et al.*, 2003). Studies in Turkey (Boz, 2004; Benlioglu *et al.*, 2004) have shown that SH conducted for 45–50 days with a soil temperature of 47.5°C effectively controlled most (99%) weeds but not horseweed (*Conyza canadensis*). Stapleton *et al.* (2005) summarized the data accumulated over 8 years of SH use in San Joaquin Valley,

California. It was demonstrated in strawberries that SH reduced both the weed number and biomass by 86–99%, being a more cost-effective weed management tool than methyl bromide.

A recent study has shown the high efficacy of SH in reducing both the vegetative infestation and persistent seedbank of the invasive plant *Acacia saligna* in natural conservation land (Cohen, 2006). The results have demonstrated for the first time that SH, as a non-chemical and non-destructive physical method, can be employed in natural conservation ecosystems, as an effective component of invasive weed management. Use of SH in natural ecosystems, where herbicides should not be used, may increase in the future.

Analysis of 197 articles focusing on soil solarization, published during the period 1982–2006, showed that 60% of the papers dealt with soil-borne pathogens and nematodes, 17% with weeds, 5% with both weeds and other soil-borne pests, and 18% with other issues such as plastic technology or biotic and abiotic factors (Fig. 11.2).

11.3 Weed Management and the Seedbank

Seedbank depletion: an important goal for weed management

Unused resources (e.g. light, water and nutrients) and propagule pressure, which is the availability of weed seed supply in the area, are two ecological factors that play an important role in determining the susceptibility of a habitat to invasion by new species (Davis *et al.*, 2000; D'Antonio *et al.*, 2001). In agroecosytems, the level of the unused resources may increase because of an increase in the supply of resources due to habitat disturbance caused by cultivation. Unfortunately, according to D'Antonio *et al.* (2001), habitat such as agricultural land may be at risk of a high probability of invasion even at relatively low levels of propagule pressure. Although weed seeds can be brought from outside by wind, water, animals and humans, the most significant source for weed propagules in agroecosystems is the seedbank – the reservoir of plant propagules in the soil (Cavers and Benoit, 1989; Radosevich *et al.*, 1997). For this reason, when we deal with sustainable management of weed control, we should consider the magnitude of the soil seedbank.

The soil seedbank is notable for its ecological importance in the persistence of plant species. It can withstand fire and disturbances such as ploughing or chemical control. Unfortunately, the seedbank of a majority of the world's most troublesome weeds is highly persistent (Holm *et al.*, 1977). For example, a single large plant of redroot pigweed (*Amaranthus retroflexus*) – one of the most widely distributed weed species in the arable crops of the world – produces more than 500,000 seeds/m^2, and some of them may be viable for as long as 40 years (Holm *et al.*, 1977), resulting in an accumulation of a huge persistent seedbank. Thus, in many cases, the key for controlling weed infestation is to reduce the persistent seedbank and eliminate the re-

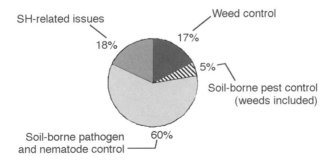

Fig. 11.2. The division of 197 SH articles published during the period 1982–2006, showing the poor representation of articles focusing on soil solarization (SH) and weed control; SH-related issues deal with polythene technology, microbial activity, etc.

establishment of seedlings (Altieri and Liebman, 1988). This assumption has been validated by a computer simulation model (Kebreab and Murdoch, 2001; Grenz *et al.*, 2005) for broomrape management. Cultural control methods such as hand-weeding, trap/catch crops, or delayed planting were very effective when combined with seedbank reduction. Therefore, any long-term integrated weed management scheme, especially for annual summer weeds (Davis, 2006), should aim at the reduction of the weed seedbank (Altieri and Liebman, 1988; Kennedy, 1999; Menalled *et al.*, 2001). Soil solarization could be an effective tool in this regard.

Seedbank characteristics

The majority of the seedbank is concentrated in the upper 15 cm of the soil profile, depending on soil cultivation history (Wilson, 1998). Most weed species, however, germinate from 0.5 to 2 cm soil depth (Holt, 1988). The depletion of the seedbank in the upper soil layer is faster than in the deeper layers due to more germination, predation, and exposure to harsh or fluctuating climatic and edaphic conditions (Dekker, 1999). When no weed management was practised, the density of weed seeds in soil after one cycle of a four-crop rotation increased by 25,000–51,000 seeds/m^2 in the upper 15 cm layer (Hill *et al.*, 1989). Wilson (1998) detected more than 130,000 seeds/m^2 in this layer, depending on the soil cultivation history.

Thompson *et al.* (1998) studied the correlation between ecological factors and seed persistence in soil in the north-western European flora. They found that annual and biennial weeds often have more persistent seeds than perennial weeds. They also found that disturbed habitats, such as agroecosystems, were often characterized by persistent seeds of species with small seeds. Most agricultural weeds are annuals, which rely on seed dormancy (delayed germination) for survival under unfavourable conditions. According to Grime (1981), the accumulation of dormant seeds in the soil seedbank plays a major role in ensuring their persistence in the soil and is possibly the most important strategy for the success of weeds.

There are several mechanisms of seed dormancy (Baskin and Baskin, 2004):

- Physical dormancy (PY) is linked with the presence of water-impermeable layer(s) of palisade cells in the seed coat. It is associated with long persistence in soil and is frequently found in legumes (Bibbery, 1947; Quinlivan, 1971; Rolston, 1978; Bradbeer, 1988; Baskin *et al.*, 2000).
- Physiological dormancy (PD) is induced by factors inhibiting embryonic activity.
- Combined dormancy (PY + PD), the seed coat is water-impermeable and the embryo is physiologically inactive.
- Morphological dormancy (MD) is caused by an undeveloped embryo.
- Morpho-physiological dormancy (MPD) is a combination of MD and PD.

In light of the above, one can postulate the following:

- Decreasing the seedbank size in the upper soil layer (e.g. by SH) will reduce the weed infestation problem, at least in the short term.
- In order to reduce the seedbank for more than one season, one should influence the weed seeds located in the deeper soil layer.

The latter can be achieved by two possible strategies: either by raising the seed from the deep soil layer to an upper layer (e.g. by cultivation), or by exposing the seeds in the lower soil layer to the same management tool applied to the upper layer. One of the advantages of SH is that its thermal effect 'leaches' to deeper layers by heat conduction through soil moisture.

Deterioration dynamics of a persistent seedbank

Schafer and Chilcote (1970) suggested dividing a buried seed population in three categories: non-dormant, dormant and non-viable. Seeds in the first two categories can exchange status or may become non-viable. Also, non-dormant seeds that germinate could either emerge from the soil surface or commit 'suicidal germination' without being able to complete their journey to the soil surface. Based on the model of Schafer and Chilcote (1969), we propose a simple model that describes the state of a seedbank at a given time:

$$S = Sd + Snd + Snv$$

where S represents the total amount or number of seeds in a seedbank at a given time (100%), Sd, dormant seeds, Snd, non-dormant seeds, and Snv, non-viable seeds. Persistence of a seedbank is determined by the relationship between Sd and Snd and the seedbank deterioration processes can be expressed by the transition directly from Sd to the Snv component or indirectly from Sd through Snd to Snv. These direct and indirect transitions may be affected by SH treatment.

11.4 Effect of SH on Weed Control: The Underlying Mechanism

How SH controls weeds

Little is known about the underlying mechanism(s) by which SH controls weeds. In many SH studies correlations between SH and number of weeds and their emergence kinetics have been compared to non-SH treatments (Horowitz *et al.*, 1983; Sauerborn *et al.*, 1989; Al-Masoom *et al.*, 1993; Abu-Irmaileh and Thahabi, 1997; Saghir, 1997; Boz, 2004). Other laboratory studies have improved our understanding of the thermal killing mechanism (Egley, 1983, 1990; Kebreab and Murdoch, 1999; Stapleton *et al.*, 2000b; Mas and Verdu, 2002; Verdu and Mas, 2004). A database describing the response of weeds to thermal killing could be useful in making decisions regarding mulching technique selection (Stapleton, 2000a). However, only a few qualitative studies combine field trial data with detailed laboratory experiments to investigate the mechanism(s) involved in weed control by SH (Rubin and Benjmain, 1984; Standifer *et al.*, 1984; Egley, 1990; Economou *et al.*, 1997; Stapleton *et al.*, 2002). Results of laboratory experiments that dealt with models of the thermal killing mechanism sometimes conflict with field data. For example, Stapleton *et al.* (2000b) indicated that thermal death of tumble pigweed (*Amaranthus albus*) occurred following heat treatment of 50°C for 113 h (temperature × duration). However, field experiments have demonstrated that thermal death was achieved even after exposure to 50°C for 1 h (Stapleton *et al.*, 2002). Two explanations

have been proposed: first, in the laboratory experiments the seeds were exposed to a constant temperature, whereas in the field the soil temperature fluctuated. Second, the soil used in the laboratory experiment was sterile, lacking ecological factors such as changes in the soil chemistry and biotic activities that may contribute to seed deterioration in the field (Stapleton *et al.*, 2002). We propose that thermal killing by SH involves interactions between physical, chemical and biological factors which facilitate deterioration of weed seedbanks.

Soil temperature and seedbank deterioration

Direct and indirect thermal killing

The main factor involved in the disinfestation by SH is thermal killing (Katan and DeVay, 1991). Organisms vary in their heat sensitivity. For example, a temperature of 50°C would be considered critically high when referring to multicellular organisms, but is considered moderate for thermophilic bacteria (Brock, 1978). This variation is determined by the inherent traits of organisms and is affected by environmental factors (Hutchison, 1976). SH represents a unique case of an extreme environment characterized by a high moist temperature which directly affects the viability of organisms that cannot tolerate it. In addition, the SH effect may also induce indirect killing by modifying the soil environment and its biological activity (Katan, 1981).

The thermal effect may cause a direct transition of the weed seed in the seedbank from the Sd to the Snv state, or from Snd to Snv as described above, or an indirect transition from the Sd state to Snd and than to the Snv state. Commonly, when PY seeds are killed, the effect of SH is indirect. Egley (1990) described two phases of the deterioration process. In the first phase, sublethal temperatures promoted germination by breaking the dormancy of hard-seeded species such as prickly sida (*Sida spinosa*), velvetleaf (*Abutilon theophrasti*), spurred anoda and pitted morning glory (*Ipomoea lacunosa*). The next phase involves direct thermal killing of seedlings. Cohen (2006) and O. Cohen *et al.* (unpublished data) have shown that direct exposure of *Acacia saligna* seeds to 50°C for 24 h resulted in direct killing of 7% of the seeds. However, the number of germinating seeds was six times higher than in the untreated control. It was also noted that acid-scarified *A. saligna* seeds completely lost their viability when exposed to a similar thermal treatment. These results implied that sublethal heating breaks the dormancy of *A. saligna* seeds making them more heat-sensitive. The elevated temperature can impose a 'weakening effect' that induces changes in the seed coats and/or metabolic processes that are not sufficient to induce germination but increase the seed's sensitivity to phytotoxic volatiles and microbial attack.

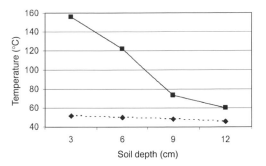

Fig. 11.3. Temperature gradient in soil profile exposed to soil solarization (SH) treatment (black diamonds, dotted line) and fire (black squares, solid line). Measurements were conducted in midsummer of 2003 by Cohen (2006). Values represent averages of ten transect measurements of soil profiles at 5 h after the fire started (09:30 hours).

SH and fire: similarities and differences

Thermal killing may occur not only in SH but also under natural conditions; wildfire is an extreme case of a thermal killing. Understanding the mechanism by which aboveground fire reduces the below-ground seedbank can be used to develop SH management practices, to achieve the same purpose. Factors involved in determining seed survival following fire are a combination of temperature intensity and duration, soil moisture, seed thermotolerance and seed location in the vertical soil profile, and the maximum depth from which a seed can emerge (Shea *et al.*, 1979; Keley, 1987). The above-mentioned factors also determine the efficacy of thermal killing by SH. However, there are following distinct differences between fire and SH in soil temperature elevations (Fig. 11.3):

- The short-lasting effect of thermal killing under fire conditions is achieved at high temperature levels, usually >500°C at the soil surface, for short durations (minutes to hours). In contrast, SH treatment produces a moderate temperature level (35–60°C) for longer durations (2–4 h per day over several weeks).
- Fire produces a relatively constant temperature, while SH is characterized by daily cyclical temperature fluctuations.
- The temperature gradient in the vertical soil profile is significantly steeper in the fire condition compared with SH (Fig. 11.3).

In spite of these notable differences, these two processes have a similar effect on seed viability. This similarity is achieved through the control of the duration of SH. The longer duration at moderate temperatures compensates for the shorter duration at high temperatures. For example, exposing *Acacia saligna* seeds to moderate dry heat of 50°C for 72 h increased germination to 78% (Cohen, 2006; O. Cohen *et al.*, unpublished data), which was equivalent to exposure of seeds to 80°C for 30 s (Jeffery *et al.*, 1988).

SH also has some added value over fire due to the effects of fluctuating temperature and higher soil moisture. Fluctuating soil temperature is more favourable for seed germination than a constant temperature (Lonsdale, 1993; Baskin and Baskin, 1998), and germinated seeds become more vulnerable to high soil temperature because of their decreased thermotolerance. Additionally, unlike fire, SH treatment is usually performed in moist soil, where weeds are more sensitive to heat than in dry soil (Delouche and Baskin, 1973; Egley, 1990; Mickelson and Grey, 2006). For example, Kebreab and Murdoch (1999) found that moist heating adversely affected the longevity of broomrape seeds, more than dry heating. Egley (1990) indicated that imbibed seeds are susceptible to high temperature due to their greater metabolic activity. Warcup (1980) found that exposure of a forest soil layer to the sun in summer was sufficient to induce elevated germination of several plant species. Mickelson and Grey (2006) found that wild oat (*Avena fatua*) seedbank decline is faster in moist than in dry soils, suggesting that management practices that increase or conserve soil moisture will also increase the rate of seedbank decline. Hence, moderate moist soil temperature may reduce the seedbank persistence either by a direct transition from *Sd* to *Snv*, or indirectly through *Snd* by breaking the dormancy and exposing the seed/seedlings to the harsh SH conditions (Schafer and Chilcote, 1970).

Soil chemical changes and seedbank deterioration

There is no direct evidence that chemical changes occurring during SH are involved in weed seedbank deterioration. Haidar *et al.* (1999) proposed that these SH induced changes are involved in the transition from *Sd* to *Snv* directly or indirectly. In addition, chemical changes in the soil atmosphere may be involved in the 'weakening effect' described above.

CO₂ concentration

CO_2 concentration in the soil atmosphere rapidly increases in the mulched soil, reaching a peak of 3.1% within 6 days (Horowitz *et al.*, 1983; Rubin and Benjamin, 1984). Several studies have indicated that CO_2 concentration levels of 2–5% in the soil atmosphere might promote seed germination (Baskin and Baskin, 1998). Hence, the transition of seeds from the *Sd* to *Snd* phase induced by changes in CO_2 concentrations could increase their vulnerability to SH-induced harsh conditions.

Organic matter decomposition and biofumigation

Organic matter (manure and plant biomass) decomposition may result in the generation of phytotoxic and volatile compounds, which may vary according to the organic matter used (Chou and Patrick, 1976; Wainwright *et al.*, 1986; Wheatley *et al.*, 1996) especially when high soil temperature is employed (Gamliel and Stapleton,

1993a,b; Gamliel *et al.*, 2000). This is called 'biofumigation' (Stapleton *et al.*, 2000a). Gamliel *et al.* (2000) have shown that, depending on the plant residues or the manure incorporated into solarized soil, generation of measurable amounts of volatile compounds such as ammonia, methanethiol, dimethyl sulphide, allylisothiocyanates, phenylisothiocyanates and aldehydes may be detected in the soil atmosphere. These compounds accumulate under the polythene to above a threshold level which is toxic to soil flora and fauna. The elevated soil temperature also increases the sensitivity of soil pests (including weed seeds) to the toxic effect of the trapped volatiles (Gamliel *et al.*, 2000), further deteriorating the seedbank persistency (Lynch, 1980; Petersen *et al.*, 2001). For example, Petersen *et al.* (2001) indicated that isothiocyanates released by turnip-rape mulch (*Brassica rapa*) suppress weed infestation in the field. A high level of isothiocyanates in the soil was found to be a strong suppressant of germination in several weeds and crops, such as scentless mayweed (*Matricaria inodora*), smooth pigweed, barnyardgrass (*Echinochloa crus-galli*), blackgrass (*Alopecurus myosuroides*) and wheat (*Triticum aestivum*). Possible mechanisms for the enhanced weed seed deterioration by SH as a result of increasing organic matter degradation include:

- *Biofumigation effect*: direct killing by transition from *Sd* and *Snd* to *Snv* as a result of exposure to above-threshold levels of toxic compounds.
- *Breaking dormancy*: indirect killing by transition from *Sd* to *Snd* due to high CO_2 concentrations.
- *Temperature elevation*: an additional increase of the solarized soil temperature by 1–3°C due to the exothermic degradation of organic matter increases the transition from *Sd* to *Snd* and from *Snd* to *Snv*.
- *Increasing soil moisture*: an increase in soil moisture increases the sensitivity of pest propagules to chemical, physical and biological killing mechanisms (Gamliel *et al.*, 2000), resulting in transition from *Sd* to *Snd* and from *Snd* to *Snv*.
- *Increasing biotic attack*: increasing soil thermophilic microbial activity may facilitate transition from *Sd* to *Snd* and from *Snd* to *Snv*.

Soil biotic changes and seedbank deterioration

Soil microorganisms play a significant role in the effect of SH on soil-borne pathogens (DeVay and Katan, 1991), but there is no sufficient evidence concerning their involvement in weed control by SH. However, it is well established that soil microorganisms affect weed seedbank deterioration processes (Halloin, 1983; Harman, 1983; Mills, 1983) that can be exploited for weed management (Kremer, 1993; Kennedy, 1999; Chee-Sanford et al., 2006). The seedbank is a major source of nutrition for microorganisms (Mills, 1983; Kremer and Schulte, 1989). The seed coat, with its dense palisade layer, is a physical barrier not only for water but is also important in protection against microbial attack (Halloin, 1983; Kremer et al., 1984; Kremer, 1986). Moreover, seeds may contain or exude chemicals that are toxic to the microorganisms (Broekaert et al., 1995). The palisade layers and the chalazal area of velvetleaf seeds contain antimicrobial substances such as phenolic compounds which diffuse from seeds to the surrounding environment, reducing the activity of seed decomposers (Kremer et al., 1984; Kremer, 1993). In addition, velvetleaf seeds accommodate antagonistic bacteria within and on its seeds which inhibit the penetration of microorganisms (Kremer et al., 1984; Chee-Sanford et al., 2006). When the seed coat is punctured or cracked, the deterioration caused by thermophilic microorganisms increases. Thus, SH, which augments the activity of thermophilic microorganisms, facilitates destruction of the seed coat and could be an effective management tool for depleting weed seedbanks in natural or agricultural ecosystems. Moreover, the increase in soil moisture and organic decomposition could further enhance the activity of these microorganisms, resulting in reduction of seed viability (Kremer, 1993; Mickelson and Grey, 2006).

What makes some weeds more tolerant to SH?

Not all weeds are equally controlled by SH treatment; some exhibit high levels of resistance (e.g.

milkvetch (*Astragalus boeticus*), *Scorpiurus muricatus*, sweetclover (*Melilotus sulcatus*), *Lavatera cretica*, bull mallow (*Malva nicaeensis*) and *Leontice leontopetalum*) (Table 11.1). Reinfestation by several weeds was stimulated in response to SH (e.g. purple nutsedge (Egley, 1983), trailing crownvetch (*Coronilla scorpioides*), *Scorpiurus muricatus* (Sauerborn et al., 1989), *Bunium elegans*, wild carrot (*Daucus* spp.), geranium (*Geranium tuberosum*) and grape hyacinth (*Muscari racemosum*)) (Linke, 1994). The important question is what traits characterize a SH-resistant weed?

The ability of annual plants to survive desiccation or other unfavourable conditions is an important evolutionary advantage for survival, especially in disturbed and unstable habitats. Annual weed seeds tend to form huge persistent sandbanks. Elmore (1990, 1991) concluded that summer annuals are more resistant to SH due to their high thermotolerance compared with winter annuals. These weed species grow during the summer months, and require a higher temperature for a longer duration for germination and in order to break seed dormancy. For example, common purslane (*Portulaca oleracea*) will be controlled only if the soil temperature reaches 60°C during SH if the seeds are located near the soil surface (Verdu and Mas, 2004). Similar results were reported for redroot pigweed (Mas and Verdu, 2002). The thermal death of tumble pigweed occurred at a temperature of 50°C for 4.7 days, or 45°C for 13 days (Stapleton et al., 2000b). However, most summer annual weeds are controlled under strong SH conditions (Elmore, 1991).

Some perennial weeds, such as purple nutsedge, exhibit a high level of tolerance to SH treatment, whereas other perennials such as johnsongrass are quite sensitive. The survival of perennial weeds against SH may be explained by the nature of their underground vegetative structures such as tubers, rhizomes or bulbs, which allow them to survive and sprout from the deep soil profile (Horowitz et al., 1983; Elmore et al. 1993). For example, only 26% of yellow nutsedge (*Cyperus esculentus*) tubers were killed by SH treatment (Hejazi et al., 1980) whereas good control of bermudagrass (Elmore et al., 1993) and johnsongrass (Sauerborn et al., 1989) was reported. Since most perennials can rapidly re-establish from underground propagules, they are poorly controlled or even stimulated by SH (Kumar *et al.*, 1993).

PY is a common trait among SH-uncontrolled seeds. This is supported by studies employing sweetclover, milkvetch and *Scorpiurus muricatus* (Rubin and Benjmain, 1983; Sauerborn *et al.*, 1989), sterile wild oat (*Avena sterilis*) (Standifer *et al.*, 1984), spurred anoda (Egley 1990) and *Coronilla scorpioides* (Linke, 1994). It could be assumed that the high thermotolerance of PY seeds is achieved by the water impermeability of the seed coat. In fact, even when the seeds are in a moist soil, the effect of heat treatment on the embryo looks similar to that under dry heating conditions. As discussed above, the thermal killing efficacy of dry heating is lower than that of moist heating. In addition, the seed coat is a physical barrier against microbial invasion and it also decreases the effect of phytotoxic solutes and volatile compounds. Therefore, the combined effect of SH on seed deterioration is assumed to be less effective in PY seeds than in other dormant seeds. Hence, the success or failure of SH in controlling PY and other dormant seeds is determined by the time when the seeds become non-dormant during SH. Seeds that become non-dormant in the early stage of the SH are better controlled and vice versa. Germination of weed seeds that become non-dormant in the late stage of the SH might even be stimulated (Fig. 11.4).

The thermal killing induced by SH decreases significantly with soil depth (Rubin and Benjamin, 1983; Arora and Yaduraju 1998). Therefore, the position of weed propagules in the vertical soil profile significantly affects seed survival following SH. Compact weed seeds (small size and rounded shape) tend to be buried into the deeper soil layers. Theoretically, these seeds could escape thermal killing and survive in the deeper soil profile. On the other hand, they cannot emerge from the soil surface due to their low nutrient content. Thus, without deep soil disturbance (e.g. tillage) after the SH treatment, the fraction of buried *Sd* seeds may remain dormant or eventually become *Snv* due to predation or decay. In addition, the fraction of buried *Snd* seeds may possibly lose their viability due to a 'suicidal germination' process, or may be transformed to *Sd* by a secondary dormancy without interfering with the crop. This, however, is not always the case and

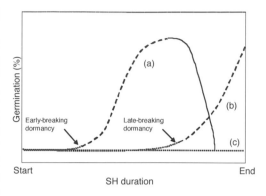

Fig. 11.4. The success or failure of soil solarization (SH) in control of PY seeds: (a) *Acacia saligna* – example of a controlled weed that completes the transition from *Sd* to *Snd* phase and from *Snd* to *Snv* phase during the duration of SH (Cohen, 2006); (b) *Coronilla scorpioides* – example of a stimulated weed where a large proportion of the seeds complete only the transition from *Sd* to *Snd* during the SH (Sauerborn, 1989; Linke, 1994); (c) *Cuscuta campestris* – example of an uncontrolled weed where seeds remain in *Sd* phase during the duration of SH. *Sd* (dotted line), *Snd* (dashed line), *Snv* (solid line) (Nir *et al.*, 1996).

there are some exceptions. Witchweed (*Striga asiatica*), for example, exhibits PD and becomes dormant in response to high levels of soil moisture (Mohamed *et al.*, 1998). It was assumed that the re-infestation with the parasite after SH treatment is caused by seeds that after-ripen and germinate and attach to the host roots from deeper layers where solarization was not effective (Osman *et al.*, 1991). Unlike small seeds, large seeds contain a large reservoir of nutrients enabling them to emerge from a deeper soil profile. Consequently, they are more difficult to control by SH than the small seeds.

11.5 Advantages and Limitations of SH

The important issue of cost–benefit analysis of SH as a weed control method is much wider than we are able to discuss in this chapter. Moreover, the real cost of SH varies from country to country and is highly dependent on the local land-use programmes (e.g. organic versus conventional farming) and the cost and availability of alternatives. Here we present some of the benefits and limitations of SH.

Economically, SH as a non-chemical tool for weed management was proven to be more cost-effective and profitable than methyl bromide (Stapleton *et al.*, 2005) and certain other treatments (Boz, 2004), especially in high-value crops (Abdul-Razik *et al.*, 1988; Yaron *et al.*, 1991; Vizantinopoulos and Katranis, 1993). Beyond the efficacy of SH as a sustainable IPM technique, it also results in increased plant growth response (IGR), especially in cases where dissolved organic matter (DOM) is available (Chen *et al.*, 2004). Additional factors such as the long-term effect on seedbank reduction and contribution to nature conservation are difficult to translate into economic models. Technological innovations, such as mulching the soil with sprayable polymer or using a wide variety of polythene sheets or other mulch techniques (Gamliel and Becker, 1996; Al Kayssi and Al-Karaghouli, 2002) will facilitate the use of SH in agriculture. These facilitations could reduce mulch longevity, increase the geographical range of usage, broaden the range of controlled weeds, improve the durability of polythene sheets, decrease polythene pollution and achieve a significant reduction in the cost of mulching. Unfortunately, as well as the favourable effects of SH, there are some limitations:

- It can only be used in regions where the climate is suitable and when the soil is free of crops (Katan, 1981).
- It is less recommended for low-income crops (e.g. arable crops) that are unable to bear the cost of the treatment (Yaron *et al.*, 1991). However, it can provide effective weed control in strawberries at a much lower cost than methyl bromide (Abdul-Razik *et al.*, 1988; Stapleton *et al.*, 2005).
- It is difficult to protect polythene sheets from damage by wind and animals, which reduces the efficacy of solarization by increasing the dissipation of heat, water and volatile substances.
- At present there is no satisfactory environmentally acceptable solution for the use or disposal of the used polythene, resulting in a pollution problem.
- Several weeds (mostly perennials and legumes) have been shown to be highly

tolerant of or are even stimulated by SH (Table 11.1). Furthermore, the high infestation of troublesome legume weeds in temperate Australian and Mediterranean ecosystems (Rubin and Benjamin, 1984; Paynter *et al.*, 2003; Emms *et al.*, 2005) limits the use of SH despite favourable climatic conditions.

Elmore (1991) described the behaviour of various weeds in response to SH. In Table 11.1 we enlarge the 'database' and summarize the response of numerous weeds to SH reported in the literature. It is interesting to note that at least 16 species listed amongst the 'world's worst weeds' (Holm *et al.*, 1977) are controlled by SH, indicating the potential of SH as a practical weed management tool.

11.6 Conclusions

SH is an attractive and environmentally friendly IPM option for agricultural systems (Stapleton, 2000a). Its effectiveness in controlling weeds and some soil-borne pests has been demonstrated in a variety of agroecosystems (Table 11.1), not only in regions with high solar radiation but also under cloudy weather conditions (Peachey *et al.*, 2001). Reduction of the seedbanks is an important goal in any long-term weed management programme. Hence, understanding the dynamics of seedbank deterioration processes in soil profile is vital before making any management decisions. The factors involved in success or failure of SH are summarized in Fig. 11.5, which illustrates correlations between various factors and their influence on the response of SH-treated seedbank (tolerance or susceptibility). Factors are divided into two categories: SH factors which determine the soil environmental conditions affecting the seedbank deterioration and factors involved with the intrinsic characteristics of the seedbank, which determine seedbank persistence upon SH (Fig. 11.5).

SH factors

Soil temperature affects weed control by SH either directly by the thermal killing, or indirectly by the 'weakening effect'. A rise in the concentration of phytotoxic volatile compounds due to

Seedbank factors

SH factors

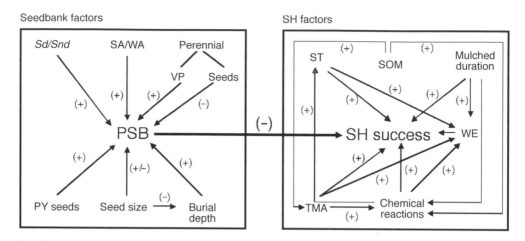

Fig. 11.5. Positive (+) and negative (−) relationships that determine the success or failure of soil solarization (SH) in weed control. PSB, the persistency of seedbank under SH conditions; *Sd/Snd*, the ratio between dormant and non-dormant seeds; VP, vegetative propagules; SA/WA, the ratio between summer and winter annual weed seeds; PY, physical dormancy; ST, soil temperature; SOM, soil organic matter; TMA, thermophilic microorganism activity; WE, weakening effect.

enhanced organic matter decomposition further accelerates seedbank deterioration. An increase in the soil organic matter is accompanied by an enhancement of thermophilic microbial activity, resulting in additional soil temperature elevation and the release of phytotoxic compounds. Thus, organic matter could advance the seed bank deterioration due to the interactions between physical, chemical and biological changes and their combined effect on the 'weakened' seeds.

Seedbank factors

Some perennial weeds could survive SH by means of their underground vegetative propagules which can emerge from the deeper soil layers. Summer and winter annual weed seeds are effectively controlled by SH. However, summer annuals are more resistant than winter annuals to SH treatment. SH induces a 'suicidal germination' of non-dormant seeds in moist soils under the polythene mulch, excluding them from the seedbank. SH also damages dormant seeds if they are able to imbibe water, due to embryonic degradation. Indeed, it is clear that water-impermeable seeds (PY and combinational (PY + PD) dormant seeds) are more tolerant to SH. Successful SH occurs when the dormancy of PY seeds is broken in the early stage of the treatment. The success or failure of SH against small seeds depends on soil disturbance. Soil cultivation after the SH may exhume small seeds that were not exposed to the thermal killing effect in the deep soil profile. Large seeds are able to emerge from deeper soil layers even without soil disturbance.

Future research should aim to improve solarization technology and develop cheaper and more environmentally acceptable mulches to enhance soil temperature transmission in the vertical soil profile. These improvements should make SH suitable for marginal climatic regions and for less profitable crops, expand the spectrum of the weeds controlled, and reduce the duration of the SH process.

11.7 References

Abdel-Rahim, M.F., Satour, M.M., Mickail, K.Y., El-Eruki, S.A., Grinstein, A., Chen, Y. and Katan, J. (1988) Effectiveness of soil solarization in furrow irrigated Egyptian soils. *Plant Disease* 72, 143–146.

Abdul-Razik, A., Grinstein, A., Zeydan, O., Rubin, B., Tal, A. and Katan, J. (1988) Soil solarization and fumigation of strawberry plots. *Acta Horticulturae* 265, 586–590.

Abu-Irmaileh, B.E. (1991a) Soil solarization controls broomrapes (*Orobanche* spp.) in host vegetable crops in the Jordan valley. *Weed Technology* 5, 575–581.

Abu-Irmaileh, B.E. (1991b) Weed control in squash and tomato fields by soil solarization in the Jordan valley. *Weed Research* 31, 125–133.

Abu-Irmaileh, B.E. and Thahabi, S. (1997) Comparative solarization effects on seed germination of *Cuscuta* and *Orobanche* species. In: Stapleton, J.J., DeVay, J.E. and Elmore, C.L. (eds) *Proceedings of the Second International Conference on Soil Solarization and Integrated Management of Soil-borne Pests, Aleppo, Syria.* FAO Plant Protection and Production Paper 147, FAO, Rome, Italy, pp. 227–235.

Al Kayssi, A.W. and Al-Karaghouli, A. (2002) A new approach for soil solarization by using paraffin-wax emulsion as a mulching material. *Renewable Energy* 26, 637–648.

Al-Masoom, A.A., Saghir, A.R. and Itani, S. (1993) Soil solarization for weed management. *Weed Technology* 7, 507–510.

Altieri, M.A. and Liebman, M. (1988) Weed management: ecological guidelines. In: Altieri, M.A. and Liebman, M. (eds) *Weed Management in Agroecosystems: Ecological Approaches.* CRC Press, Boca Raton, FL, USA.

Altieri, M.A. (1992) Agroecological foundations of alternative agriculture in California. *Agriculture, Ecosystems and Environment* 39, 23–53.

Arora, A. and Yaduraju, N.T. (1998) High-temperature effects on germination and viability of weed seeds in soil. *Journal of Agronomy and Crop Science* 181, 35–43.

Avissar, R., Mahrer, Y., Margulies, L. and Katan, J. (1986a) Field aging to transparent polyethylene mulches. I. Photometric properties. *Soil Science Society of America Journal* 50, 202–205.

Avissar, R., Naot, O., Mahrer, Y. and Katan, J. (1986b) Field aging of transparent polyethylene mulches. II. Influence on the effectiveness of soil heating. *Soil Science Society of America Journal* 50, 205–209.

Baskin, C.C. and Baskin, J.M. (1998) *Seeds, Ecology, Biogeography, and Evolution of Dormancy and Germination.* Academic Press, San Diego, CA, USA.

Baskin, J.M. and Baskin, C.C. (2004) A classification system for seed dormancy. *Seed Science Research* 14, 1–16.

Baskin, J.M., Baskin, C.C. and Li, X. (2000) Taxonomy, anatomy and evolution of physical dormancy in seeds. *Plant Species Biology* 15, 139–152.

Benlioglu, S., Boz, O., Kaskavalci, Y.G. and Benlioglu, K. (2004) Alternative soil solarization treatments for the control of soil-borne diseases and weeds of strawberry in the western Anatolia of Turkey. *Journal of Phytopathology* 153, 423–430.

Ben-Yaphet, Y., Stapleton, J.J., Wakeman, R.J. and DeVay, J.A. (1987) Comparative effects of soil solarization with single and double layers of polyethylene film on survival of *Fusarium oxysporum* f.sp. *vasinfectum*. *Phytoparasitica* 15, 181–185.

Bibbery, R.B. (1947) Physiological studies of weed seed germination. *Plant Physiology* 23, 467–484.

Boz, O. (2004) Efficacy and profitability of solarization for weed control in strawberry. *Asian Journal of Plant Sciences* 3, 731–735.

Bradbeer J.W. (1988) *Seed Dormancy and Germination.* Chapman and Hall, London.

Brock, T.D. (1978) *Thermophilic Microorganisms and Life at High Temperatures.* Springer-Verlag, New York.

Broekaert, W.F., Terras, F.R.G., Camue, B.P.A. and Osborn, R.W. (1995) Plant defenses: novel antimicrobial peptides as components of host defense system. *Plant Physiology* 108, 1353–1358.

Caussanel, J-P., Trouvelot, A., Vivant, J. and Gianinazzi, S. (1997) Effects of soil solarization on weed infestation and mycorrhiza management. In: Stapleton, J.J., DeVay, J.E. and Elmore, C.L. (eds) *Proceedings of the Second International Conference on Soil Solarization and Integrated Management of Soil-borne Pests,* Aleppo, Syria. FAO Plant Protection and Production Paper No. 147, FAO, Rome, Italy, pp. 212–225.

Cavers, P.B. and Benoit, D.L. (1989) Seedbanks in arable land. In: Leck, M.A., Parker, V.T. and Simpson, R.L. (eds) *Ecology of Soil Seedbanks.* Academic Press, New York, pp. 309–328.

Chapin, F.S., III, Zavaleta, E.S., Eviner, V.T., Naylor, R.L., Vitousek, P.M., Reynolds, H.L., Hooper, D.U., Lavorel, S., Sala, O.E., Hobbie, S.E., Mack, M.C. and Díaz, S. (2000) Consequences of changing biodiversity. *Nature* 405, 234–242.

Chase, C.A., Sinclair, J.R. and Locascio, S.N. (1999) Effect of soil temperature and tuber depth on *Cyperus* spp. control. *Weed Science* 47, 467–472.

Chauhan, Y.S., Nene, Y.L., Johansen, C., Haware, M.P., Saxena, N.P., Singh, S., Sharma, S.B., Sahrawat, K.L., Burford, J.R., Rupela, O.P., Kumar Rao, J.V.D.K. and Sithanantham, S. (1988) *Effects of Soil Solarization on Pigeonpea and Chickpea*. Research Bulletin No. 11, International Crop Research Institute for the Semi-Arid Tropics, Andhra Pradesh, India.

Chee-Sanford, C.J., Williams, M.M., II, Davis, A.S. and Sims, G.K. (2006) Do microorganisms influence seedbank dynamics? *Weed Science* 54, 575–587.

Chen, Y., Katan, J., Gamliel, A., Aviad, T. and Schnitzer, M. (2004) Involvement of soluble organic matter in increased plant growth in solarized soils. *Biology and Fertility of Soils* 32, 28–34.

Chou, C.H. and Patrick, Z.A. (1976) Identification and phytotoxic activity of compounds produced during decomposition of corn and rye residues in soil. *Journal of Chemical Ecology* 2, 369–387.

Cohen, O. (2006) Understanding the seedbank eco-biology of plant invader (*Acacia saligna*) in natural and management conditions. PhD thesis. Ben-Gurion University of the Negev, Israel.

D'Antonio, C., Levine, J. and Thomsen, M. (2001) Ecosystem resistance to invasion and the role of propagule supply: a California perspective. *Journal of Mediterranean Ecology* 2, 233–246.

Davis, M.A., Grime, J.P. and Thompson, K. (2000) Fluctuating resources in plant communities: a general theory of invisibility. *Journal of Ecology* 88, 528–534.

Davis, S.A. (2006) When does it make sense to target the weed seed bank? *Weed Science* 54, 558–565.

Dekker, J. (1999) Soil weed seedbanks and weed management. In: Buhler, D.D. (ed.) *Expanding the Context of Weed Management*. Food Products Press, New York.

Delouche, J.C. and Baskin, C.C. (1973) Accelerated aging temperatures for predicting the relative storability of seed lots. *Seed Science Technology* 1, 427–452.

DeVay, J.E. and Katan, J. (1991) Mechanisms of pathogen control in solarized soil. In: Katan, J and DeVay. J.E. (eds) *Soil Solarization*. CRC Publications, Boca Raton, FL, USA, pp. 87–101.

Economou, G., Mavrogiannopoulos, G. and Paspatis, E.A. (1997) Weed seed responsiveness to thermal degree hours under laboratory conditions and soil solarization in greenhouse. In: Stapleton, J.J., DeVay, J.E. and Elmore, C.L. (eds) *Proceedings of the Second International Conference on Soil Solarization and Integrated Management of Soil-Borne Pests, Aleppo, Syria*. FAO Plant Protection and Production Paper No. 147, FAO, Rome, Italy, pp. 246–263.

Egley, G.H. (1983) Weed seed and seedling reductions by soil solarization with transparent polyethylene sheets. *Weed Science* 31, 404–409.

Egley, G.H. (1990) High-temperature effects on germination and survival of weed seeds in soil. *Weed Science* 38, 429–435.

Elmore, C.L. (1990) Use of solarization for weed control. In: DeVay, J.E., Stapleton, J.J. and Elmore, C.L. (eds) *Proceedings of the First International Conference on Soil Solarization*. Amman, Jordan, pp. 19–25.

Elmore, C.L. (1991) Weed control by soil solarization. In: Katan, J. and DeVay, J.E. (eds) *Soil Solarization*. CRC Publications, Boca Raton, FL, USA, pp. 61–72.

Elmore, C.L., Roncoroni, J.A. and Giraud, D.D. (1993) Perennial weeds respond to control by soil solarization. *California Agriculture* 47, 19–22.

Emms, J., Virtue, J.G., Preston, C. and Bellotti, W.D. (2005) Legumes in temperate Australia: a survey of naturalization and impact in natural ecosystems. *Biological Conservation* 125, 323–333.

FAO (1995) *Taller Regional Solarizacion del Suelo*. División de Producción y Protección Vegetal, FAO, Rome, Italy, 68 pp.

Gamliel, A. and Becker, E. (1996) *A Method for Applying Plastic Mulch*. Israel Patent No. 118787; USA Patent No. 6,270,291 B2.

Gamliel, A. and Stapleton, J.J. (1993a) Effect of chicken compost or ammonium phosphate and solarization in pathogen control, rhizosphere microorganisms, and lettuce growth. *Plant Disease* 77, 886–891.

Gamliel, A. and Stapleton, J.J. (1993b) Characterization of antifungal volatile compounds evolved from solarized soil amended with cabbage residues. *Phytopathology* 83, 899–905.

Gamliel, A., Austeraweil, M. and Kritzman, M. (2000) Non-chemical approach to soilborne pest management: organic amendments. *Crop Protection* 19, 847–853.

Goldewijk, K.K. (2001) Estimating global land use change over the past 300 years: the HYDE database. *Global Biogeochemical Cycles* 15, 417–434.

Grenz, J.H., Manschadi, A.M., Meinke, H. and Sauerborn, J. (2005) Assessing strategies for *Orobanche* sp. control using a combined seedbank and competition model. *Agronomy Journal* 97, 1551–1559.

Grime, J.P. (1981) The role of seed dormancy in vegetation dynamics. *Annals of Applied Biology* 98, 555–558.

Grinstein, A., Katan, J. Abdul-Razik, A., Zeydan, O. and Elad, Y. (1979) Control of *Sclerotium rolfsii* and weeds in peanuts by solar heating of the soil. *Plant Disease Reporter* 63, 1056–1059.

Haidar, M.A. and Iskandarani, N. (1997) Soil solarization for control of dodder (*Cuscuta* spp.) and other weeds in cabbage. In: Stapleton, J.J., DeVay, J.E. and Elmore, C.L. (eds) *Proceedings of the Second International Conference on Soil Solarization and Integrated Management of Soil-borne Pests, Aleppo, Syria*. FAO Plant Protection and Production Paper No. 147, FAO, Rome, Italy, pp. 264–275.

Haidar, M.A., Iskandarani, N., Sidahmed, R. and Baalbki, R. (1999) Response of field dodder (*Cuscuta campestris*) seeds to soil solarization and chicken manure. *Crop Science* 18, 253–258.

Halloin, J.M. (1983) Deterioration resistance mechanisms in seeds. *Phytopathology* 73, 335–339.

Harman, G.E. (1983) Mechanisms of seed infection and pathogenesis. *Phytopathology* 73, 325–329.

Hejazi, M.J., Kastler, J.D. and Norris, R.F. (1980) Control of yellow nutsedge by tarping the soil with clear polyethylene plastic. *Proceedings of the Western Society of Weed Science* 33, 120–126.

Hill, N.M., Patriquin, D.M. and Vander Kloet, S.P. (1989) Weed seedbank and vegetation at the beginning and end of the first cycle of a 4-course crop rotation with minimal weed control. *Journal of Applied Ecology* 26, 233–246.

Holm, L.G., Plucknett, D.L., Pancho, J.V. and Herberger, J.P. (1977) *The World's Worst Weeds: Distribution and Biology*. University of Hawaii, Honolulu, Hawaii, USA.

Holt, J.S. (1988) Ecological and physiological characteristics of weeds. In: Altieri, M.A. and Liebman, M. (eds) *Weed Management in Agroecosystems: Ecological Approaches*. CRC Press, Boca Raton, FL, USA, pp. 7–23.

Horowitz, M. (1980) Weed research in Israel. *Weed Science* 28, 457–460.

Horowitz, M., Regev, Y. and Herzlinger, G. (1983) Solarization for weed control. *Weed Science* 31, 170–179.

Hutchison, V.H. (1976) Factors influencing thermal tolerances of individual organisms. In: *Thermal Ecology II, Proceedings of a Symposium, August, Georgia, USA*. CONF-750425, 1976, pp. 10–26.

Ioannou, N. (2000) Soil solarization as a substitute for methyl bromide fumigation in greenhouse tomato production in Cyprus. *Phytoparasitica* 28, 1–9.

Jacobsohn, R., Greensberger, A., Katan, J., Levi, M. and Alon, H. 1980. Control of Egyptian broomrape (*Orobanche aegyptiaca*) and other weeds by means of solar heating of the soil by polyethylene mulching. *Weed Science* 28, 312–316.

Jeffery, D.J., Holmes, P.M. and Rebelo, A.G. (1988) Effects of dry heat on seed germination in selected indigenous and alien legume species in South Africa. *South African Journal of Botany* 54, 28–34.

Katan, J. (1981) Solarization heating (solarization) of soil for control of soilborne pest. *Annual Review of Phytopathology* 19, 211–236.

Katan, J. (1987) Soil solarization. In: Chet, I. (ed.) *Innovative Approaches to Plant Disease Control*. John Wiley and Sons, New York, pp. 77–105.

Katan, J. (1999) The methyl bromide issue: problems and potential solutions. *Journal of Plant Pathology* 81, 153–159.

Katan, J. (2000) Physical and cultural methods for the management of soil-borne pathogens. *Crop Science* 19, 725–731.

Katan, J. and DeVay, J.E. (1991) Soil solarization: historical perspectives, principles, and uses. In: Katan, J. and DeVay, J.E. (eds) *Soil Solarization*. CRC Publications, Boca Raton, pp. 23–37.

Katan, J., Greenberger, A., Alon, H. and Grinstein, A. (1976) Solar heating by polyethylene mulching for the control of diseases caused by soil-borne pathogens. *Phytopathology* 66, 683–688.

Kebreab, E. and Murdoch, A.J. (1999) Effect of temperature and humidity on the longevity of *Orobanche* seeds. *Weed Research* 39, 199–211.

Kebreab, E. and Murdoch, A.J. (2001) Simulation of integrated control strategies for *Orobanche* spp. based on a life cycle model. *Experimental Agriculture* 37, 37–51.

Keley, J.E. (1987) Role of fire in seed germination of woody taxa in California chaparral. *Ecology* 68, 434–443.

Kennedy, A.C. (1999) Soil microorganisms for weed management. In: Buhler, D.D. (ed.) *Expending the Context of Weed Management*. Food Products Press, New York, pp. 123–138.

Khan, A.R., Srivasava, R.C., Ghoral, A.K. and Singh, S.R. (2003) Efficient soil solarization for weed control in the rain-fed upland rice ecosystem. *International Agrophysics* 17, 99–103.

Kremer, R.J. (1986) Antimicrobial activity of velvetleaf (*Abutilon theophrasti*) seeds. *Weed Science* 34, 617–622.

Kremer, R.J. (1993) Management of weed seed banks with microorganisms. *Ecological Application* 3, 42–52.

Kremer, R.J. and Schulte, L.K. (1989) Influence of chemical treatment and *Fusarium oxysporum* on velvetleaf. *Weed Technology* 3, 369–374.

Kremer, R.J., Hughes, L.B., Jer, L.B. and Aldrich, R.J. (1984) Examination of microorganisms and deterioration resistance mechanisms associated with velvetleaf seed. *Agronomy Journal* 76, 745–749.

Kumar, B., Yaduraju, N.T, Ahuja, K.N. and Prasad, D. (1993) Effect of soil solarization on weeds and nematodes under tropical Indian conditions. *Weed Research* 33, 423–429.

Kumar, R. and Sharma, J. (2005) Effect of soil solarization on true potato (*Solanum tuberosum* L.) seed germination, seedling growth, weed population and tuber yield. *Potato Research* 48, 15–23.

Linke, K.H. (1994) Effects of soil solarization on arable weeds under Mediterranean conditions: control, lack of response or stimulation. *Crop Protection* 13, 114–120.

Lira-Saldivar, R.H., Salas, M.A., Cruz, J., Coronado, F.D., Guerrero, E. and Gallegos, G. (2004) Solarization and goat manure on weed management and melon yield. *Phyton* 53, 205–211.

Lonsdale, W.M. (1993) Losses from the seedbank *Mimosa pigra*: soil microorganisms vs. temperature fluctuations. *Journal of Applied Ecology* 30, 654–660.

Lynch, J.M. (1980) Effect of organic acids on the germination of seeds and growth of seedlings. *Plant, Cell and Environment* 3, 255–259.

Mahrer, Y., Naot, O., Rawitz, E. and Katan, J. (1984) Temperature and moisture regimes in soils mulched with transparent polyethylene. *Soil Science Society of America Journal* 48, 362–367.

Marenco, R.A. and Lustosa, D.C. (2000) Soil solarization for weed control in carrot. *Pesquisa Agropecuária Brasileira* 35, 2025–2032.

Mas, M.T. and Verdu, A.M.C. (2002) Effects of thermal shocks on the germination of *Amarantus retroflexus*: use of the Excel solver tool to model cumulative germination. *Seed Science and Technology* 30, 299–310.

Mauromicale, G., La-Monaco, A., Longo, A.M.G. and Restuccia, A. (2005) Soil solarization, a non-chemical method to control branched broomrape (*Orobanche ramose*) and improve the yield of greenhouse tomato. *Weed Science* 53, 877–883.

McGovern, R.J., McSorley, R. and Wang, K.J. (2004) Optimizing bed orientation and number of plastic layers for solarization in Florida. *Annual Proceedings of the Soil Science Society of Florida* 63, 92–95.

Menalled, F.D., Gross, K.L. and Hammond, M. (2001) Weed aboveground and seedbank community responses to agricultural management systems. *Ecological Applications* 11, 1586–1601.

Mickelson, J.A. and Grey, W.E. (2006) Effect of soil water content on wild oat (*Avena fatua*) seed mortality and seedling emergence. *Weed Science* 54, 255–262.

Mills, J.T. (1983) Insect–fungus associations influencing seed deterioration. *Phytopathology* 73, 330–335.

Mohamed, A.H., Ejeta, G., Butler, L.G. and Housley, T.L. (1998) Moisture content and dormancy in *Striga asiatica* seeds. *Weed Research* 38, 257–265.

Moya, M. and Furukawa, G. (2000) Use of solar energy (solarization) for weed control in greenhouse soil for ornamental crops. *New Zealand Plant Protection* 53, 34–37.

Mudalagiriyappa, Nanjappa, H.V. and Ramachandrappa, B.K. (1996) Effect of soil solarization on weed growth and yield of kharif groundnut (*Arachis hypogaea*). *Indian Journal of Agronomy* 44, 396–399.

Nir, E., Rubin, B. and Zharasov, S.W. (1996) On the biology and selective control of field dodder (*Cuscuta campestris*). In: Moreno, M.T., Cuberu, J.I., Berner, D., Joel, D., Musselman, L.J. and Parker, C. (eds) *Advances in Parasitic Plant Research*. Proceedings of the 6th International Symposium on Parasitic Weeds, Cordoba, Spain, pp. 809–816.

Osman, M.A., Raju, P.S. and Peacock, J.M. (1991) The effect of soil temperature, moisture and nitrogen on *Striga asiatica* (L.) kuntze seed germination, viability and emergence on sorghum (*Sorghum bicolor* L. Moench) roots under field conditions. *Plant and Soil* 131, 265–273.

Paynter, Q., Csurhes, S.M., Heard, T.A., Ireson, J., Julien, M.H., Lloyd, J., Lonsdale, W.M., Palmer, W.A., Sheppard, A.W., Klinken, R.D.V. and van Klinken, R.D. (2003) Worth the risk? Introduction of legumes can cause more harm than good: an Australian perspective. *Australian Systematic Botany* 16, 81–88.

Peachey, R.E., Pinkerton, J.N., Ivors, K.L., Miller, M.L. and Moore, L.W. (2001) Effect of soil solarization, cover crops and metham on field emergence and survival of buried annual bluegrass (*Poa annua*). *Weed Technology* 15, 81–88.

Petersen, J., Belz, R., Walker, F. and Hurle, K. (2001) Weed suppression by release of isothiocyanates from turnip–rape mulch. *Agronomy Journal* 93, 37–43.

Pimentel, D., Acquay, H., Biltonen, M., Rice, P., Sliva, N., Nelson, J., Lipner, V., Giordano, S., Horowitz, A. and D'Amore, M. (1992) Environmental and economic cost of pesticide use. *Bioscience* 42, 750–760.

Quinlivan, B.J. (1971) Seed coat impermeability in legumes. *Journal of the Australian Institute of Agricultural Science* 37, 283–295.

Radosevich, S., Holt, J. and Ghersa, C. (1997) *Weed Ecology; Implications for Management*. John Wiley and Sons, New York.

Ricci, M.S.F., De Almeida, D.L., Riberio, R.D.L.D., Dequino, A.M., Pereira, J.C., De-Polli, H., Reis, V.M. and Eklund, C.R. (1999) *Cyperus rotundus* control by solarization. *Biological Agriculture and Horticulture* 17, 151–157.

Ristaino, J.B. and Thomas, W. (1997) Agriculture, methyl bromide and the environment. *Plant Disease* 81, 964–978.

Rolston, M.P. (1978) Water impermeable seed dormancy. *Botanical Review* 44, 365–396.

Rubin, B. and Benjamin, A. (1983) Solar heating of the soil: effect on soil incorporated herbicides and on weed control. *Weed Science* 31, 819–825.

Rubin, B. and Benjamin, A. (1984) Solar heating of the soil: involvement of environmental factors in the weed control process. *Weed Science* 32, 138–144.

Saghir, A.R. (1997) Soil solarization: an alternative technique for weed management in hot climates. In: Stapleton, J.J., DeVay, J.E. and Elmore, C.L. (eds) *Proceedings of the Second International Conference on Soil Solarization and Integrated Management of Soil-Borne Pests, Aleppo, Syria*. FAO Plant Protection and Production Paper No. 147, FAO, Rome, Italy, pp. 206–211.

Sauerborn, J., Linke, K.-H., Saxena, M.C. and Koch, W. (1989) Solarization; a physical control method for weeds and parasitic plants (*Orobanche* spp.) in Mediterranean agriculture. *Weed Research* 29, 391–397.

Schafer, D.E. and Chilcote, D.O. (1969) Factors influencing persistence and depletion in buried seed populations: a model for analysis of parameters of buried seeds persistence and depletion. *Crop Science* 9, 417–418.

Schafer, D.E. and Chilcote, D.O. (1970) Factors influencing persistence and depletion in buried seed populations. II. The effect of soil temperature and moisture. *Crop Science* 10, 342–345.

Shea, S.R., McMormick, J. and Portlock, C.C. (1979) The effect of fires on regeneration of leguminous species in the northern Jarrah (*Eucalyptus maginata* Sm) forest of Western Australia. *Australian Journal of Ecology* 4, 195–205.

Shlevin, E., Mahrer, Y. and Katan, J. (2004) Effect of moisture on thermal inactivation of soilborne pathogens under structural solarization. *Phytopathology* 94, 132–137.

Singh, R. (2006) Use of soil solarization in weed management on soybean under Indian condition. *Tropical Science* 46, 70–73.

Singh, V.P., Dixit, A., Mishra, J.S. and Yaduraju, N.T. (2004) Effect of period of soil solarization and weed-control measures on weed growth, and productivity of soybean (*Glycine max*). *Indian Journal of Agricultural Sciences* 74, 324–328.

Standifer, L.C., Wilson, P.W. and Sorbet, R.P. (1984) Effect of solarization on soil weed seed populations. *Weed Science* 32, 569–573.

Stapleton, J.J. (1997) Soil solarization: an alternative soil disinfestation strategy comes of age. *UC Plant Protection Quarterly* 7, 1–5.

Stapleton, J.J. (2000a) Soil solarization in various agricultural production systems. *Crop Science* 37, 837–841.

Stapleton, J.J. (2000b) Developing alternative heat treatments for disinfestation of soil and planting media. *International Plant Propagators' Society: Combined Proceedings of Annual Meetings* 50, 561–563.

Stapleton, J.J. and DeVay, J.E. (1986) Soil solarization: a non-chemical approach for management of plant pathogens and pest. *Crop Protection* 5, 190–198.

Stapleton, J.J., Elmore, C.L. and DeVay, J.E. (2000a) Solarization and biofumigation help disinfest soil. *California Agriculture* 54, 42–45.

Stapleton, J.J., Parther, T.S. and Dahlquist, R.M. (2000b) Implementation and validation of a thermal death database to predict efficacy of solarization for weed management in California. *UC Plant Protect Quarterly* 10, 9–10.

Stapleton, J.J., Parther, T.S., Mallek, S.B., Ruiz, T.S. and Elmore, C.L. (2002) High temperature solarization for production of weed-free container soils and potting mixes. *HortTechnology* 12, 697–700.

Stapleton, J.J., Molinar, R.H., Lynn-Patterson, K., McFeeters, S.K. and Shrestha, A. (2005) Soil solarization provides weed control for limited-resource and organic growers in warmer climates. *California Agriculture* 59, 84–89.

Stevens, C., Khan, V.A., Brown, J.E., Hochmuth, G., Splittstoesser, W.E. and Granberry, D.M. (1991) Plastic chemistry and technology as related to plasticulture and solar heating of soil. In: Katan, J. and DeVay, J.E. (eds) *Soil Solarization*. CRC Publications, Boca Raton, FL, USA, pp. 151–158.

Stevens, C., Khan, V.A., Okoronkwo, T., Tang, A.Y. and Wilson, M. (1989) Evaluation of pre-plant applications of clear polyethylene mulch for controlling weeds in central Alabama. *Proceedings of the National Agricultural Plastics Congress* 20, 65–70.

Tekin, Y., Kadyoolu, Y. and Uremip, Y. (1997) Studies on soil solarization against root-knot nematode and weeds in vegetable greenhouses in the Mediterranean region of Turkey. In: Stapleton, J.J., DeVay, J.E. and Elmore, C.L. (eds) *Proceedings of the Second International Conference on Soil Solarization and Integrated Management of Soil-borne Pests, Aleppo, Syria*. FAO Plant Protection and Production Paper No. 147, FAO, Rome, Italy, pp. 604–615.

Thompson, K., Bakker, J.P., Bakker, R.M. and Hodgson, J.G. (1998) Ecological correlates of seed persistence in soil in the north-west European flora. *Journal of Ecology* 86, 163–169.

Tilman, D., Fargione, J., Wolff, B., D'Antonio, C., Dobson, A., Howarth, R., Schindler, D., Schlesinger, W.H., Simberloff, D. and Swackhamer, D. (2001) Forecasting agriculturally driven global environmental change. *Science* 292, 281–284.

Verdu, A.M.C. and Mas, M.T. (2004) Modeling the effects of thermal shocks varying in temperature and duration on cumulative germination of *Porulaca oleracea* L. *Seed Science and Technology* 32, 297–308.

Vitousek, P., Aber, J.D., Howarth, R.W., Likens, G.E., Malson, P.A., Schindler, D.W., Schlesinger, W.H. and Tilman, D.G. (1997) Technical report: human alteration of the global nitrogen cycle: sources and consequences. *Ecological Applications* 3, 737–750.

Vizantinopoulos, S. and Katranis, N. (1993) Soil solarization in Greece. *Weed Research* 33, 225–230.

Wainwright, M., Nevell, W. and Grayston, S.J. (1986) Effects of organic matter on sulphur oxidation in soil and influence of sulphur oxidation on soil nitrification. *Plant and Soil* 96, 369–376.

Warcup, J.H. (1980) Effect of Heat treatment of forest soil on germination of buried seed. *Australian Journal of Botany* 28, 567–571.

Wheatley, R.H., Millar, S.E. and Griffiths, D.W. (1996) The production of volatile organic compounds during nitrogen transformations in soils. *Plant and Soil* 181, 163–167.

Wilson, R.J. (1998) Biology of weed seeds in the soil. In: Altieri, M.A. and Liebman, M. (eds) *Weed Management in Agroecosystems: Ecological Approaches.* CRC Press, Boca Raton, FL, USA, pp. 25–39.

Yaron, D., Regev, A. and Spector, R. (1991) Economic evaluation of soil solarization and disinfestation. In: Katan, J. and DeVay, J.E. (eds) *Soil Solarization.* CRC Publications, Boca Raton, FL, USA, pp. 171–190.

12 Non-chemical Weed Management: Synopsis, Integration and the Future

M.K. Upadhyaya[1] and R.E. Blackshaw[2]

[1]*Faculty of Land and Food Systems, University of British Columbia, Vancouver, BC, Canada, V6T 1Z4;* [2]*Agriculture and Agri-Food Canada, Lethbridge, AB, Canada, T1J 4B1*

12.1 Introduction

Weeds cause significant loss to agricultural as well as non-agricultural ecosystems. In the past 60 years most weed control research has focused on the development of the technology to control weeds rather than using a holistic approach to learn about weeds in complex ecosystems and to employ this knowledge in order to develop weed management strategies which consider economic, ecological and social factors simultaneously.

The availability and acceptance of highly effective and selective herbicides after World War II shifted the focus of weed management from non-chemical options, which had been practised for centuries, to weed control using herbicides. Weeds were not considered as components of agroecosystems comprising a complex web of interactions, and so sustainability issues were easily ignored and preventive or suppressive approaches to weed management were sidelined. The ecological and social consequences of herbicide use were either ignored or played down. Research on environmental impact, toxicology, and weed biology and ecology took a back seat.

Excessive reliance on the use of synthetic organic herbicides, with a narrow short-term focus on the maximization of control, prevented the improvement of other weed management strategies. Cultural weed management options were ignored and research on other non-chemical weed management methods was adversely affected. Several potentially useful weed management options were labelled 'uneconomical' or 'impractical', and their technological development was discontinued. The lack of research on non-chemical options for weed management has made weeds a serious problem in organic farming. Environmental concerns, development of herbicide resistance in weed populations, and a growing demand for pesticide-free produce have renewed interest in non-chemical weed management.

This chapter summarizes the various non-chemical weed management strategies covered in this book, and discusses their integration and future potential, keeping economic, ecological and social factors in mind.

12.2 Non-chemical Weed Management Tool-box

Preventive strategies

A sound prevention strategy is an essential, but often overlooked, component of any integrated weed management (IWM) strategy (Thill and Mallory-Smith, 1997). The saying 'An ounce of prevention is better than a pound of cure' is indeed very applicable to weed management.

Many serious weed problems around the world have been traced to inadvertent weed introductions. Ignorance or delayed response to new weed problems has allowed many of these unintentional introductions to become serious problems. Prompt action to control or eliminate, if possible, new weed problems is therefore essential.

Weeds can be disseminated by many mechanisms, including crop seeds, vegetative propagules, straw and hay, soil, manure, wind, water, animals, machinery and vehicles (Thill and Mallory-Smith, 1997), and weed introductions can be either deliberate (e.g. ornamental and agricultural use) or unintentional. Preventive management at the farm and at the landscape/ecosystem level requires knowledge of the processes and practices that contribute to species introduction, proliferation and dispersal. A sound understanding of persistence and dissemination strategies for specific weeds is essential in order to develop preventive strategies.

Weed prevention strategies aim at preventing: (i) initial introduction; (ii) infestation development; and (iii) dispersal of weeds and their propagules. Because of their role in reproduction and dissemination and their ability to withstand extreme environments, seeds represent an important stage in the life cycle of many weeds. Weed management strategies could be developed at the level of an individual farm, region, district, state or province, country or continent. In this regard it is important to note, when implementing a strategy, that the movement of weeds over a region extending beyond the area for which the strategy was developed must be considered – think globally while acting locally. The formulation and practice of good weed prevention strategies may involve individual and group responsibilities as well as government-enacted laws to prevent the introduction and dissemination of weed propagules (e.g. a Quarantine Act or a Seed Act).

Exploiting weed–crop interaction to manage weeds

In order to effectively manage weed problems, an understanding of weed–crop interaction is essential (Zimdahl, 2004). Assessing the merits (including the economics) of any weed management practice requires a clear understanding of the impact of weeds on a given crop or on a non-crop environment. It is necessary to have a holistic understanding of how different weeds, present at different densities during various stages of crop growth, influence crop performance, as well as knowing when weed control is justified. Knowledge of the economic threshold (the minimum weed density at which weed control is economically justified) and critical period of weed interference – when weeds must be controlled to obtain maximum yield – is essential. A sound knowledge of weed population dynamics and how it is affected by different weed management strategies is important in developing an optimum crop management strategy.

Several experimental approaches have been developed to study plant–plant interactions and their underlying mechanisms (Weigelt and Jolliffe, 2003; Furness et al., 2005). These include: additive and replacement series experimental designs, inverse density models, allometric analysis, neighbourhood analysis, size structure analysis, and the tracking of competition over time. These approaches differ in their strengths and limitations with regard to the information they provide about plant–plant interaction and their mechanistic underpinnings. Studies involving a crop and single weed species grown at unrealistic densities and conditions may provide some useful information but are of little value in developing a holistic understanding of plant–plant interaction in complex agroecosystems.

It is important that any knowledge and understanding developed using these approaches is not confined to scientific journals or books but is made available to producers in a form that they can use to make weed management decisions in their specific situations.

Cultural options

Globally, cultural control has been one of the most widely used weed control options for centuries. The introduction of effective, selective and inexpensive herbicides has diverted emphasis away from cultural control during the past 50 years. However, cultural control has now begun to regain its importance within IWM systems in recent years.

Cultural weed control options include: crop rotation, increasing the competitive ability of a crop, delayed or early seeding, flooding, inclusion of green manure and cover crops, and intercropping (Bond and Grundy, 2001; Shrestha *et al.*, 2004). While, taken individually, many of these options may not provide the desired level of weed control, the degree and consistency of control can be increased significantly by integrating several of these options in a multi-year weed-management strategy.

The ability of crops to compete against weeds could be increased by selecting the right crops and cultivars, considering the weeds present as well as the climate, ensuring rapid and uniform crop emergence through proper seedbed preparation, and by using the right seed and seeding depth, increasing planting density and adapting planting patterns wherever possible to crowd out weeds, adequate and localized resource (water, fertilizer) application, and optimum management of the crop, including insect pest and disease management. Plant breeders can play a significant role in this regard by developing canopy architectures that maximize a crop's ability to compete with weeds.

Allelochemicals, released by either the living cover crop or its dead residues, can also influence weed growth and weed seed germination. Cover crops not only reduce weed growth during their life cycle but also lower weed pressure in subsequent crops by reducing soil-borne seed banks and the below-ground food reserves of perennial weeds, due to competition in the year when the cover crop was grown. In addition to a direct reduction in weed seed production due to competition for resources, it has also been suggested that cover crops can reduce the size of soil seed banks by increasing the activities of soil predators.

In addition to controlling weeds, cover crops also reduce soil erosion by wind and water, increase soil organic matter content, improve soil structure, and influence the soil's nutrient status, nutrient cycling, soil biology, and insect pests and diseases (Blackshaw *et al.*, 2005). While the presence of a cover crop with a cash crop may adversely affect the latter because of competition for resources, a cash crop grown following a leguminous cover crop may benefit from the nitrogen fixed by the legume. Furthermore, the presence of a cover crop, compared with fallow land, may also enhance soil mycorrhizal potential, which in turn may benefit a subsequent mycorrhizal cash crop.

Cover crops

Cover crops (e.g. rye, hairy vetch, red clover, sweetclover, velvetbean, cowpea) are grown for their various ecological benefits in an agroecosystem, including weed suppression (Akemo *et al.*, 2000; Blackshaw *et al.*, 2001; Ross *et al.*, 2001) and not as cash crops. They can be grown in rotation, during a fallow period, during an off-season winter period (a more acceptable approach for many farmers), or simultaneously during part or all of the life cycle of a cash crop. Depending upon their specific objectives, they have been referred to as smother crops, green manure crops, living mulches, and catch crops.

Cover crops control weeds mainly by absorbing photosynthetically active radiation and by lowering the red : far-red ratio of transmitted light, which in turn influences the germination of light-requiring weed seeds.

Allelopathic interactions

Many plant species (e.g. rye, sorghum, mustards, velvetbean, black walnut) are known to release chemicals which can influence associated species either directly by influencing their growth and/or seed germination, or indirectly by affecting soil biology (e.g. by inhibiting mycorrhizal inoculation potential) (Inderjit and Keating, 1999; Weston and Duke, 2003). This phenomenon, called allelopathy, could be used to suppress weeds by using companion or rotational crops, mulching with plant residues, applying plant extracts, or by incorporating allelopathic potential in crop cultivars using plant improvement techniques (Einhellig and Leather, 1988; Weston, 1996, 2005; Inderjit and Bhowmik, 2002). While allelopathy seems to offer interesting potential for IWM systems, more research is needed before this potential can be fully exploited under field conditions.

The short persistence of allelochemicals in the environment and their high specificity may limit their usefulness for weed control. More research is needed in order to understand the regulation of production of allelochemicals and their mechanism of action, the genetics of allelopathy, the flow of genes responsible for allelopathy in field populations and its implications, the specificity and fate of allelochemicals in the soil, factors affecting the susceptibility of weeds to allelochemicals and their influence on soil biology, as well as interactions between allelochemicals.

Biological control using arthropods

Biological control, particularly using phytophagous insects, offers an attractive option for the control of introduced weedy species in certain situations (Julien and Griffiths, 1998; McFadyen, 1998). Natural enemies of the weed are introduced from the weed's native region to re-establish the natural control that was disrupted when the weed was brought to the new area. This environmentally friendly option offers several advantages. Unlike tillage it does not cause an abrupt disruption in ecosystems, damage the soil structure, or make the soil vulnerable to erosion. No synthetic herbicides are added to the environment and the control is long-lasting and less expensive in the long run. Weed control using this option in difficult-to-reach (e.g. steep or rough terrains) or environmentally sensitive (e.g. river or lakeside) areas is easier and/or more acceptable than mechanical and chemical options.

The establishment of a classical biological control programme is a long and expensive process. An extensive analysis of the weed problem is necessary before making a decision to implement biological control for a specific weed. A study of the nature and location of weed infestation, a survey of the area infested, survival and persistence strategy of the weed (mode of reproduction, fecundity, seed bank dynamics, seed germination behaviour), and analyses of economic impact (both current and future projections) of the weed, conflicts of interest, and impact of climate and any anticipated pesticide use, particularly insecticides, is necessary. After determining where the weed was introduced from, its natural enemies are surveyed, suitable host-specific agents identified, host specificity re-confirmed, and selected agents are introduced, multiplied and released. It is important to ensure that no undesirable pests or pathogens are co-introduced inadvertently. Both insect populations and weed infestations are then regularly monitored to determine the success of the weed control programme. If an acceptable level of control is not achieved, it may be necessary to introduce additional agents, preferably focusing on a different aspect of weed biology.

Biological control using phytophagous insects can be a very useful option for perennial ecosystems, particularly with a physically continuous stand of a single weedy species (e.g. rangelands or a non-agricultural system). However, classical biological control has several limitations in cultivated crops. The presence of several weedy species, taxonomically related to economically important crops, in cultivated agroecosystems requires the introduction of many host-specific biotic agents, which increases the cost of weed control programmes as well as the risk of inadvertently introducing undesirable agents. Some biotic agents are also vulnerable to sudden and/or drastic changes that occur with crop harvesting, crop rotation, tillage, and fallowing of land. Moreover, achieving an acceptable level of weed control during the critical period of weed interference may not be possible if the necessary level of insect population during this period is not achieved and maintained. The use of insecticide(s) to control crop insect pests, if needed, may harm the biotic agent(s) used. When weed infestation is spotty or discontinuous and the biotic agent used not very mobile, there may be a problem. For these reasons, biological control using insects has not been very practical in annual cropping systems. More research is needed to develop this option for weed management in these systems.

Bioherbicides

The use of bioherbicides for weed control involves overwhelming weeds with single or multiple applications of a pathogen (Hoagland, 2001). When the organism used is a fungal pathogen, it is called a mycoherbicide. Pathogens selected for bioherbicide development are generally indigenous and, because of their environmental benefits, registration of bioherbicides is generally a shorter and less expensive process compared with synthetic herbicides.

Despite their enormous potential for use in IWM systems, only a few bioherbicides have been registered to date. In order to fully realize the potential of these environmentally friendly weed management tools, further research to identify bioherbicides for important weeds and to improve their formulation, production, application technology, efficacy, and compatibility with other agricultural practices is needed. Pathogens with low virulence but other desirable characteristics (e.g. specificity, desirable epidemiology) could be genetically engineered to increase their virulence, thus allowing some of the pathogens discovered previously to be used as bioherbicides in the future.

Mechanical weed management

Mechanical weed control involves tillage as well as the cutting and pulling of weeds and is probably the oldest weed management tool (Wicks *et al.*, 1995; Cloutier and Leblanc, 2001). Many mechanical operations that farmers around the world have traditionally practised have evolved, at least in part, because of their weed control benefit. Like many other non-chemical options, the availability and acceptability of herbicides has diverted attention away from research on mechanical weed management, hampering technology development in this area.

A variety of tillage operations have been developed with different objectives. Tillage not only controls weeds, it also breaks up the soil to prepare the seedbed, facilitating rapid and uniform germination and root penetration, increases soil aeration and rainfall penetration, and provides the soil-surface topography needed for specific crops (raised beds, furrows, etc).

Tillage also has some negative effects on agroecosystems. It breaks and exposes the soil, making it vulnerable to wind and water erosion, reduces soil moisture and organic matter contents, influences seed movement in the soil profile, induces germination of some weed seeds, and may damage soil structure and crop roots and compact the subsoil. Weed control in

crop rows, problems in crop residue management, and weather dependency are major challenges for weed control by tillage.

In many parts of the world where soil erosion is a problem, conventional tillage practices have been questioned and crop production involving reduced tillage is being adapted. A reduction in tillage could influence the spectrum of weeds present and make the control of volunteer seedlings difficult. The presence of a trash cover may intercept herbicide spray, if used, sheltering weed seedlings below the trash. Interception of light by trash may also influence the germination of light-requiring weed seeds.

Since tillage moves weed seeds in the soil profile and may stimulate their germination either by exposing light-sensitive seeds to solar radiation or by oxygenating the soil air, tillage practices have a strong influence on soil-borne seed banks. More research is needed on the effects of various tillage practices on the distribution of weed seeds in the soil profile, the size of seed banks, seed longevity, and the overall dynamics of seed banks.

Reduction of seed germination by night-time tillage has suggested some interesting but yet to be exploited opportunities for non-chemical weed management (Hartmann and Nezadal, 1990; Scopel *et al.*, 1994). However, before the potential of night-time tillage can be fully exploited, several questions need to be answered. Whether night tillage selects in favour of species that do not require light for germination, the eventual fate of seeds that do not germinate, the practicality, economics and effectiveness of this practice, the light threshold for seed germination, and the influence of smoothing of the soil surface following tillage and of supplemental light during the daytime cultivation requires further investigation.

Rapid advances in tillage technology, such as automatic guidance systems (mechanical or electronic) to allow the development of self-guided, self-propelled, autonomous equipment to control weeds with precision and minimal operator intervention; the use of real-time image acquisition and analysis and global positioning systems to map the field, crop and weeds; the use of compressed air to blow small weeds out of the crop rows; intelligent weeders that use sensors to distinguish weeds from crops and selectively remove weeds from crops, may make this option a more effective weed management tool in future.

Mowing, which reduces the leaf area of weeds, slows weed growth, and decreases or prevents weed seed production, is particularly effective for upright annual weeds in some ecosystems (e.g. turf). Many weeds produce multiple shoots following mowing due to the release of apical dominance. These weeds may be controlled by repeated mowing and cutting at a lower height. More research on the optimum timing and frequency of mowing for specific weeds is needed. This is particularly true for perennial weeds, where one of the goals of mowing, in addition to reducing competition and seed production, is to deplete underground food reserves in order to weaken the ability of the weed to regrow. With each cutting, the below-ground food reserves are depleted as new shoots are produced. Mowing repeatedly at carefully determined intervals, to prevent the replenishment of below-ground reserves by the shoot, weakens the weed and decreases its ability to compete.

Non-living mulches

Both natural (organic or inorganic) and synthetic non-living mulches are used for weed control in agricultural as well as non-agricultural systems (Bond and Grundy, 2001). Depending on their shape, these mulches can be divided into various categories: sheet mulches (e.g. black, clear or coloured polythene, geotextiles, paper, needle-punched fabrics, and carpets) or particulate mulches (e.g. straw and hay, grass clippings, leaf mould, industrial crop waste, coffee grounds, dry fruit shells, shredded and chipped bark or wood, sawdust, crushed rock, and gravel). These mulches control weeds by inhibiting germination of light-requiring seeds, reducing weed growth by partially or completely absorbing light (dark mulches), solar heating (clear plastic mulches), and/or by physically interfering with weed seedling growth.

Depending on the material used, non-living mulches may also protect soil from wind and water erosion; add organic matter to soil; help conserve soil moisture; increase or decrease rainfall penetration, nutrient leaching and soil

oxygen content, depending on the material used; influence soil properties; improve produce quality by separating it from the soil; and reduce transmission of soil-borne pathogens to shoots due to raindrop splashing. Natural organic mulches (e.g. straw, grass clippings, sawdust) may alter the soil carbon/nitrogen ratio, which may influence nutrient availability.

While non-living mulches have been used for a long time, their cost (material, laying, lifting, disposal), availability, susceptibility to weather (wetting, solar ultraviolet radiation), blowing away by wind and ripping have limited their use. Further research on the development of mulches suitable for a variety of agricultural and non-agricultural systems, their influence on soil properties and crop performance, environmental impacts, biodegradability, reflection and transmission characteristics, impregnation with agrochemicals, foam and spray-on mulches, and seed mate systems that incorporate seeds into perforated biodegradable mulches may increase the usefulness of this option. Certain mulches may introduce weed seeds, phytotoxins or allelochemicals, and pathogens to an area. Care should be taken to avoid these introductions.

Thermal options

Several options for controlling weeds using heat have been developed (Bond and Grundy, 2001; Laguë et al., 2001; Shrestha et al., 2004). These include: fire, directed flaming, hot water, steam, microwave, infrared, ultraviolet radiation, electrocution and freezing. Heating results in the coagulation of proteins and bursting of protoplasm due to expansion, which kills the tissue. Weeds can also be killed by exposure to very low temperature, e.g. by exposing aquatic weeds to low air temperature by removing water from a pond or lake or by freezing terrestrial weeds using dry ice or liquid nitrogen. Some of these options are currently used commercially (e.g. flaming and steam), some need further research and technology development before their adaptation to commercial agriculture, and some appear to be impractical at present (e.g. lasers and electrocution). Before the potential of thermal weed control options is fully realized, a comparative cost–benefit analysis with other weed control options, technology development to improve efficacy, effect on soil biology, and integration with other agroecosystem practices must be thoroughly analysed. It is interesting that some of the options previously considered 'impractical' or 'interesting ideas' are now being reconsidered due to public concerns about environment and health, the development of herbicide resistance, and increased social pressure to regulate herbicides.

Thermal weed control options offer several advantages. They generally do not disturb buried seeds, leave dead biomass on the soil surface, which offers protection against erosion and moisture loss, and may kill some insect pests and pathogens and, most importantly, do not pollute the environment with synthetic herbicides. Their high costs, lack of residual control, risk of starting a fire, and/or concern for applicator safety, however, have limited their use. While some of these options are occasionally promoted as highly environmentally friendly, it must be noted that they involve the use of a large amount of fossil-fuel energy and are not entirely non-polluting.

Solarization or solar heating is also a thermal weed control option which involves the laying of a clear plastic film on well-prepared soil to allow solar radiation to pass through, but which prevents heat from escaping. This increases surface soil and air temperatures below the film which kills weed shoots and some below-ground vegetative parts, weed seeds, nematodes, and soil-borne pathogens and insect pests (Katan, 1981, 1987; Stapleton and DeVay, 1986). The addition of organic matter, particularly chopped crucifer biomass and certain manures (e.g. chicken manure) to soil has been reported to increase the effectiveness of solar heating by releasing antimicrobial volatiles. This procedure is called biofumigation (Stapleton *et al.*, 2000).

Solarization can be an effective and environmentally friendly weed management option in regions where abundant solar heat is available. It has been used for a variety of crops (e.g. tomato, peppers, lettuce, cucumber and aubergine) in many parts of the world. In addition to benefiting the crop by controlling weeds and some soil-borne organisms, solarization also increases the availability of soil nutrients and improves some soil properties. The plastic cover used for solarization also helps in conserving soil moisture.

Despite its several positive attributes, solarization has not been widely used globally. Low availability of heat (duration and intensity), the cost of the plastic film and labour, the removal and disposal of films, as well as the unavailability of land during solarization, have limited its use. For many regions, the adequacy of the available heat to control prevalent weeds by solarization has not been determined.

12.3 Integration of Non-chemical Weed Management Options

An integrated weed management (IWM) strategy involves selection, integration and implementation of weed management options based on economic, ecological and social principles. Weed management decisions (i.e. what, how much, when, and how to control) should be based on optimizing, but not necessarily maximizing, weed control.

IWM involves a holistic consideration of weeds in a complex ecosystem. The development of an IWM strategy requires a sound knowledge of the biology and ecology of the weeds involved, tools of weed management, as well as an out-of-the-box thinking to develop novel options which may exploit natural regulating forces. The weed manager must realize that what happens in a given field influences and is influenced by an area extending beyond that field, and weed management in any one year impacts the agroecosystem in subsequent years. Accordingly, both spatial and temporal aspects must be evaluated. Development of improved weed monitoring techniques, capability to predict losses by modelling weed density–crop loss relationships, establishment of economic threshold(s), and a knowledge of critical period of weed interference are central to this approach. Effective weed management information has to be made available to producers in an understandable, acceptable and usable form.

No one weed control option is a universal panacea for achieving weed management goals in complex ecosystems. Weeds are components of agroecosystems consisting of a complex web of intra- and inter-ecosystem interactions.

Weeds, as well as their control options, have a variety of impacts, both desirable and undesirable, on agroecosystems. The integration of available weed management options with a long-term strategy to achieve a desired level of control, discourage the build-up of specific weeds, increase weed diversity, and minimize negative impacts is therefore necessary. Weed distribution patterns in the field should be taken into consideration in making weed management and technology development decisions. Weed management research and technology development therefore needs interdisciplinary cooperation with inputs from all stakeholders.

Weedy species are not entirely our enemies, and therefore not only targets for elimination, but represent part of an agroecosystem with some positive attributes. They may benefit the crop by either diverting insect pests away or by supporting predators of harmful pests. There may also be some beneficial effects on associated non-crop communities (e.g. soil organisms including mycorrhizas) and the physical environment of the ecosystem. Therefore, the level at which they can be tolerated must be carefully determined.

Integration of different weed management options has been shown to increase species diversity, which has a variety of benefits that have been shown in studies with unmanaged communities (Clements *et al.*, 1994). Species diversity could affect weed–crop interactions, non-plant communities, as well as the physical environment of the ecosystem. Monoculture and reliance on a single weed control option has been shown to reduce species diversity in agroecosystems, making them less stable (Clements *et al.*, 1994). More research on the influence of IWM on various components of species diversity, disturbance and re-colonization, inter- and intra-community (both plant and non-plant) interactions, dynamics of seed banks, and impacts on the physical environment of the ecosystem is needed. A sound understanding of the influence of weed management on species diversity may allow management decisions to achieve more desirable community composition and sustainable agroecosystems.

A sound weed management plan should have a strategy to prevent the introduction and dissemination of weeds, enhance the ability of crops to compete with weeds, and combine a variety of weed management options to prevent weeds from adapting to any one control practice. Weeds should be managed to some acceptable level, considering their positive as well as their negative attributes, and should not be simply considered as targets for elimination.

12.4 References

Akemo, M.C., Regnier, E.E. and Bennett, M.A. (2000) Weed suppression in spring-sown rye–pea cover crop mixes. *Weed Technology* 14, 545–549.

Blackshaw, R.E., Moyer, J.R., Doram, R.C. and Boswell, A.L. (2001) Yellow sweetclover, green manure, and its residues effectively suppress weeds during fallow. *Weed Science* 49, 406–413.

Blackshaw, R.E., Moyer, J.R. and Huang, H.C. (2005) Beneficial effects of cover crops on soil health and crop management. *Recent Research Developments in Soil Science* 1, 15–35.

Bond, W. and Grundy, A.C. (2001) Non-chemical weed management in organic farming systems. *Weed Research* 41, 383–405.

Clements, D.R., Weise, S.F. and Swanton, C.J. (1994) Integrated weed management and weed species diversity. *Phytoprotection* 75, 1–18.

Cloutier, D.C. and Leblanc, M.L. (2001) Mechanical weed control in agriculture. In: Vincent, C., Panneton, B. and Fleurat-Lessard, F. (eds) *Physical Control Methods in Plant Protection*. Springer-Verlag, Berlin, Germany, pp. 191–204.

Einhellig, F.A. and Leather, G.R. (1988) Potential for exploiting allelopathy to enhance crop production. *Journal of Chemical Ecology* 14, 1829–1844.

Furness, N.H., Jolliffe, P.A. and Upadhyaya, M.K. (2005) Experimental approaches to studying effects of UV-B radiation on plant competitive interactions. *Phytochemistry and Photobiology* 81, 1026–1037.

Hartmann, K.M. and Nezadal, W. (1990) Photocontrol of weeds without herbicides. *Naturwissenschaften* 77, 158–163.

Hoagland, R.E. (2001) Microbial allelochemicals and pathogens as bioherbicidal agents. *Weed Technology* 15, 835–857.

Inderjit and Bhowmik, P.C. (2002) Importance of allelochemical in weed invasiveness and their natural control. In: Inderjit and Mallik, A.U. (eds) *Chemical Ecology of Plants: Allelopathy in Aquatic and Terrestrial Ecosystems*. Birkhauser-Verlag, Basel, Switzerland, pp. 188–197.

Inderjit and Keating, K.I. (1999) Allelopathy: principles, procedures, processes, and promises for biological control. *Advances in Agronomy* 67, 141–231.

Julien, M.H. and Griffiths, M.W. (1998) Biological control of weeds: a world catalogue of agents and their target weeds. CABI Publishing, Wallingford, UK.

Katan, J. (1981) Solarization heating (solarization) of soil for control of soilborne pest. *Annual Review of Phytopathology* 19, 211–236.

Katan, J. (1987) Soil solarization. In: Chet, I. (ed.) *Innovative Approaches to Plant Disease Control*. John Wiley and Sons, New York, pp. 77–105.

Laguë, C., Gill, J. and Péloquin, G. (2001) Thermal control in plant protection. In: Vincent, C., Panneton, B. and Fleurat-Lessard, F. (eds) *Physical Control Methods in Plant Protection*. Springer-Verlag, Berlin, Germany, pp. 35–46.

McFadyen, R.E.C. (1998) Biological control of weeds. *Annual Review of Entomology* 43, 369–393.

Ross, S.M., King, J.R., Izaurralde, R.C. and O'Donovan, J.T. (2001) Weed suppression by seven clover species. *Agronomy Journal* 93, 820–827.

Scopel, A.L., Ballare, C.L. and Radosevich, S.R. (1994) Photostimulation of seed germination during soil tillage. *New Phytologist* 126, 145–152.

Shrestha, A., Clements, D.R. and Upadhyaya, M.K. (2004) Weed management in agroecosystems. towards a holistic approach. *Recent Research Developments in Crop Science* 1, 451–477.

Stapleton, J.J. and DeVay, J.E. (1986) Soil solarization: a non-chemical approach for management of plant pathogens and pests. *Crop Protection* 5, 190–198.

Stapleton, J.J., Elmore, C.L. and DeVay, J.E. (2000) Solarization and biofumigation help disinfest soil. *California Agriculture* 54, 42–45.

Thill, D.C. and Mallory-Smith, C.A. (1997) The nature and consequence of weed spread in cropping systems. *Weed Science* 45, 337–342.

Weigelt, A. and Jolliffe, P. (2003) Indices of plant competition. *Journal of Ecology* 91, 707–720.

Weston, L.A. (1996) Utilization of allelopathy for weed management in agroecosystems. *Agronomy Journal* 88, 860–866.

Weston, L.A. (2005) History and current trends in the use of allelopathy for weed management. In: Harper, J.D., An, M., Wu, H. and Kent, J.H. (eds) *Allelopathy: Establishing the Scientific Base: Proceedings of the 4th World Congress on Allelopathy*, Wagga Wagga, Australia, pp. 15–21.

Weston, L.A. and Duke, S.O. (2003) Weed and crop allelopathy. *Critical Reviews in Plant Sciences* 22, 367–389.

Wicks, G.A., Burnside, O.C. and Warwick, L.F. (1995) Mechanical weed management. In: Smith, A.E. (ed.) *Handbook of Weed Management Systems*. Marcel Dekker, New York, pp. 51–99.

Zimdahl, R.L. (2004) *Weed–Crop Competition*, 2nd edn. Blackwell Publishing, Oxford, UK.

Index

Abutilon theophrasti 6, 9, 15, 29, 32, 68, 96, 110,
 189, 198
Acacia saligna 192, 196
Azolla filiculoides 83
Adonis aestivalis 180
Aegilops cylindrica 5, 14, 31, 35, 45
Aerial dispersal of pathogens 107
Aerosolization 97
Aeschynomene 94, 98, 102, 104
 indica 94
 virginica 106–107, 109
Agents
 bioherbicidal 74, 108, 209
 biological 93, 109
 emulsifying 101
 stabilizing 97
Ageratum conyzoides 3
Agropyron repens 62
Agrostemma githago 3
Ailanthis altissima 74
Ailanthone 69
Ailanthus altissima 69, 73, 74–75
Ailanthus glandulosa 73
Aizoaceae 185
Alginate
 bead formulations 97
 gels 107
Alhagi maurorum 180
Alkaloids 69
Allelochemical degradation 66
Allelochemicals 65, 68–69, 71–75, 203–204, 209
Allelopathic 39–41, 44, 59, 65–69, 71–76, 148,
 151, 203
Allelopathy 47, 65–67, 71, 73–5, 151, 203–204, 209
Alleycropping 150
Alliaria petiolata 85, 89

Allium
 cepa 146
 porrum 41
Allylisothiocyanates 190
Alm 14
Alopecurus myosuroides 191
Alpha-acid levels 145
Alsike 51
Alternanthera philoxeroides 84
Alternaria 96, 99–101, 106, 107, 109, 110
Alvarenga 62
Amaranthus 9–10, 41, 57, 68–70, 103, 110, 149,
 180, 187, 188
Amarolide 69
Ambrosia
 artemisiifolia 146
 trifida 68
2-aminophenoxazin-3-one 71
Ammannia coccinea 66
Anagallis coerulea 180
Anaphase stages 71
Anchus aggregata 180
Anecdotal evidence 86
Animal's intestinal tract 8
Anna Karenina principle 90
Annual
 ryegrass 16
 sowthistle 184
Annual bluegrass 100, 169, 183, 185, 199
Anoda cristata 96, 110, 180
Anthracnose 107
Antifungal volatile compounds 197
Antimicrobial 191, 198
Antioxidants 97
Apiaceae 182, 184
Apium graveolens 41

Aposphaeria amaranthi 104
Apple 104, 109, 148, 150
Applied electrostatic fields 173
Aqueous
    production systems 72
    spore suspensions 101–102
    spray mixture 104
    suspensions 101–102
Aquifoliaceae 66
Arachis hypogaea 198
Aristolochia maurorum 180
Artemisia
    annua 69, 73–74
    vulgaris 73
Artemisinin 69, 73
    production 74
Arthropods 204
    grassland 175
Arum italicum 180
Asclepias syriaca 14
Asparagus 72
Aspectos gerais 15
Aspergillus species 108
Asphodelus tenuipholius 180
Asteraceae 181, 183–185
Astragalus 16, 180, 191
Atmospheric aerosols 97
Aubergine 102, 207
Autotrophic microbes 157
Avena
    fatua 4, 15, 30–31, 35, 44–46, 190, 198
        interference 32
    sativa 40, 50, 65
    sterilis 192
    strigosa Schreb 58
Azolla filiculoides 83, 90

Bacteria 73, 77, 108, 110, 157
    antagonistic 191
    thermophilic 188
Bahiagrass 146
Balloonvine 4
Band
    steamer 166
    steaming 165, 166, 171, 174
Banded treatment 170
Bark 66, 69, 136, 139, 142, 149
    chipped 138, 140, 206
    mulches 138–139, 146–147, 149, 152
Barley 22, 32, 37, 39–41, 44, 46, 54, 62–63, 65,
        67, 146, 151
    intercropping 45
    plant density 32
    silage 39
Barnyardgrass 10, 30, 32, 66, 75, 182, 191

Beans 41, 45, 52, 117, 123–124, 157
    green 63
    jack 50–52, 59
Beauveria bassiana 102
Bed
    shaping, initial 143
    system, permanent 151
Bed-making machines 143
Beet, red 118, 124
Beetles 162
    carabid 162, 173
    leaf 88
Bellevalia sp 180
Below-ground 170, 207
    food reserves 203, 206
    meristems 159
Belowground herbivory 88
Bermudagrass 52, 185, 192
Beta vulgaris 32, 173
Between-row mowing 132
Bidens spp 8
Bindweed 181
Bioactive 66, 71
    compounds 66–67
    long-chain hydroquinones 70
    root exudates 67
Bioaerosols 97
Bioassays 72
    hydroponic 71
    leaf disc 102
    modified soil 72
Biocompostable polymer 144
Biocontrol 82, 86–87, 90, 107, 110
    agents 85–89, 103–105, 109–110, 150
        classical 94
        fungal 110
        herbivorous 83
        potential 81, 85, 94
        seed-feeding 83
        weed 86–87, 91
    benefits 150
    classical 94, 103
    fungi 96, 99, 107
Biodegradable
    mulch films 153
    mulches 144
    non-plastic products 144
    plastic products 144
    polymers 144
    products 144
Biofumigation 190, 207
Bioherbicide 76, 93–94, 96–97, 98, 102–103,
        105–108, 204–205
    bacterial 104
    broad-spectrum 106
    solid-state 97
Bioherbicide Formulation 97, 100
Biological Conservation 197

Biological control 74, 78–79, 81, 83, 85–91, 98,
    106–110, 204, 209
  agents 77–78, 86, 88–91, 98, 106, 108–109
  classical 78, 80, 89, 204
Biological Invasions 88–89, 91
BioMal 99
Biomass 18–19, 40, 42, 50, 54, 56, 66–67, 69, 84,
    146, 148, 170, 186
  above-ground 12
  chopped crucifer 207
  dead 207
Biopesticides 94, 106, 109–110
Biotic 39, 186
  agents 204
  factors 77
  interactions 29
  resistance framework 79
Biotransformation 71
Bipolaris sorghicola 96, 101
Birds 3, 77
  predatory 148
Bitter principle 74
Black-grass 191
Blackspot 147
  spores splashing 147
Blue pimpernel 180
Blueberries 152, 157
Botrytis
  cinerea 95
    spores 109
  group 146
Branched broomrape 198
Brassica 40, 50, 63
  campestris 146
  napus 39
  nigra 180
  oleracea 41, 53, 146
  parachinensis 146
  rapa 35, 190
Brazilian
  jack bean 52
  pepper 146
Broccoli 62, 173
Bromus
  diandrus Roth 46
  secalinus 42, 45
  tectorum 35, 44, 46, 85
Broomrape 183, 185
Buckwheat 51, 65, 67, 148, 151
Bud stage 9
Buddingh basket weeder Model 122
Buddingh Model 124
Bull mallow 191
Bunium elegans 191
Burner
  angle 160
  position 161

power 161
  settings 159
  type 161
Burners 156–157, 160–161
  covered 160–161
  gas-phase 160
  infrared radiant gas 162
  open 157, 160
  standard open 160
Burning 139, 156, 163, 171, 175
  controlled 156
  low-intensity 156
Bursaphelenchus xylophilus 147
Butyric acids 146

Cabbage 53, 124, 138, 146, 197
  residues 197
  white 62
Cactoblastis cactorum 86
Cajanus cajan 59
Calendula arvensis 181
Calopogonium mucunoides 53
Camelina sativa 4
Chamomilla suaveolens 159
Canada thistle 14, 52, 54–56, 181
Canal muck 135
Canavalia
  brasiliensis 52
  ensiformis 50
Cannabis
  inoculum 108
  sativa 95
Canola 37, 39, 41–42, 45
  rotating 35
  rotations 44
Canopy 67, 158, 169
  architectures, developing 203
  closure 36–37, 139
  development 42
  penetration 164, 167
    leaf 164
Capacity 58, 124, 161, 166
  spore-carrying 109
  water-holding 156
Capsella
  bursa-pastoris 169
  rubella 181
Carabid activity 162
Carbon 56, 95
  dioxide 142
    snow 173
  emissions 161
Cardboard 138
Cardiospermum halicacabum 4
Carduus nutans 83
Carpet 138, 206

Carrier 98, 101, 107, 123, 150
Carrots 113, 115, 118, 122, 124–125, 151, 157,
        162, 165, 198
    organic 133
    wild 191
Carthamus
    flavescens 181
    syriacum 181
    tinctorius 30
Cash crops 49, 54, 56–57, 59–61, 203
    annual 56
    competitive 55
    early-harvested 54
    early spring 50
    mycorrhizal 203
Cassava 40, 47
    systems 41
Cassia
    obtusifolia 110
    occidentalis 100
    rotundifolia 53
Casuarina equisetifolia 146
Catmint 67, 72–73
Cattle 4, 7, 8, 14–16
    manure 8, 15
Cauliflower 146
Cavara 89
C-curved 119
    shanks 120
Cells 155
    rupture 157
    ruptures 167
    rupturing 158
Cellulose
    sheeting 150
    sheets deteriorate, heavy 136
Cenchrus echinatus 8
Centaurea
    iberica 181
    maculosa 10, 80, 87, 89
    solstitialis 94
Centrifugal phylogenetic method 85
Centrosema pubescens 53
Cercospora dubia 110
Cereal 5, 31, 46, 52, 54–55, 63, 66, 67, 95, 117,
        119, 139, 145
    temperate 156
    undersown 54
Charcoal, activated 103
Chenopodium
    album 8, 10, 31, 69, 110, 159, 173, 181
    murale 181
    pumila 181
Chewings fescue 67
Chickpea 196
Chicory 118, 126
    fields 115

Chinampa
    reconstructed 153
    system 135
Chinese cabbage 138
Chisel-ploughing 5
Chlamydospores 99
    live 98
Chloris gayana 181
Chlorothalonil 103
Chromolaena odorata 83
Chrozophora tinctoria 181
Chrysanthemum
    coronarium 181
Chrysomelidae 88
Cichorium intybus 181
Cinnemethylin 68
Cirsium arvense 52, 54, 62, 64, 83, 181
Clear
    polyethylene mulch 200
    polythene sheeting 137
Clover 50–51, 54, 56, 62, 63, 209
    berseem 51
    crimson 50
    plants 55
    red 54–55, 203
    subterranean 50, 53
    white 51, 53–55
CO_2 172
    concentration levels 190
    concentrations 190
    laser 132, 170, 174
        energy 170
    snow 170
Coccodes 101, 104
Cocklebur 96, 102
    common 109
Cocoa
    bean mulch 151
    shell mulch 145
    shells 140
Coffea arabica 8
Coffee 8, 140
Cofrancescor 88
Cogongrass 41, 52, 60, 66
    suppression 44, 52
Coleoptera 88–89, 156
Collego 97, 102, 106
    application 103
    product 98
Colletotrichum 100–101
    coccodes 110
    gloeosporioides 94–96, 98, 104, 109
        hydrophilic fungus 99
    malvarum 96
    truncatum 95, 104, 107–109
        conidia 107
        microsclerotia 107–108

Colorado potato beetle 157, 173
Coloured
  mulch technology 153
  polythene sheeting 137
  products 137
Combined solid substrate 96
Combustion
  by-products 172
  surface 162
Commelina communis 181
Common
  cocklebur 185
  dayflower 181
  groundsel 184
  lambsquarters 181
  purslane 100, 108, 184, 192
Compatibility 103–104, 205
  of Bioherbicides 103
Competition 18, 24, 27, 29–33, 35, 36, 42, 44–46,
    52–56, 63–64, 77, 93, 103, 109, 130, 158,
    202–203
  interspecific 33, 45
  intraspecific 21
  model 197
  two-species 24
Competitive
  abilities, relative 19
  crop canopies 42
Competitiveness 12, 36–37, 45
Compost 8, 9, 14, 145–146, 150, 152
  chicken 197
  domestic 151
  effects 152
  immature 146
  maturity 146
  mulch effects 152
  mushroom 139
  piles 7
  windrow 9
Composted
  green waste 148
  materials 7, 145
  mulches 147
Composting 7, 9, 13–14
  manure 8
  process 146
  windrow 16
Conductivity 188
  improved thermal 178
Cones 124
  opposite 124
  truncated steel 124
Conicus 83, 86
Conidia 94–95, 97, 101, 108
  gloeosporioides 101
  virulent 95
Conidial germination 109–110
  inhibited 103

Conservation
  tillage
    corn production systems 132
    practices 93
    systems 46, 56
Control
  agents 80–87
    classical biological 91
    disperse 80
    effective biological 89
    evaluating potential biological 88
    potential 81, 85
      biological 80, 107, 109
    bolters 173
  methods
    cultural 18, 187
    non-chemical weed 161
    non-conventional weed 131
Convolvulus
  althaeoiedes 181
  arvensis 9, 69, 185
Conyza
  bonarinsis 181
  canadensis 2, 15, 186
Corn 30–32, 61, 62, 74, 95, 97, 131–133, 164,
    173–174, 196
Corn/cassava systems 44
Cornmeal 97, 99
Coronilla scorpioides 191–192
Coronopus didymus 181
Cotton 104, 157, 159, 163, 175
Cottonseed flour 95
Cover-crop residues 149
  decaying 40
Cover-cropping systems 52, 62
Cover crops 39, 49, 51, 53, 55, 57, 59, 61, 63
Covered flamers 158, 160–161, 163
  standard 161
Coverings 7, 165
  green 137
  tissue 169
Cowpea 41, 50, 53, 63, 153, 203
  intercrop 44
Critical period of competition concept 12
Crop
  allelopathy 75, 209
  canopy 140
    development 46
  competition 12, 19, 36–37, 55, 209
  competitiveness 22, 37, 45
  contamination 136
  damage 122, 124, 126
  density 22–24, 27–28, 30, 37, 47
  detection system 128
  development 59, 62, 112
  ecosystem 20
  emergence 37, 113, 115–117, 157–158, 160–161

Crop *continued*
  establishment 53, 56, 67, 130, 140
    optimal 36
    rapid 138
    uniform 37
  fertilization 38–39
  interactions 17–18, 21, 23–25, 28–29, 63, 202, 208
  interference 31
  losses 31
  model 29
  mulches 63, 152
  performance 167, 206
  planting 20, 40, 56, 115, 135–139
  pollinators 24
  population 22, 57, 61, 168
  quality 136, 145–146
  residues 7, 40, 50, 52–53, 59, 61, 63–64, 129, 136, 148
  root pruning 119
  rotation 18, 35, 44–45, 53, 58, 60, 171, 203–204
  rows 115–116, 118–121, 123–126, 128–131, 147, 157–158, 160, 205–206
  seeding 113
  seedling growth 169
  seedlings 119
  seeds 3, 4, 14, 113, 117, 129, 138, 202
  straw 156
  stubble 129
  suppression 53, 61
  systems 54, 58
    forage 75
Cropping 31, 49, 55, 61, 63, 135, 143–144
  alley 74
  continuous 36
  double 152
  management
    strategy 131
    systems 131
  organic 93
  patterns 12
  systems 16, 25, 28, 38, 42–45, 55–56, 65, 103, 111, 115, 128–129, 164, 209
    non-inversion 128
    organic arable 62
    soil-inversion-based 111–112
    transgenic 105
Crops
  allelopathic 72
  catch 49
  competitive 27, 36, 42
  green manure 19, 55, 203
  heat-tolerant 157, 159, 162
  rotational 203
  trap/catch 187
  undersown 54, 62
  weed-seed-contaminated 13

Crotalaria
  juncea 52
  spectabilis 102
Crupina vulgaris 16
Crushed rock 139, 206
Crustaceans 77
Cucumber 67, 151, 207
Cucumis sativus 70, 151
Cucurbita 95
  texana 100, 106
Cultivation 4–6, 15, 52, 115–117, 122, 124, 126, 131, 135, 152, 155, 161, 163, 172, 186–187
  blind 116
  daytime 205
  depth 113, 117
  inter-row 118–119, 124, 129
  intra row 125
  late 118
  multiple 12, 118
  plough 15
  pre-crop emergence 117, 129
  precision 120
  raised bed 112, 115
  stubble 55
Cultivators 112, 115–116, 118–121, 123–124, 126, 131
  broadcast 115
  heavy 112, 120
  higher-residue 58
  intra-row 115, 123
  light-duty 119
  precision 125–126
  rear-mounted 123
  rolling 119–120
  rotary tilling 120
Cultural
  methods 133, 174, 197
  practices 25, 35, 42–43
  strategies 43, 62
  weed control practices 44
Cultural weed management 35, 37, 39, 41, 43, 45, 47, 201
Culture 62, 109
  agar 95
  liquid 96
    shake 99
  medium 95
  submerged 96
Curcurbitia texana 110
Cuscuta 195
  campestris 192, 197–198
  monogyna 181
Cyanogenic glucosides 66
Cynodon dactylon 3, 5, 7, 8, 14, 52, 185
Cyperus 52, 182
  esculentus 104, 107, 146, 181, 192
  rotundus 7, 83, 103, 110, 180–181, 199
Cypress spurge 89
Cytisus scoparius 90

Dactylaria higginsii 103, 110
Dactyloctenium aegyptium 182
Dairy manure 14
  composted 9
Danish S-tines 120
Datura stramonium 96, 100, 107
Daucus 182, 191
  aureus 182
Decomposers, seed 191
Decomposition 145, 148, 190, 196
  anaerobic 142
  organic 191
  processes 50
  slow 59, 139
Decontaminating 131
Decreased weed seed production 27
Deer 67
  mule 16
Defoliators 84
Degradation 63, 73–74, 137–138, 144
  increasing organic matter 191
Delay senescence 53, 57
Delayed
  control 62
  planting date 43
  response 202
  seeding of winter wheat 38
Demographic modelling 81
Denaturation, protein 155
Denitrification 156–157
Density
  planting 203
  insect 91
  jointed goatgrass 31, 36
  microbial population 151
  thresholds 26
Density-dependent pathogen mortality 103
Deplete underground food reserves 206
Descurainia sophia 20, 182
Desiccation 111, 115, 157–158, 175, 191
Design
  additive 18, 21, 33
    experiment 24
  experiments 81
    standard additive 28
  full-factorial 19
Desmodium tortuosum 8, 101, 107
D-glucopyranosyl-4-hydroxy 74
Dibber drill 129
Dielectric heating 168
Different-coloured plastic mulches 144
Digera arvensis 182
Digitaria 3, 68
  sanguinalis 8, 70, 182
2-dihydroailanthone 69
4-dihydroxyhydrocinnamic acids 72
4-Dihydroxyphenyl 75

4-dihydroxyphenylalanine 69, 70, 75
Dimethyl sulphide 190
Diplotaxis erucoides 182
Direct-sown 145
Discs 112–113, 119–121, 126
  notched 121
  spike 113–115
Disease
  cycles, secondary 105
  development 38, 102
    suppressed 103
  resistance 37, 53, 57
  severity 101, 104
  spores 147
Diseases 36, 59, 75, 105, 136, 146–147, 158, 195,
    198, 203
  insect-vectored viral 151
  rust-induced 107
  virus 147
Dispersal 2, 3, 5, 7, 11, 14–15, 18, 77, 202
  adaptations 2
  long-distance 2, 6
  natural 94
  slow 79
  weed 2, 3, 11–12, 14, 202
Dispersion 5, 7, 8
  human 3
  long-distance 3
  promoted 5
  propagule 2
  short-distance 3
  weed 3, 8, 11
Dissemination 2, 3, 14, 202
  rapid 82
  strategies 202
  weed 3, 5, 208
  of weed seeds 15–16
Dissolved organic matter 193
Dodder 181
Dominance 18, 25
  apical 206
Dormancy 187, 189–190, 195, 198
  breaking 38
  cycles 148
  physical 194–195
  secondary 193
Dormant 188, 193–194
  seeds 187–188, 192, 195
Dose-response models 173
Double-tent
  method technique 179
  technique 179
Downy brome 35–36, 41, 44, 46
Droplet size 101–102
Drought 67, 71
  conditions 71, 142
  stress 72
Duck-salad 66

Ear tree 146
Echinochloa 3, 4, 68
  colona 182
  crus-galli 10, 30, 32, 68, 75, 182, 191
Ecological approach 30, 32, 44, 47, 62, 132, 195,
      197, 200
Ecology 14, 17–18, 32, 81, 88, 90–91, 108, 195,
      199, 201, 207
  of Fire 175
  of Intercropping 47
  of Soil Seedbanks 196
  weed competition/population 32
Economic
  evaluation 90
    of soil solarization 200
  threshold approach 33
  thresholds 32–33, 43, 46, 77, 202, 207
    long-term 26
Economics 24, 26, 31, 43, 94, 136, 143, 174, 202,
      205
  of Cultural Weed Control Practices 43
  of mechanical weed management 131
  of Non-living Mulches 142
Ecosystem
  function 86
  processes 85, 87
  resistance 196
  threat 108
Ecosystems 2, 3, 14, 31, 63, 85, 89, 91, 175, 195,
      204, 206, 208
  agricultural 191
  closed 135
  complex 201, 207
  low-input 103
  managed 94
  native 87
  natural 186, 197
    conservation 186
  non-agricultural 201
Effectiveness 50, 77, 87, 132, 164, 167–168, 180,
      194–195, 205, 207
  of biological control agents 91
  cost- 179
  of soil solarization 195
Efficacy 94, 97–98, 103, 105, 109–110, 115, 136,
      158, 179, 189, 196, 200, 205, 207
  bioherbicidal 103
Egyptian broomrape 179, 183, 197
Eichornia crassipes 83
EIL 24–26, 28
  concept 25, 28
    single-season 25
  estimates 29
Electrical
  infrared emitters 163
  methods 168
    of killing plants 173

seed treatment 175
  systems 123
Electrocution 155, 167–168, 172, 206
  weed 168, 175
Electrothermal 173
Eleucine indica 182
Elymus repens 46
Elytrigia repens 52, 136
Embryonic degradation processes 195
Emergence 14, 20, 22–23, 29–32, 37, 40, 46, 52,
      57, 111–113, 115, 149, 151, 158, 185, 199
  of annual weeds 133, 136
  crop's 12
  early 57
  kinetics 188
  parameter 22
  patterns 38
  rapid 36
  reduction 180
  variability 44
Empirical
  models 29, 30
  observations 17
Energy 128, 155, 163–164, 167–171, 174
  absorption 168
  conversion factors 171
  converting electromagnetic 168
  efficiency 161, 171–172
  electrical 167, 172
  fossil-fuel 207
  inputs 155, 161, 163, 168, 171–172, 175
  losses 164, 169
  microwave 169, 174
  requirements 163, 170–171, 173
  solar 177, 198
Energy-use
  efficiencies 163, 167, 171
Ensiling 10, 13–14, 16
  process 10
Enterolobium cyclocarpum 146
Enyinnia 47
Epidemiological mechanisms of mycoherbicides
      effectiveness 110
Epidemiology 106, 205
Ericaceae 66
Erosion 54, 58, 93, 128, 138, 150, 155, 172, 204,
      207
  accelerated 156
  control purposes 138
  reducing 60
Erwinia carotovora 104
Escola Superior 15
Eucalyptus maginata 199
Euphorbia
  cyparissia 89
  esula 80, 83, 89
Exclusion 81, 135
  principle, competitive 18

Exotic 78, 94, 145
  invasion 89
  plant invasions 90
Exotic Arthropod Biological Control Agents 91
Experimental designs 18–20, 82, 202
  additive 21
  Regression Techniques 18
Experiments
  additive 21
  long-term cropping systems 31
  multiple-choice feeding preference 90
  organic farming crop rotation 63
Exposure
  microwave 168
  repeated 165
  time 155–156, 164
  to UV radiation 169

Fabaceae 183–184
Fabrics 137, 143, 151
  needle-punched 138, 206
  strong 139
  weed 139
  woven 137
Fagopyrum
  esculentum 151
  tataricum 32
Fallow 45, 59, 62, 208
  land 203
  periods 52, 57, 59 61, 203
Farming 2, 13, 46, 55, 57, 111, 193
  organic 54, 57, 63, 133, 152, 157, 161,
      171–172, 201
  system 13
Farms, organic 54, 57, 113
Fate of Allelochemicals in Soil 71
Fatty acids 96
Fennel 115, 126
Fermentation 10
  processes 7, 10
Fertigation 142
Fertility 53, 57, 60, 67, 112, 141, 196
Fertilization 12, 21, 39, 153, 203
  strategies 45
Fertilizer 2, 38, 42, 58, 112, 140, 203
  application method 42, 44
  applications 142
    annual 142
  autumn-applied 39
  banded 43
  nitrogen requirements 57
  placement 42, 46
  timing 39
Festuca rubra 67
Fibre production 3

Fibres 3, 139, 150
  cellulose 150
  mesh 150
  natural 138, 143
  recycled 138
Field
  assessment 90
  cage 86
  dodder 181, 197–198
  emergence 199
  enclosures 85
  exposure 144
  flaming 162
  incineration 143
Films 137, 143–144, 150, 207
  biodegradable starch-based 144
  blue 140
  polyethylene 196
  white 140
  white-on-black 140
Finger weeders 124
First-principle model outcomes 27
First principles of weed ecology 17 18
Fitness
  density-dependent 109
  frequency-dependent 109
Flame
  burners 163
    covered 163
    ordinary gas 162
  deflects 160
  duration 162
  heat 158
  intensity 162
  treatment, single 159
  weeders 124, 160–163
    covered 160, 163
  weeding 155, 157–158, 161–162, 170–175
    banded 158
    extreme LPG 162
    selective 157, 159, 162
Flamers 156, 160–161
Flaming 25, 27, 113, 155–164, 170–173, 175, 206
  band 160
  directed 206
  equipment 161
  moderate 158
  selective 160
    post-emergence 157
  technique 158
  technology 159
Flax 4, 6, 138
Flooding 2, 38, 46, 77, 203
Flowering 58, 74, 156
  periods 35
  stage 10
Foam mulching system 151

Foams 149–150, 206
    biodegradable 164
Forages 9, 10, 61, 133
Formulation
    additives 102
    composition 97
    ingredients 97–98
Foxtail
    barley 10
    green 9, 10, 30, 38
Freeze-killing 170
Fruits, winged 2
Fuel 58, 112, 156, 159, 161, 171
    consumption 160–161
    converting 169, 171
    diesel 163, 166, 168–169
    fossil 155, 161, 171–172
    input 158, 161
    liquid 157
    pressure 160
Fungal 90, 106, 147, 156, 162, 167, 177, 197, 204
    biomass 96
    disease 146, 147, 152
    effective 108
    entomopathogens 101
    mycelium 96, 99
        culturing 96
    spores 94, 145
    sporulation 95
    spray mixture 101
Fungi 93–96, 99, 101–104, 108, 110, 147, 157, 162
    plant
        parasitic 109
        pathogenic 95
    pluvivorous 107
    rust 104
Fungicides 103, 108
Fungicides, copper-based 145
Fungus associations 198
Fur 2, 3, 11, 14
    animal's 8
Furrows 123, 129, 195, 205
Fusarium
    arthrosporioides 97
    lateritium 109–110
    oxysporum 95, 108, 198
        f.sp 109, 196
    solani 95, 99, 106, 110

Galerucella
    calmariensis 88
    pusilla 91
Gangs 117, 119–121
    inter-row 120
    of wheels 120–121
Geotextiles 137–138, 206

Germination 16, 40, 44, 57, 70, 77, 99, 110,
        115, 120, 145, 148, 150–151, 189–190,
        195–196, 205
    conidium 108
    cumulative 198, 200
    delayed 187
    elevated 190
    growth response 75
    increased 37, 190
    inhibiting 206
    promoted 189
Glandular trichomes 69
Glass 145
    crushed 139
Gloeocercospora sorghi 96
    johnsongrass bioherbicide 109
Gloeosporioides 102
Glycine max 4, 33, 37, 52, 109, 132–133, 199
Goats 15, 90
Golden loosestrife beetle 91
Gould competition 46
GPS 42, 131
GPS-controlled patch spraying 31
Grain
    legumes 36
    production 3
    sorghum 65, 75
Granular formulations 99, 109
Granular Pesta formulation 109
Granules 97–99
    kaolin 107
    non-disintegrable 97
    producing 99
    uniform-sized 99
    wettable 97
Grape hyacinth 191
Grass
    carp 77, 88
    clippings 138–139, 206
Gravel 139, 164, 206
Grazing 77
    goat 91
Green
    foxtail levels, reduced 39
    manure 39, 49, 54–55, 58–59, 62, 64–66, 203,
        208
        annual 54
Green Manure and Cover Crops 39
Grey mucuna 52
Groundcover, woven 137
Growth
    media 95–96
        concentrated 95
    medium 95–96
        development 96
        residual liquid 99
    suppression 52
        weed 36, 69

Guidance systems 122, 131
  automatic 131, 205
  manual 121
Guide
  weeding tools 128
  wheels 123

Habitat
  disturbance 186
Habitats 3, 11, 79, 82, 85, 186
  contrasting 73
  disturbed 187
  favourable 50, 147
  ideal 2, 148
  open 175
  sensitive natural 93
  unstable 192
Hairy vetch 50, 53, 56–57, 62, 63, 64, 67, 203
  desiccated 64
  mulch 57
  residue 47, 57, 64
Hand-weeding 41, 67, 137, 161–162, 165, 166, 187
Hanta virus 86
Harrowing 59, 117, 133–134
  aggressive 117
  weed 54, 117, 132–133
Harrows 112–113, 115–117, 124, 133
  chain 116
  flex-tine 113, 115–116
  weed 132
Harvesters 4, 5, 14–15
Harvesting 4, 5, 18, 42, 59
  problems 147
Hazards 105, 164, 170
  associated 97
  connected health 172
  environmental 145, 179
  high-intensity wild-fire 156
  intrinsic 105
  potential 145
Health 37, 91, 110, 155, 172, 182, 207
  human 66, 177
  public 65
  risks 97, 145
Heat 137, 148, 156, 158, 161–163, 166–169, 177, 179, 190, 206–207
  absorption 139
  conductance 143
  conserved 179
  convection 160, 163
  damage 158
  dissipation 164
  dry 190, 197
  injury 155, 159
  losses 161
  sensitivity 188
  solar 207

tolerance 158
transfer 156, 161, 168, 173
  improved 161
treatment 156, 188, 192, 200
units 156
Helianthus annuus 6, 35, 60, 96
Heliotropium sp 182
Hemp 138
  broomrape 185
  sesbania 95, 99, 101–102, 104, 107
    control 95, 99, 101, 103–104
    weed 108
  sunn 52, 60
Hemp-nettle 33
Herbicidal
  activity 75
    potential 69
  development 65
  effects 74
  inputs, minimizing 67
  strategies 64
Herbivore 77–81, 83–86, 88
  aboveground 88
  combinations 84
  exclusion 84
  exotic 91
  feeding 81
  generalist 78
  multiple 82
  preferences 80
  released insect 85
  releases 88
  specialist invertebrate 78
  specialized 85
  systems 79
  insect 88
Heteranthera limosa 66, 73
Heterotrophic microbes 157
Hilling 63, 126
Hoeing 120, 122
  manual 125
  time 60
Hops 145, 150
Hordeum vulgare 37, 45–46, 54, 65, 146
Horses 8, 14
Horticultural systems, organic 151
Host 86, 94, 102, 104, 147, 195
  defense system 196
  plant
    abundance 79
    individuals 84
    populations 79
  preference 101
  range 85, 105
    alteration 106
    fundamental 85
    limited 102
    realized 85

Host *continued*
 roots 193
 selectivity 102
 specificity 81–83, 85–86, 88, 91, 94, 102, 107, 204
  of control agents 81
Host specificity
 realized 83
 screening 86
Host-specificity investigations 82
Host-specificity tests 90
Hot water 155, 163–164, 167, 172, 174, 206
 application 163–164
  equipment 164
 applicator, narrow 164
 treatment 156, 171
Hydrilla 95, 105, 109
 verticillata 95, 105
  submersed aquatic macrophyte 109
Hydrolysed-corn-derived materials 95
Hydroxamic acids 68, 75
2-hydroxy-5-methoxy-3 70
4-hydroxybenzoic 72
4-hydroxyhydrocinnamic 72
Hydroxyphenylpyruvate dioxygenase 68
Hylobius transversovittatus 88–89
Hypericum 91
 perforatum 83
Hyperparasitism 103

Ice, dry 170, 206
Imperata cylindrica 41, 52
Impermeable mulches 147
Implementation 22, 27, 80, 93, 107, 200, 207
 of SH 179
 of Soil Solarization for Weed Control 178
Impregnated mulches 149
Inclusion of silage crops 39
Incorporation of bioherbicide propagules 97
Increased risk of fungal diseases 147
Indices of plant competition 33, 209
Inert
 carrier material 98
 rehydrating osmoticum 98
Infective units 94, 102
Infrared
 cotton defoliation 175
 light 137
 radiation techniques 173
Inhibition of weed seed germination 173
Inoculum 95–96
 concentrations 103
  reduced 95
 density 95
  high 95
 desiccation-tolerant 95
Insect 24, 77–78, 82, 88–90, 147, 156, 162, 164,
  168, 198, 204

behaviour 90
fauna 81
groups 147, 156, 162
harmful 155
introductions 87
manipulation 30
mobility 156
pests 147, 155, 161, 172, 203, 207
 control crop 204
 diverting 208
 repelling 147
phytophagous 204
responses 175
vectors 147
Integrated Management 45, 195–197, 199, 200
Integrated pest management 15, 107, 155
 sustainable 177
Integration
 of agronomic practices 44
 of biological control agents 109
 of cover crops 62
Inter-cropping 59
Inter-crops 41
Inter-ecosystem interactions 208
Inter-ridge 129
Inter-row 119–121, 124
 area 119, 122
 cultivators 115, 118–120, 123–124, 126–127,
  129
 hoeing 165
 spacing 129
 width 119
Intercropping 40–41, 44, 47, 59, 61, 203
 kale 45
 leeks 44
 strategies 45
 systems 40, 61
Intercrops 40–41
 annual 47
 cash 53
 lentil 44
Interference 30, 44–46, 56, 58, 61, 63, 66, 78
 plant allelochemical 74
Interplant interactions 21
Interseeding 63
Intra-row 125, 174
 band, treated 165
 weeders 123, 131
  intelligent 128
 weeding 124, 126, 132, 157
  action 126
Invasibility 73, 90
Invasive
 plants 3, 78, 80–83, 85–86, 88–90, 156
  managing 89
Invasive Alien Species 90
Invasiveness 73, 89
 weed 74, 209

Invert 101
emulsion 102, 106–107, 109
droplet size 108
emulsions 107
formulation 104
Ipomoea 180, 182
lacunosa 182, 189
lacunose 69
IR
radiators 162–163
weeders 171
IR-burner heats 162
IR-radiator 162
Irrigation 2, 11–12
canals 7, 11
drip 178
foliar 147
trickle 142, 147
Isothiocyanates 190, 199
Italian ryegrass 54, 62, 163
IWM 60, 201, 207–208
approach 60
programme 60
strategy 207

Johnsongrass 15, 96, 101, 185, 192
control 15, 16
infestation 15
reduced 38
Jointed goatgrass 5, 14, 31, 35–36, 42, 45
Juglans 75
nigra 66, 74
Juglone 66, 75, 140
Jussiae decurrens 102
Jussiaea decurrens 106

Kale 41
Kaolin clay 99
Kochia scoparia 2, 11
Kudzu 40, 53, 98, 106, 108

Lablab purpureus 40
Lactuca sativa 146
Lagonychium farctuma 182
Lamium amplexicaule 182
Landraces 67
Landscape 2, 11, 65, 67, 75
Lantana camara 83
Large
blue alkanet 180
crabgrass 182
Larval
acceptance 86
mobility 85

Laser wavebands 155
Lasers 130, 132, 170–172, 174, 206
infrared regime 170
Lateritium 96, 99, 104
Lavatera cretica 191
Lavorazione 133
Laws 1, 3, 4, 12, 202
seed 3
purity 4
Layers
single polythene 179
stratospheric ozone 177
L-DOPA 69, 70, 72, 74–75
Leaf
canopy 49, 52, 164
litter 137
sheaths 158
stage 31, 158–159
Leaf area 29, 31, 36, 46, 130, 206
Leaf-beetles 89
Leafy spurge 80, 89
Leek 41, 128, 165–166
Legumes 14, 36, 45, 53, 57, 63, 65, 197, 199, 203
annual 53, 57, 61–62
living 57
tropical 70
Lens culinaris 3, 36
Lentils 3, 36, 41
Leontice leontopetalum 191
Lepidium
draba 9
sativum 69, 146
Leptinotarsa decemlineata 157
Leptospermone 68
Lesser swinecress 181
Lettuce 75, 113, 128, 133, 146, 207
crop 131
floating tropical water 84
growth 197
Life
cycles 35, 42, 202–203
weed 27, 50, 148
expectancy 135, 137, 144
Life cycle
analysis 172, 174
Light
absorption 170
colour 140
cultivators 112
interception 46, 128
mulch materials 136
reflectance 147
spectrums 145
threshold 205
transmission 149
high 149
transmittance 64

Light-requiring weed seeds 203, 205
Linaria
    chalepensis 183
    vulgaris 3
Linseed meal 95
Linum 183
    usitatissimum 4
Liquid
    culture production 108
    culturing of microsclerotia of Mycoleptodiscus
        terrestris 109
    formulations 97, 99, 101
    spray formulations 99
Lithospermum arvense 5
Livestock 3, 7, 8, 11
    farming 2
    feedstuffs 10
Living
    mulch system 62
    roots of rice plants 69
Lolium
    multiflorum 54
    perenne 67
    persicum 3, 45
    rigidum 16, 39, 44–46, 183
        competition 45
        seedlings 14
London rocket 184
Long-term
    application of first principles 25
    cover-cropping strategies 61
    rotation 36
Longevity 12, 14, 65–66, 94–95, 136, 140, 190, 198
    seed 205
Loss
    models 23
    predictions 28
Low-fertility
    conditions 59
    system 135
LPG 157, 159, 162
    gas burners 143
    rates 158
Lucerne 6, 139, 157–158
Lupin 46, 58, 60
Lupinus 58
    angustifolia 46
Lycopersicon esculentum 102
Lygodium microphyllum 105, 107
Lygus lineolaris 108
Lythrum salicaria 83, 85, 88–91

Macroconidia 99
Macroptilium atropurpureum 53
Macroscopic solids 97

Maize 5, 10, 23, 29, 35, 40–43, 45, 47, 53, 57,
    59–62, 68, 117, 126–127, 157, 162
    emergence 41
    rows 130
    starch 144
    transplanting 46
Malva 183
    nicaeensis 191
        Bull 183
    parviflora 183
    pusilla 10, 99
    sylvestris 183
Management
    alternatives 23–25
        weed 17
    biorational pest 73
    broomrape 187
    of cover crops 57, 61
    crop residue 36, 42, 205
    decision aids 28
    fungal pest 156
    invasive plant 77
        species 88
    mycorrhiza 196
    of plant pathogens 199, 209
    practices
        cultural weed 35, 44
        weed 13, 17, 26–27
    preventive 2, 202
    strategies 18, 45
        integrated non-chemical weed 174
        non-chemical weed 22, 201
        weed 18, 201–202
    stubble 37
    sustainable 58, 186
        pest 74
        systems 62, 73
        improved weed 35
        integrated weed 38, 46
        non-chemical weed 28
        non-inversion soil 129
        orchard groundcover 152
        sustainable weed 35
    weed community 38
    of weed
        seedbanks 153, 198
Manihot esculenta 41
Manure 4, 8, 9, 13–15, 112, 190, 202, 207
    animal 178
    application 13
    beef feedlot 9
    chicken 197, 207
    composted beef cattle feedlot 15
    composting 8
    cow 13
    goat 198
    tonne of 4, 8
    windrow composting 14

Manuring, green 54, 62
Mass
  production
    of microorganisms 107
    techniques 82
  rearing 88, 89
Matricaria
  chamomilla 183
  inodora 190
Mechanical
  control methods 132
  disturbance 55
  in-row cultivation 131
  management of weeds 133
  removal of weeds 58
  roller-crimper 62
  weed
    control 55, 113, 119, 131–134, 161, 205, 208
    management 111–113, 115, 117, 119, 121,
      123, 125, 127, 129, 131, 133–134, 205, 209
    pullers 130
Media 109
  commercial 8
  selection 95
  solid 96, 109
Medic, annual 53
Medicago 41, 53
  arabica 183
  polymorpha 183
  sativa 6, 69
Medium 96
  basal 99
  selection 96
  submerged-liquid sporulation 109
  vegetable-juice-based liquid 96
Melaleuca leucadendron 146
Melilotus
  officinalis 40, 50
  sulcatus 191
Melucella laevis 183
Mercurialis annua 183
Mesotrione 68, 75
Metabolic
  activity 178
  processes 189–190
Metabolism 75
  cellular 97
  rapid 66
Metabolites 72, 97, 131
  secondary 105
Methanethiol 190
Methyl bromide 177, 186, 193, 199
Methylxanthine toxicosis 151
Microbial
  activity 139, 142, 153, 186
    exothermic 178
    thermophilic 191, 194
  agent 97

allelochemicals 74, 108, 209
attack 189, 191
biomass 142
community structure 174
degradation 71
invasion 192
pesticides 107
recolonization 174
Microconidia 99
Microorganisms 35, 97, 105, 107, 146, 164, 191,
  198
  airborne 109
  preexisting 98
  rhizosphere 197
  thermophilic 191
Microplate assays 96
Microsclerotia 95, 109
Microsclerotial inoculum 107
Microwave
  action 168
  application equipment 169
Microwaves 155, 168–169, 171, 173, 175, 206
Milkvetch 180, 191–192
  disperse clover 16
Mimosa pigra 90
Minimum-tillage 57–58
  corn production 64
  systems 57
Miscanthus sinensis 70
Mites 77–78, 175
Models 23, 26–33, 79, 82, 94, 107, 116–117, 120,
  133, 149, 188, 198, 199
  crop
    competition 23
    interference 28
  demographic 84, 89
  dynamic 32
  economic 193
  first-principle 26–27
  high-residue 117
  inverse density 202
  life cycle 198
  long-term threshold 26
  matrix 91
  mechanistic 29
  multispecies canopy 32
  natural herbicide 74
  neighborhood 31, 33
  non-linear regression 21
  oriented 31
  plant growth 30
  predictive 3, 84
  regression 23
  simplified weed population projection 26
  soybean decision 28
  structured population 90
  two-parameter 149
  weed competition 31

Moist heating 185, 190, 192
Momilactone 69, 70, 74
Monitoring protocols 82, 87
　developing 87
　standardized 82
Montia perfoliata 183
Morrenia odorata 98
Mortality, weed 30
Movement, seed 205
Mowing 9, 52, 55–56, 61, 77, 130, 132, 206
　frequent 55
　repeated 206
　timely 67
MSW 146
　compost mulch 141
　composted 152
　ethylene oxide 146
Mucuna 50
　cochinchinensis 40
　pruriens 52, 65, 70, 75
Mulch
　ages 149
　colours 145
　decomposes 138
　degradation 152
　effect 60, 152
　formation 59
　layer 136
　　non-living 136
　mass 149
　materials 136, 138, 142, 147–150
　mats 138
　rates, low 149
　species 44
　techniques 193
　tenders 142
Mulches 59, 129, 135–152, 195, 206
　aluminium-painted 147
　biosolid/bark compost 141
　cereals 152
　coarse 136
　compost 151
　dark 206
　degradable 151
　durable 143
　effective 60, 149
　film 135
　foam 144
　gravel 139
　hay 147
　impermeable 142, 145
　landscape fabric 135
　living 45–46, 49, 53, 57, 203
　non-combustible 140
　novel 150
　paper/polymerized vegetable oil 152
　papier mache-type 150

particulate 140, 206
physical properties of 64, 153
plant 153
red 146, 151–152
rubber 140
ryegrass 143
sawdust 139, 152
sheet 136, 206
silver 140
spray-on 206
stone 139, 147
surface 53, 151
synthetic 150
turnip-rape 190
weed-suppressive 66
white 146
wood 138, 147
woodchip 139
woven polypropylene 135
Mulching 59, 135, 137, 142–143, 152–153,
　　177–178, 180, 193, 203
　characteristics 144
　common polythene 179
　duration 180
　film 144
　material 136, 179, 195
　polyethylene 197–198
　properties 149
　stubble 142
　technique, common 179
　techniques 179
Muscari
　racemosum 183, 191
Mustard 40, 50–51, 130, 180, 203
　black 65
　garlic 85, 89
　white 163
Myagrum perfoliatum 183
Mycelial
　formulations 94
　growth 95, 103, 110
　　vegetative 95
Mycoherbicidal
　activity 108
　performance 101
Mycoherbicide 93–98, 101–103, 106–110, 205
　applying 106
　delivery system 99
　effectiveness 103
　effectiveness 110
　formulations 99
　　liquid 99
　　solid substrate 99
　oil-based 97
　pre-emergence 99
　registered 97, 99

Mycoinsecticides 107
Mycoleptodiscus terrestris 95
Mycorrhizal
    inoculation, inhibiting 203
    mutualisms 85
Myriophyllum spicatum 170
Myrothecium verrucaria 106–108, 110
    spore viability 104

Neighbourhood
    analysis 202
    conditions 21
    designs 20
    factors 20
    models 21
    types 28
Nematode 58, 77, 107, 147, 151, 164, 177, 186,
        198, 207
    entomogenous 107
    root-knot 200
Nepeta 67, 73
Nicandra physalodes 3
Niche assembly theory 18
Niches 19, 59, 93
Nicotiana tabacum 102
Nitrogen 31, 38–39, 44–45, 53, 55–57, 63, 95,
        138, 140, 152, 199, 203
    accumulation 64
    cycling 157
    dynamics 91, 134
    fertilizer 38–39, 44, 141, 172
        application method 40
        liquid 39
        timing 39, 44
    grain 63
    liquid 170, 173, 206
    losses 63
    management 64
    mineral 45
    nutrition benefits 152
    requirement, high 57
    sources 95, 110, 152
        complex 96
    transformations 200
    uptake 39, 44
        lower 39
No-till 42–43, 63
    planter 129
    systems 36, 58, 60
No-tillage 58, 60, 62, 64
    corn 62
    culture 75
    permanent 58
    practices 60
    production 58
    soybean 63
    system 57–58, 61–62

Nodding broomrape 185
Non-chemical
    disinfestation practice 177
    weed management 17–18, 23–25, 27, 30, 35,
        49, 65, 71, 77, 93, 111, 135, 155, 201
Non-invasive behaviour 79
Non-inversion system 129
Non-linear regression 19
    analysis 26
Non-living mulches 135–137, 139, 141–143, 145,
        147, 149–151, 153, 206
    synthetic 206
Non-mycorrhizal 40
Non-selective flame weeding pre-emergence 160
Non-target Effects of Biological Control 90
Non-target plant injury 105
Non-volatile sesquiterpenoid lactone 69
Nontarget
    effects 90
    feeding of leaf-beetles 89
Northern jointvetch 94, 98, 102–104, 106–107, 109
    control 98
    controlled 102
Novel antimicrobial peptides 196
Nutsedge 182
Nutsedge, purple 66, 103, 110, 180, 185, 191–192
Nymphaea spp 170

Oats 37, 40, 51, 60, 65
    black 58–59
    sterile 45
Onions 113, 118, 122, 146, 157, 159, 166, 173
    bulb 174
    seeded 124, 127
    single-line-sown 165
    spring-sown 117
Open flamers 161
    standard 161
Operator safety 167, 169, 171
Opposite cones connect 124
Optimizing
    bed orientation 198
    nutritional conditions 108
Optimum
    crop
        management strategy 202
        mixtures 20
    superheating 167
Opuntia 83
    cacti 84
    stricta 3
Orchards 40, 88, 130, 136–137, 143, 163
    fruit 8
    prune 140
Organic
    acid MSW 146

Organic *continued*
  farming systems 175, 208
  matter
    content 112, 138, 142, 203, 205
    decomposition 190
      enhanced 194
    mulch 61, 136, 139–142, 145–148
      natural 206
      non-composted 146
    no-till 133
Organisms
  mycoherbicidal 106
  potential biocontrol 94
  soil-surface-inhabiting 162
Organs
  above-ground 52
  perennial storage 38, 54
Ornamental goldenrod 67
Ornithogalum narbonense 183
Orobanche 183, 195, 197, 199
  aegyptiaca 180, 183, 197
  cernua 185
  crenata Crenate 183
  ramosa 185
    Hemp 183
  ramose 198
  seeds 198
Orthoceras 109
  microconidia 98
Oryza sativa 3, 38, 73, 108–109
Osmoconditioning 37
  effect, seed 46
Osmopriming 37
  seed 44
Oxalis cormiculata 183

Panax quinquefolius 152
Panicum miliaceum 3, 14, 132
Papaver
  dubium 183
  rhoeas 183
Papaveraceae 183
  horned poppy 184
Parasitoids 82
Particle mulches 135–138, 142
  coarse-textured 147
  potential 138
Paspalum notatum 146
Pasture
  species 3
    tropical 15
  systems, rotating 10
Pastures 8, 10, 15–16, 130, 135, 139
  annual 91
  clear grass 151

Patches 6, 16–17, 42
  dense weed 42
  initial 5
  isolated weed 2
  weed 17
Patchiness 24
  weed 28
Pathogen
  interactions 94
  virulence 101
    maintaining 95
Pathogenesis 110, 197
Pathogenicity 103, 108
Pathogens 49, 50, 57, 74, 77–78, 93, 96, 99,
    101–103, 105–108, 110, 155, 164–165,
    172, 177, 197–198, 204–207
  aggressive 105
  bacterial 177
  fungal 106, 109
  indigenous 94
  native 103
  rust 94
  soilborne 199
  viral 90
  virulent 94
  weed 97–98, 102, 108
P-benzoquinone 70
Pear cacti 83
Pearl millet 58–59
Peas 39, 41, 45–46, 61, 117, 133, 208
Peat moss 139
Pectinophora gossypiella 162
Pellet processing 14
Penicillium bilaiae 108
Pennisetum americanum 58
2-pentadecatriene 70
Peppers 143, 207
Perennial
  forages 42
  grass 136
  groundcovers, selected 73
  production systems 72
  root systems 12
  ryegrass 67
  weeds 12, 14, 52, 55, 60, 111–112, 115, 130,
    136–137, 155, 159, 185, 187, 192, 194, 196
    creeping 55
    suppressing 52
Perennials 12, 40, 52, 55–56, 61, 63, 66, 112, 137,
    180, 185, 192
Perforated biodegradable film overlay 150
Persian darnel 45
Persistence 18, 71, 150, 187–188, 194, 199, 202
  strategy 204
Persistent sandbank 192
Pest
  complexes 68

control 136
  chemical 177
  improving 178
  process 177
  predators 24
  problems 135, 146
  resistance 67
  suppression 72, 147
    enhanced 150
  tolerance 67
Pesta 98–100, 107
  formulations 98
    amended 98
  granules 98–99, 107
Pests 49, 57, 59, 62, 65, 67, 98, 136, 145, 147–148,
    168, 177–178, 186, 194–197, 199, 200, 204
  exotic 93
  hibernating 148
  weed 94
Phalaris 183
  brachystachys 183
  paradoxa 183
Phaseolus vulgaris 41, 74, 95, 146
Phenolic acids 69, 75, 146
Phenylisothiocyanates 190
Phomopsis amaranthicola 103
Phompsis amaranthicola 110
Photocontrol of weeds 112, 132, 209
Photodegradable
  films 144
  polyolefin polymer 144
Photostimulation of seed germination 209
Phragmites australis 88, 91
P-hydroxybenzoic acid 72
Phyllanthus fraternus 183
Phylloplane 99
Phytopathogens 106
Phytotoxic
  fatty acids 148
  volatile compounds 194
  volatiles 189
Phytotoxicity, phenolic acid 73
Phytotoxins 50, 53, 56, 61, 136, 145, 206
  releasing 50, 53
Pigeonpea 51, 196
  dwarf 59
Pine
  needles 139
Pistia stratiotes 3
Pisum sativum 39, 133
Plantago spp 183
Plantains 139
  false horn 152
Planting
  date 36, 42–43
  delayed 113, 187
  distances 66

media 199
pattern 37, 42, 203
season 12
winter wheat 42
Plastic
  film 142, 144, 207
    clear 144, 207
  layers 198
  mulches 137, 140, 142–143, 147, 149, 151–152,
    179
    clear 141, 148, 206
    coloured 144
    green 149
    sheeted 140
  mulching 178
  sheeting 136, 177
  technology 186
Plastics 140, 143, 149, 151, 200
  clear polyethylene 197
  degradable 143
Plasticuture 200
Ploughing 55, 59, 112–113, 187
  mouldboard 132
Ploughs 111, 129
  chisel 112
  disc 112
  mouldboard 112
  powered rotary 112
  stubble mulch blade 129
  two-layer 129, 133
  wide-blade sweep 129
Pneumat 126, 127
  weeder 126
Poa annua 70, 159, 169, 183, 185, 199
Poly-tunnels 137
Polycultures 20
Polyethylene
  co-polymer 144
  glycol 37
  mulches 151–152
    black 151
    transparent 198
Polygonum 68, 96, 159
  aviculare 159
  convolvulus 4
  equisetiforme Horsetail 183
  persicaria 183
  polyspermum 184
Polystigma rubrum subsp 101, 109
Polythene 178–179, 190
  aged 179
  black 135, 137–138, 142, 148–149, 179
  blue 140
  clear 185
  coloured 206
  films 144
  mulches 141, 147, 149, 195
    black 141–142
    clear 148

Polythene *continued*
  pollution 179, 193
  products 140
  properties 178–179
  sheets 177–179, 193
    single 179
    transparent 179
  technology 178–179, 186
  transparent 178–179
  type 179
Populations 20–21, 25–26, 32, 35, 53, 60, 62, 73,
    80, 84, 86–87, 91, 153, 167
  control agent 82
  earthworm 148
  herbivore 81
  host-plant 87
  insect 147, 156, 204
  native prey 87
  neighborhood 32
  nematode 147
  pest 58
  self-sustaining 84, 86
  soil microbial 156
Portulaca oleracea 69, 184, 192
Post-emergence 99, 117, 125, 133, 157
  control 68
  foliar sprays 99
  herbicide control decisions 33
  sprays 99
  treatment 121
Potato 54, 63, 102, 117, 128, 131, 147, 157–158,
    160, 198
  haulms 157
  tops 132
Potentilla recta 10
Potting mixtures 147, 149
Power
  conversion ratio 170
  take-off 112
    tractor's 120
Powered machines 112
Pre-emergence 99, 172
  harrowing 134
  herbicides 68, 99, 185
Pre-planting 99
Precautionary principle 80
Precision Agriculture 45
Predation 61, 187, 193
Predator populations 87
  generalist 87
Predators 50, 82, 208
  seed 61
Predictions 23, 29, 30, 74, 82–83, 133, 149
Prescription burning 156
Preventative 13
  measures 145

Prevention 2, 3, 11–13, 136, 150, 191, 201
Prevention Strategies 3, 5, 7, 9, 11, 13, 15
Preventive
  actions 2
  approaches 13
  method, valid 112
  practice 2
  strategies 201–202
Prickly-Pear 89
Primary tillage 111–112, 129
Primrose Primulaceae 184
Propagule
  pressure 186
    weed 186
  reservoirs 11
  supply 196
Propagules 2, 4, 5, 7, 11, 82, 94, 97, 111, 202
  fungal 99
  underground 192
  vegetative 5, 8, 12–13, 138, 145, 194, 202
  weed 1, 5, 13, 192, 202
Propane 157, 161, 163, 171
  burners 157
  combustion 161
  consumption 161
  doses 158
  flamers 113, 163
  flames 159
Proso millet 14, 132
Prosopis farcta 185
Protectants 101, 108
Protection 58, 101, 108, 132–133, 142, 159, 165,
    191, 207
Puccinia
  canaliculata 104, 107
  jaceae 94
Pueraria
  montana 106, 108
  phaseolides 53
  phaseoloides 40, 53
Pulses 42, 45
  high-voltage 167
Punch planting 133
Punk tree 146
Purity, seed 4
Purple
  loosestrife 85, 88–91
  nutsedge 181
Purslane, common 108, 192
Pusilla 88

Quackgrass 52, 54–56, 60
  control 55
  suppressed 54
Quarantine 81–82, 85, 91, 202
Quassinoid 69, 75

Radiation
  electromagnetic 168
  incoming 49, 140
  influx, increased 179
  infrared 137, 140, 160, 162–163, 172–173
  levels 169, 173
  long-wave 179
  microwave 168, 172–173
Radish 59, 60
  wild 33
Ragweed 146
Railroad right-of-way 165
Rain splash 147
Raindrop splashing 206
Rainfall 58, 67, 145
  penetration 205–206
Raised
  bed systems, permanent 128
  beds 115, 129, 144, 165, 205
Rangelands 88, 204
Ranunculus
  arvensis 184
  repens 184
Rape 148
  forage 148
  mulch 199
  oilseed 37, 46
  spring-seeded 124
Raphanus
  raphanistrum 184
  sativus 59
Raspberries 142
Rattlebox, smooth 51
Real-time image
  acquisition 131, 205
  analysis 31
Recycled
  ground wood pallets 142
  materials 139, 150
  products 135, 138
Red
  clover undersown 54, 62
  film 140
  sorrel
Reduced-tillage techniques 112
Reducing weed seed production 60
Reduction
  of fungal disease problems 146
  of insect pests 147
  of seed germination 205
  seedling 196
  weed 52
Regression equations 32
Regression Techniques 18
Reseeding 82
  active 82
Residues 40, 50, 53, 56–59, 61–63, 65–66, 145,
  148, 208

chemical 103, 155, 161, 171–172
dead 203
decomposing 56
forest 149
fresh 50
hard woody 139
nontoxic 144
organic 140
weed 112
Resistance 36, 66, 86, 93, 116, 125, 144, 191
  biotic 78, 90
  developing 71
Rhagadiolus stellatus 184
Rhinocyllus conicus 89, 90
  seed head feeding weevil 83
Rhizoctonia solani 147
Rhizomes 5, 52, 54–55, 170, 192
Rhizosphere 70–71, 73, 153
  bacteria 73
Rice 3, 38, 46, 65–67, 72–75, 102, 107–109, 199
  accessions 66
  allelopathy 69, 74
  cultivars 15, 66, 74
    non-allelopathic 72
  cultivated 4, 66
  direct-seeded 30
  ecosystems 32
  straw 74
  transplanted 38
  transplanting 46
Ricinus communis dispersion 6
Ridges 116, 123, 129
  permanent 128
Ridging 126–127, 129, 132
Rights-of-way 164
Rigid-tine harrows 116
Risk 4, 5, 12–13, 15, 28, 38, 40, 46, 56, 84, 86,
  90–91, 97, 105, 108, 110, 145, 147,
  171–172, 199
  analysis 107
  assessment 170
  factor 105
  mitigation 105
  non-target 108
  perceived 2
  potential 145
Risk-benefit-cost analysis 91
Roadside 7, 67
Rod weeder 129
Rodents 3, 147
Rolling
  cages 121
  coulters 167
  harrow 113–115, 119, 121, 126, 133
Root
  exudates 73–74
  feeders 84

Rosa spp 95
Rotary
  cultivators 119–120
  hoes 116–117
  tillers 120
Rotating crops 35
Rotation
  combinations 35
  design 42–43
Rotations 27, 35–36, 39, 42, 44, 49, 54–55, 58–60,
    62, 88, 120, 123, 203
  cover-cropping 61
  fallow 35
  four-crop 36, 187
  traditional small-grain 42
  two-crop 36
Rottboellia exaltata 5
Row
  crop systems 165
  crops 63, 113, 118–119, 131, 157, 171, 174
  distance 119
  position 123
  spacing 36–38, 43, 45
  spacing
    effects 30
    narrow 42
    reducing 37
  width 133
Rubrum stromata 109
Rumen 10
  digestion 13–14
Rumex
  acetocella 184
  crispus Curly 184
Runner rooting 144
Russian thistle 2
Rye 37, 40, 50, 53, 56–57, 60, 62, 64–65, 68, 148,
    203
  control 64
  interseeding 56
  residues 56–57, 71, 196
  spring-sown 61, 208
  termination 56
  white clover 51
Ryegrass 16, 39, 45–46, 54, 62, 67, 139, 143, 173,
    183

Saccharum 15
  officinarum 7
Safety 86–88, 89, 91, 102, 105–106, 139
  applicator 207
  environmental 72
  feature 81
  operator's 168
Safflower 30
Sagittaria montevidensis 3
Saligna seeds 189

Salsola kali 2
Salvinia 3
  molesta 84
Satureja hortensis 146, 153
Sawdust 139, 142, 206
Scale production, large 108
Scaling responses 28
Sclerotina sclerotiorum 107
Sclerotinia sclerotiorum 106
Sclerotium rolfsii 197
Scorpiurus muricatus 184, 191–192
Scotch broom 90
Seashells 139
Secale cereale 40, 50, 63, 65–66
Secondary dispersal of seeds 15
Sedges 52
Seed
  banks 12–13, 15, 26, 39, 42–43, 46, 59, 203, 205
    dynamics of 204–205, 208
    local 82
  coat impermeability 199
  dispersal 5, 14–15
    preventing weed 11
    reducing 11
    weed 3
  dormancy 148, 187, 192, 195, 197
  dormancy, impermeable 199
  germination 37, 136, 148, 169, 190, 195,
      197–199, 203, 205, 209
    behaviour 204
    phytochrome-mediated 50
    preventing 148
    reduced 168
    reducing weed 145
    suppressing weed 149
    uniform 203
  mortality 61, 198
    enhancing weed 52
  persistence 200
  populations 61
    buried 188, 199
    weed 152, 199
  predation 62
    weed 52, 57
  survival 16, 185, 189, 192
    minimizing weed 43
    weed 35, 196
  transportation 2
Seed viability 10–12, 16, 35, 189–191
  reducing 185
Seedbank 43, 148–149, 151, 165, 186–188,
    189–191, 194–196
  below-ground 189
  characteristics 187
  depletion 12, 186
  deterioration 188, 190–191, 194
    processes 177, 188, 194
    weed 190

manipulation 148
natural 165
persistency 190, 194
persistent 186–188
reduction 187, 193
size 139
weed 60
Seedbed
  operation, false 113
  preparation 112, 115, 133, 156, 203
    final 112, 129
  technique 112–113, 133
    false 112–113, 115
  prepared 135
Seeding 5, 31, 38, 123, 129, 143
  delayed 38
  depth 203
    result, uniform 37
  direct 38, 46
  early 203
  machinery 36
  rates 21, 23–24, 30, 38, 43, 45, 54, 117
    increased
      crop 25, 27, 30
      winter wheat 42
Seedling 8, 14, 38, 50, 69, 74, 132, 134, 145, 147,
      163, 165, 187, 189, 198
  developing 72
  development 158
  emergence 35, 44, 133, 165, 171, 198
    monitored 168
    weed 15, 36, 135, 148, 151, 165–166, 174
  establishment 43
    improved 156
    row crop 75
    weed 42
  germinating 150
  growth 198
    weed 206
  nursery 8
  production 8
  stages, early 159
  transplant 8
  volunteer cereal 145
Seedmat systems 150
Seeds 2–5, 7–13, 15–16, 35, 50, 52, 62, 69, 70, 88,
      111, 150, 168, 187–193, 195, 197–198,
      205–206
  blown-in 139
  buried 172, 193, 207
  dispersing 3
  germinated 190
  germinating 189
  imbibed 165, 168, 190
  immature 5
  light-requiring 206
  light-sensitive 205
  moist 178

monitored 11
non-dormant 188, 194–195
non-viable 188
persistent 187
saved 4, 8
water-impermeable 195
weakened 194
wind-blown 145
wind-dispersed 138
Selection
  of culture medium 95
  mulching technique 188
  pressures 79, 93
  procedures 85
  seed 12
  strain 105
Selective
  heating 168–169
  intra-row flame weeding 159
  phytotoxin 73
  post-emergence intra-row flaming 159
  treatment 159
Self-regenerating intercrops 11
Senecio
  jacobaea 83
  vernalis 184
  vulgaris 159, 169, 184
Senna obtusifolia 96, 110
Septoria polygonorum 96
Sergoleone 70
Sesbania
  exaltata 95, 107–108
    biocontrol 107
  punicea 83
Setaria
  glauca 184
  viridis 9, 10, 30, 38
SH 177–180, 185–186, 188–195
  duration of 180, 189, 192
  efficacy 178, 180, 193
  factors 194
  failure of 192, 194–195
SH killing process 188
SH-resistant 191
SH thermal
  effect 189
  killing 189
SH-treated seedbank 194
Shanks 119–120, 129
  short 116
  vibrating 119–120
Sheep 8–9, 15–16, 90–91
  pasture 4
  reported 8
Sheets 99, 137, 179
  flat 135
  of plastic 136
  transparent polyethylene 196

Shells 145
  almond 140
  dry fruit 206
Shelter belts 139
Shields 2, 15, 129
Shinjudilactone 74
Showy
  crotalaria 102
  rattlebox 59
Shredded newspaper 140, 152
Sicklepod 96, 100, 101, 104, 109–110
Sicklepod control 99
Sida 96, 102, 109–110, 189
  spinosa 96, 110, 184, 189
Silage 5, 10, 15, 39, 45
  crops 39
  maize 113, 116, 134
  producing 10
Silene noctiflora seeds 5
Silybum marianum 91
Simulation
  models 29, 31, 33
    computer 187
    physiological growth 29
Sinapis
  alba 163
    control 163
  arvensis 170, 184
Sisymbrium spp 184
Slugs 77, 147
Smother crops 41, 49, 63, 203
  effective 50
Snails 77, 147
Snap peas 62, 146
Sodium alginate 99, 110
Soil
  acidification 156
  active 66, 71
  aeration 205
  aggregates 156
  application 163
  atmosphere 180, 190
  bacteria 166
  bare 58, 137, 148, 162
  biology 63, 203–204, 207
  carbon/nitrogen ratio 206
  chemistry 169
  compaction 140
    confining 115
    decreased 129
    traffic-induced 172
  conditions 5, 111, 113, 125
    anaerobic 142
    dry 147
    optimal 143
    salty 67
  conservation 140, 142, 205

contours 120, 124
cooler 53, 56
crust 115–117, 119
cultivation 55, 119, 178, 195
depth 141, 166, 178, 180, 187, 192
disinfestation 175, 177, 199
  strategy 199
disturbance 37, 52, 193, 195
drier 167
dry 38, 115, 178, 190
erosion 135, 145, 150, 156, 203
fertility 53–54, 57–58, 135, 152
flora 190
fungi 89
heat 175
heating 195
  process 179
inversion 111, 128–129
juglone 74
microbes 40, 157
microbial
  activity 142
  biomass 162, 175
microflora 162
microorganisms 191, 198
moisture 37, 47, 50, 57, 61, 64, 140, 142–143,
    148, 150, 156, 158, 178, 188–191, 193,
    205
  conservation 50, 137, 142
  conserve 190, 206
  conserving 207
  preserving 180
  retention of 53, 136
mulched 190
mulching 178–179
  duration 178
mycorrhizal 203
nitrification 200
nitrogen 39, 62, 141
  levels 39
nutrients 153, 207
organisms 156, 162, 166, 179, 208
oxygen content 206
pathogen control 185
predators 203
preparation 5, 178
respiration 142
rhizosphere 72
scarification 175
solarization 147, 177–179, 181, 183, 186–187,
    189, 191, 193, 195–200, 209
  failure of 192, 194
  techniques 187
solarized 190, 196–197
steamed 166
steaming 165, 171, 174
  mobile 165
  technology 165

structure 115, 123, 141, 158, 203–204
    improved 136
surface
    litter 148
    smooth 178
surface, uneven 111
temperature 64, 140–141, 148, 156, 163,
        165–166, 168, 178–179, 186, 188, 192,
        194, 199
    solarized 178
tillage 59, 131, 209
Soil Seedbanks 148, 186–187, 196
Soilborne pest 197, 209
management 197
Solanum
    luteum 184
    melonegra 102
    nigrum 170, 184
    ptycanthum 102
    tuberosum 54, 198
    viarum 104
Solar
    heating 177, 197–200, 206–207
    radiation 101, 108, 169, 177, 179, 205, 207
Solarization 137, 148, 152, 180, 185, 193,
        196–200, 207, 209
    heating 197, 209
    single-layer 179
    structural 199
    technology 195
Solid substrate fermentation 95
Solidago 67
    altissima 70
Sonchus
    arvensis 52, 54, 63
    oleraceus 184
Sorghum 6, 41, 44, 46, 65–67, 70, 71, 73, 74, 128,
        199, 203
    accessions 73
    bicolor 6, 50, 65, 70, 73–75, 199
    halepense 6, 14, 38, 41, 69, 96, 184, 185
    root hairs 76
    seedling 71
        growth 62
    seedlings 76
    sudanense 65
Sorghum-sudangrass 50–51
Sorgoleone 69–71, 73
    exude 70
    natural product 74–75
    production 76
Sowing 37, 54–55, 59, 93, 113, 145, 166
    date 133
    machine 119
    time 54, 113, 134
Soy-Dox 101
Soybean 4, 23, 30, 33, 37–38, 46, 51–53, 58, 60,
        62–63, 98, 103–104, 108–109, 117,
        128–130, 199

fields 5, 130
flour 95
no-till 132
Spatial
    arrangement 20, 32
    diversification 41
    dynamics 15
    pattern 45
    uniformity 38, 46–47
Spatio-temporal variation 74
Species
    alien 91
        legume 197
    annual legume 57
    aquatic 88
    assemblages 18
    associated 203
    broadleaved 8, 10
    competitive 36
    composition 148
    density of 19
    diversity 28, 208
        weed 28, 31, 208
    endangered 80, 85
    endemic 11
    herbivore 80–81
    insect 86
    interact 18
    introduction 2, 202
    invasive 2, 11, 80, 82, 90, 156, 173
    late-emerging 113
    later-germinating 38
    neighbouring 65
    non-indigenous 86–87
    non-target 86, 156
    noxious 3
Spergula fallax 184
Spiders 120, 148
Spikes 116
    rubber 124
Spinach 113, 115, 118
    organic 133
Spore 94–97, 99, 101–102, 104, 136, 145
    canaliculata 104
    cassiae 96, 104
    dried CGA 98
    germination 95, 101, 103–104
        increased 101
    production 95, 99
    reviving 98
    types 99
    volume 95
Sporulation 95–96, 103
Spotted
    burclover 183
    knapweed seed viability 16
Spraying, patch 132

Spring
  barley 33, 37, 55
    weed harrowed 46
  cereals 31, 35
  tines 112, 124–125
    steel 126
    vertical 126
  wheat 30, 32, 37–38, 44–47
    reduced 41
Spring-tine harrows 117
Spurred anoda 96, 99, 101, 110, 180, 189, 192
S-shaped 119
  logistic models 158
  shanks 120
Starthistle, yellow 94, 107
Steam 155, 163–167, 172, 206
  application 174
    equipment 167
  exposure 165
  generator 166
  sterilization 136, 175
  superheated 164, 165, 175
  temperature 165
  treatment 165–167, 175
Steaming 165–166
  technology 167
  time 165
Stellaria media 159, 184
Stones 136, 138–139
Stranglervine 109
Strategie collaudate 132
Straw 5, 7, 136, 139, 142, 145, 148–149, 202, 206
  baled sugarcane 7
  dry 140
  mulch 145, 147
Strawberries 124, 137, 145–147, 157, 186, 193, 195–196
  matted-row 153
  ripening 151
Striga 38, 41, 52
  asiatica 193, 199
    seeds 198
  hermonthica 44
    damage 46
Submerged
  culture fermentations 96
  fermentation techniques 96
  liquid culture techniques 97
Substrates, foam spray-on 150
Subsurface tiller/transplanter 129
Sugar beet 31, 32, 45, 117, 128, 132, 158, 166, 173
Sugarcane 5, 7, 11
Suicidal germination process 193
Sulphur
  cinquefoil, reduced 10
  oxides 172
Sunflower 6, 35, 42–43, 60, 65, 67, 148
  broomrape 109

mulch 151
response 45
spring-sown 148
wild 96
Surfactants 96, 99, 101, 105, 164
  non-ionic 99
  nonoxynol 101
Suspension 97, 99
  liquid 7
Sustainable IPM technique 193
Sustainable weed management 35, 44, 132
Sweeps 119–120, 129
  sharp 120
Sweet basil 151
Sweetclover 40, 191–192, 203
Salvinia molesta 83–84

Tadpole performance 89
Taraxacum officinale Weber 52
Tartary buckwheat 32
Teeth 112
  curved 117, 121
  penetration 117
Tephrosia pedicellata 53
Termites 147–148
Terrestres 15, 105
Terrestrial microorganisms 108
Texas gourd 99, 106, 110
Theobromine 145
Theory of Island Biogeography 32
Thermal 157, 170–174, 206–207
  control
    methods 172
  death 188
  defoliation 163
  degree hours 196
  effects 178, 188
  heating methods 155
  inactivation 199
  injury 155, 162, 167
  killing 177, 188, 193
    direct 189
    effect 192, 195
    efficacy 192
    indirect 188
    mechanism 188–189, 194
    process 189
  methods 168, 171–172
    sustainable 172
  properties 144, 152
  shocks 198, 200
  techniques 175
  tolerances 197
  treatment 162, 189
Thermal Weed Control 155, 157, 159, 161, 163, 165, 167, 169, 170–173, 175

Thermophilic microorganism activity 194
Thiadiazuron 104
Thistle 55
  invasive 91
  musk 90
  nodding 83
  variegated 91
Threshold 25, 28
  aesthetic 77
  concept 27–28
  economic optimum 26, 30
  level 190
  long-term 27
  single-season 26
Tillage 12, 24–25, 27, 37, 46, 58–59, 61, 111–113,
    115, 129, 131–133, 156, 193, 204–205
  daylight 113
  equipment 115
  night-time 205
  non-inversion 14, 128–129
  operations 112, 205
    final secondary 112
  reduced 57, 112, 205
  ridge 129
  secondary 111–112, 115
  shallow 115, 120
  systems 58, 64
  tertiary 115
  timing 62
  tools 131
    non-inversion 129
  zero 32, 37, 45
Time released herbicide formulations 99
Timing of Weed Control 41
Tines 15, 112, 115–117, 125–126
  flexible 112, 116
  metal 124
Tomato 57, 102, 137, 143, 146, 147, 151, 152,
    207
Torsion
  weeder action 125
  weeders 123–126
Tractor 111–113, 116, 118, 120–121, 123, 126,
    131, 168
Trailing crownvetch 191
Transmitted light 203
Transparent
  plastic degrades 137
  polyethylene mulches 195
  polythene film 177
Transplanted
  basil 138
  brassicas 138
  cabbages 124
Transplanting 38
Trianthema
  monogyna 185
  portulacastrum 185

Tribulus terrestris 185
Trichothecene mycotoxins 105
Trichothecenes 105
  macrocyclic 106
Trifolium 50
  incarnatum 50, 149
  pratense 54
  repens 53
  subterraneum 50
Triticum aestivum 4, 31, 35, 44–47, 58, 65, 146, 191
Truncatum 98–99, 101–103
  microsclerotia 98
  spores 97–98
Tumbleweeds 11
Turf 93, 130, 206
Turfgrass quality 73
Turfgrasses 67, 75
Tween-20 101

Ultra-high
  frequency electromagnetic field 173
  temperatures 173
Ultraviolet 101, 137, 155
  light 96
  radiation 101, 108, 169, 172–173, 206
Undersowing 54–55
  clover 55, 63
Uprooting 112, 115, 121, 124–125, 130, 132, 134
  effect 112
  weed 124–125
Uredospores 104
Urtica urens 159
  Burning 185
UV 101, 137, 170
  damage 102
    escaped 169
  protectants 102, 108
  protection 101, 179
    of fungal entomopathogens 101
  radiation 169, 171
    damage dose-response curves 169
UV-induced damage 169

Vaccaria pyramidata 185
V-blades 129
Vectors 147–148
Vegetable
  crops
    non-competitive 41
    transplanted 126
  production 13, 61, 128, 133, 151, 162
    systems 62
Vegetables 46, 54, 64, 115, 151
  machine-harvested 5
  planted 117
  transplanted 124

Vegetation 57, 61, 93, 105, 135–138, 150, 197
  control 175
    aquatic 77
    fire-prone 189
    management 156
    non-target 105
    weed 63
Velvetbean 40, 50, 53, 65, 69, 74–75, 203
  accessions 52
  roots 69
  summer fallow 51
Velvetleaf 6, 9, 15, 29, 32, 46, 96, 104, 109–110,
    189, 198
  control 99, 101, 104, 110
  inoculated 101
  interference 32
  seeds 191
Vermiculite 96, 99
Verrucaria 105–106
Vertical axis brush weeder 121
Verticillium
  dahliae 147
  wilt 147, 152
Vesicular-arbuscular mychorrhizae 40
Vibrating 116, 126
  teeth 126
Vicia 58
  sativa 3
  villosa 50, 63
Vigna unguiculata 41, 50, 153
Vineyards 40, 130, 141, 158
Virulence 95–96, 205
  enhanced 106
  increased 95
  low 106, 205
Viruses 77, 107, 147
Volatiles 167, 172, 200
  foliar 73
  releasing antimicrobial 207
  trapped 190
Voles 148
V-shaped blades 129

Walnut 66, 72, 139
  black 65–66, 74, 203
  hulls 140
    decomposed 140
Walnut Shells for Weed Control 151
Wastes 11, 139
  composted 136, 142, 145, 152
    municipal green 142
  decomposed green 146
  food-processing plant 11
  industrial crop 206
  trimming 146
Water-in-oil 101
Water-jet 130, 132

Water-soluble phenolics 66
Waterfern, red 90
Waterlogged conditions 142
Wavelength-selective mulches 140
Wavelength transmission 179
Wavelengths 145, 168, 170
  infrared 137
  light reflectance 140
  long 137
Wax 101, 159
  damaged foliar epicuticular 165
Waxy cuticle 99
Weaving techniques 137
Weed
  biocontrol 78–83, 86, 88–89, 91
  biological control programs 88
  communities 2, 19–21, 24, 28, 30, 32
    managed 24
  competitive interactions 38
  control 12–13, 25, 31, 43–46, 54, 57–58,
    110–113, 128–133, 135–140, 149–153,
    155–158, 160–161, 163–175, 177–179,
    185–186, 194–207
    biological 77–78, 83–84, 86, 89–91, 93,
     106–110, 152, 209
    chemical 93, 151, 161–162, 171
    classical biological 83
    complete 12
    cultural 44, 64, 131–132, 174, 203
    early 150
    effective 164, 179, 193
     annual 142
    electrical 168, 171, 174–175
  density 7, 12, 21–28, 30–31, 35, 41–43, 50, 59,
    60, 113, 126, 133, 167–168, 171, 202, 207
    decreased 115
    estimates 33
    threshold concept 27
    thresholds 24, 26, 28, 43
  dynamics 45, 49
  economic thresholds 24, 31–32
  electrocution system 167
  emergence 12, 22, 40–41, 45, 50, 56–57, 63–64,
    113, 116, 132–133, 135–136, 142,
    148–149, 153, 166
    delaying 23
    early 137
    flush of 116
    reducing 149
    time 23
  infestation
    avoiding 2
    densities 38
    preventing 12
  interactions 29, 30, 32, 41
  interference experiments 28
  introductions 12, 202
  management
    costs 24

decision models 33
  integrated 14, 30, 36, 43, 47, 60, 201, 207–208
  invasive 67, 186
  puller 130–131
  seed bank 11–14, 36, 39, 40, 43, 55, 60, 31, 115, 148, 151, 153, 185, 187–188, 191, 196–198
  seed viability 8–10, 13–16, 150, 195
  seeds 3–5, 7–16, 38, 40, 113, 117, 129, 145, 148, 150, 165–166, 168–169, 178, 187, 195–196, 205–207
    buried 155, 161
    deposited 11
    germination of light-requiring 203, 205
    suppression 38, 41–42, 45–47, 50–51, 56–57, 59–68, 136–137, 142, 144, 146, 150, 153, 199, 203, 208–209
Weed control
  microwave-based 169
  mulch 140
  non-chemical 115, 173
  non-selective 157–158, 160
    mechanical 113
  residual 155, 161, 172
  season-long 131
  selective 157
  spectrum 102, 104, 167
  strategy 148, 153
  system 164, 175
  techniques 131
    primary 130
Weed-control strategies 151
Weed-crop Interactions 19, 21, 23, 25, 29, 31, 33
Weed-Fix cultivator 122
Weed-free mulch 145
Weed harrow 55
Weed seed
  content 11
  development 12
  germination 16, 67, 112, 135, 145, 148–149, 173, 196, 203
  production 11, 13, 18, 22, 26–27, 52, 57, 60–61, 203, 206
Weed-seed-contaminated commodities 7
Weed Seed Survival in Livestock Systems 15
Weed-suppressive
  abilities 66, 72
  benefits 56
  capabilities, effective 61
Weed thresholds 33
Weeders 120, 123
  basket 119, 121
  brush 119–120
  infrared 162
  intelligent 128, 206
  mechanical 111
  spring-hoe 125
Weeding 122
  costs 143
    reduced manual 123

equipment 123
inter-row
  brush 124
  precision 115
precision inter-row 113–114
Weeds
  alien 7
  aquatic 170, 206
  exotic 2
  flooding suppresses 38
  freezing terrestrial 206
  hard-seeded 180
  invasive 65, 93–94, 106
  mow 130
  rangeland 85
  surface germinating 39
Weevil 83, 88–90
  frond-feeding 90
Weevil, vine 148
Wettable
  powders 97
  silica gel powder 99
Wheat 1, 23, 31–32, 37, 39, 41, 44–46, 58, 60, 65, 67–68, 95, 98, 99, 107, 146, 152, 191
Wild
  buckwheat 9, 10
  oat 1, 5, 10, 15, 22, 30, 32–33, 35, 38, 44–46, 190, 198
    Avena 9
    sterile 192
    interference 32
    seedlings 38
Wildfire 156, 175, 189
Winged waterprimrose 102, 106
Wood chip 138–139, 147
Wormwood, annual 69, 73
Wound inoculations 102

Xanthan gum 101
Xanthium
  pensylvanicum 185
  spinosum 185
  strumarium 8, 68, 96, 109, 185

Yeast 147
  autolysed 95
  brewers 95
Yellow
  foxtail 184
  nutsedge 107, 146, 192, 197
  sweetclover 50, 52, 62, 208

Zea mays 5, 30–32, 35, 53, 74, 132–133
Zero-tillage production systems 44